Inhomogeneous Random Evolutions and Their Applications

Inhomogeneous Random Evolutions and Their Applications

Anatoliy Swishchuk

CRC Press
Taylor & Francis Group
Boca Raton London New York

CRC Press is an imprint of the
Taylor & Francis Group, an **informa** business

A CHAPMAN & HALL BOOK

CRC Press
Taylor & Francis Group
6000 Broken Sound Parkway NW, Suite 300
Boca Raton, FL 33487-2742

First issued in paperback 2021

© 2020 by Taylor & Francis Group, LLC
CRC Press is an imprint of Taylor & Francis Group, an Informa business

No claim to original U.S. Government works

Printed on acid-free paper

ISBN-13: 978-1-138-31347-7 (hbk)
ISBN-13: 978-1-03-208229-5 (pbk)

Visit the Taylor & Francis Web site at
http://www.taylorandfrancis.com

and the CRC Press Web site at
http://www.crcpress.com

To Stefan Banach

Contents

Preface

"Mathematics is the most beautiful and the most powerful creation of the human spirit. Only those countries that cultivate mathematics can be strong and powerful."

Stefan Banach

My favourite mathematician is Stefan Banach. My favourite space (besides universal space) is a Banach space, and my favourite evolution is a random one, namely, a random evolution in a Banach space. That is why this book is devoted to Stefan Banach and to random evolutions in Banach spaces.

My first knowledge about operators appeared in lectures presented by Yurij Berezansky (a member of the National Academy of Sciences of Ukraine (NASU) in Kyiv) at the Faculty of Mathematics at Kyiv State University back in 1976, where he taught a course in functional analysis consisting of operator theory and various abstract spaces, including Banach spaces. I liked his clear, neat and comprehensive lectures and fell in love with operator theory. At the same time, being at the University, I attended all lectures on stochastic processes delivered by Anatoliy Skorokhod (also a member of the NASU in Kyiv, who passed away in 2011 in East Lansing, Michigan, where he worked as a Professor at Michigan State University). Those lectures were also very interesting and exciting. Both professors were like artists who painted their beautiful paintings with chalk on a blackboard. During these lectures I got the idea to create a process that is random and, at the same time, evolves in a Banach space. I did not yet know about the existence of random evolution created by Reuben Hersh (see [6]). R. Hersh also mentioned in one of his books [7] that the term 'random evolution' was proposed by Peter Lax, his Ph.D. supervisor. It became known to me at the Institute of Mathematics (NASU, Kyiv) during my graduate studies (1981-1984). I found and realized then that there is a beautiful and remarkable object, random evolutions (i.e., operator-valued random processes), whose dynamics are in Banach spaces. They were different from random linear operators studied in a book [10] written by A. Skorokhod: This book was devoted to random linear operators in Hilbert spaces. Of course, I really enjoyed reading this book. However, my research intention was aimed at random evolutions in Banach spaces.

Over the course of my graduate studies, random evolutions in Banach spaces were rather promising and very interesting to me. Thus, I decided to pursue my research in this area. As a result, I defended my Ph.D. thesis on limit theorems for Markov random evolutions and their applications in

1984 (under the supervision of Volodymyr S. Korolyuk - a mathematician at the Institute of Mathematics in Kyiv, and a memebr of the NASU), and my Doctor of Science (D.Sc) thesis on limit theorems for semi-Markov random evolutions and their applications in 1992 (also at the Institute of Mathematics in Kyiv). Following these years, Dr. Korolyuk and I published many papers and books, all of which were devoted to homogeneous random evolutions - the operator semi-groups and stochastic processes (Markov or semi-Markov) were homogeneous in time.

And, of course, I was eager to move on and do research in the area of inhomogeneous random evolutions, where the operator semigroups and stochastic processes are inhomogeneous in time. However, due to many circumstances in my life, this dream was postponed until I received one particular Ph.D. student at the University of Calgary, namely, Nelson Vadori (a student of mine from 2011 to 2015) who was willing to undertake this difficult task and managed to accomplish it on a high note. Together, we published many papers, and many results in this book involving inhomogeneous random evolutions and their limit theorems came from our joint papers in this area (see [13], [15], [16]). I would like to thank Nelson very much for his hard work, for his love of mathematics and of this topic, and for his dedication to many problems that seemed unsolvable at first. I was very lucky to have Nelson Vadori as my student.

I also would like to mention that during my studies as a graduate and doctorate student, I periodically (as do many other graduates in stochastics and functional analysis) attended a seminar at the Institute of Mathematics in Kyiv organized by Anatoliy Skorokhod and Yuri Daletsky (also a member of the NASU), devoted to infinite-dimensional processes. The seminar was always open-minded and light-hearted, with many interesting problems frequently raised by the attendees. After the seminar, we usually continued to talk about such topics and problems outside the Institute, frequently in nearby coffee shops (such as _Café_-a small coffee shop in Kyiv). Of course, we knew about the well-known Scottish Café (previously titled _Kawiarnia Szkocka_ in Polish) on Fedro street in Lviv (Polish: _Lwów_) back in the 1930s, and we were under the influence of its activities (initiated then by Stefan Banach, Stanislaw Ulam and Stanislaw Mazur). We were excited to read and to discuss many problems from the Scottish Book ([14]) (Polish: _Księga Szkocka_), and tried to solve them. We also studied and discussed many papers and books in this area, especially original ones, for example, Banach's own works. These remarkable and exciting memories will remain in my mind for the rest of my life.

The world-wide recognition of Banach's results came after the publication of his book in 1931 (see [3]), which in the following year was translated into French (see [4]). The most exciting fact was that this book was also translated into Ukrainian and published in 1948 (see [5]). I thoroughly enjoyed this book, and, as I remember, I devoured it in probably one week or so at that time. The topic of Banach's book was originated in 1922: Banach submitted his original work [1] for a doctorate at the Jan Kazimierz University (JKU) in

Lviv (Polish: *Lwów*; now The Ivan Franko National University of Lviv), and it was published in the third volume of *Fundamenta Mathematicae*. The same year, he obtained his Ph.D. (see [2]) at the JKU and in 1922, after completing postdoctoral work, Banach was appointed to a full professor position at the JKU. He had previously already published six papers, however, this was the first one dedicated to linear operation theory. Banach's first book [3] on linear operation theory was in Polish and was published in 1931 in Warsaw (Polish: *Warszawa*), then the extended version [4] was published in 1932 in French. The English translation [3] was published in 1987 by North-Holland. The Ukrainian translation [5] was published in 1948 in Kyiv.

I would like to mention, the book [8] is a concise overview of Banach's pioneering work, and includes all the materials in a way that makes matters more easily understandable.

I would like to finish this Preface with the words presented by Mark Kac (who considered Banach "probably the greatest Polish mathematician of all times") and Hugo Steinhaus, dedicated to Stefan Banach:

- "In the short fifty-three years of his life (...) he succeeded in combining an overwhelming flow of brilliant ideas with a style of high living that few men could sustain" - M. Kac ([9]);

- "He combined within himself a spark of genius with an astonishing internal urge, which addressed him incessantly in the words of the poet: "there is only one thing: the ardent glory of one's craft ("Iln'ya que la gliore ardent du métier"-Verlaine)"-and mathematicians well know that their craft consists of the same secret as the poet's craft."- H. Steinhaus ([11]).

Anatoliy Swishchuk
Calgary, Alberta, Canada
April 26, 2019

Bibliography

[1] Banach, S.S. Sur les opérations dans les ensembles abstraits et leur application aux équations intégrals. *Fundamenta Mathematicae*, v.3, 1920. (In English: 'On operators defined on abstract sets and their applications to integral equations').

[2] Banach, S. Sur les opérations dans les ensembles abstraits et leur application aux équations intégrals. *Ph.D. Thesis*, Lvov University, 1922. (In English: 'On operators defined on abstract sets and their applications to integral equations').

[3] Banach, S.S. *Teoria Operacyj. Tom 1. Operacje Linjowe.* Kasa im. Mianowskiego, Warszawa, 1931. (English Translation: *Theory of Linear Operations*, North-Holland, Amsterdam, New-York-Oxford-Tokyo, 1987).

[4] Banach, S.S. *Théorie des Opérations Linéares.* v. 1. Monograpfie Matematyczne, Fundusz Kultury Narodowej, Warszawa, 1932.

[5] Banach, S. S. *Kurs Funktsional'nogo Analizu.* Radianska Shkola, Kyiv, 1948 (In Ukrainian). (In English: 'A Course in Functional Analysis').

[6] Hersh, R. Random evolutions: a survey of results and problems. *Rocky Mount. Math. J.*, 1974, v. 4, 443-477.

[7] Hersh, R. *Peter Lax, Mathematician: An Illustrated Memoir.* AMS, 2015.

[8] Jakimowicz, E. and Miranowicz, A. *STEFAN BANACH. Remarkable Life, Brilliant Mathematics.* Gdánsk University Press. 2011.

[9] Kac, M. *Enigma of Chance. An Autobiography*, Harper&Row, NY, 1985.

[10] Skorokhod, A. *Random Linear Operators.* D. Reidel Publ. Comp., Dordrecht, Holland, 1984. (original ©1978 by Naukova Dumka, Kyiv).

[11] Steinhaus, H. Stefan Banach. An Address Delivered at the Stefan Banach Memorial Conference. *Wiadomosci Matematyczne*, 1961, 4, 251-259.

[12] Swishchuk A. V. and Wu J. *Evolutions of Biological Systems in Random Media. Limit Theorems and Stability.* Kluwer AP, Dordrecht, The Netherlands, 218p. Serie: Mathematical Modelling: Theory and Applications, 2003, v. 18.

[13] Swishchuk, A. and Vadori, N. A semi-Markovian modelling of limit order books. *SIAM J. Fin. Math.*, 2017, 8(1), 240-273.

[14] Ulam, S. *The Scottish Book: a Collection of Problems.* Los Alamos, 1957.

[15] Vadori, N. and Swishchuk, A. Convergence of Random Bounded Linear Operators in the Skorokhod Space. *Random Operators and Stochastic Equations*, 2019, 27 (3), 1-13 (https://doi.org/10.1515/rose-2019-2011).

[16] Vadori, N. and Swishchuk, A. Inhomogeneous Random Evolutions: Limit Theorems and Financial Applications. *Mathematics*, 2019, 7, 447; doi:10.3390/math7050447.

Acknowledgments

I would like to thank David Grubbs (Editor, CRC Press/Taylor & Francis Group) who reminded me many times about writing a book that I promised him, and finally, we did it.

My thanks also go to Shashi Kumar (Cenveo, Stanford, Connecticut) who helped me a lot with the LaTEX preparation of all the chapters in Krantz style suitable for the CRC Press. His help was so valuable and on time that I submitted my book well before the deadline.

I would also like to thank all my graduate students and participants of our finance seminars 'Lunch at the Lab', 'Lobster' and 'Hawks', for their curiosity and dedication to the seminars' activities. As I have already mentioned in the Preface, I would like specifically to thank to my ex-Ph.D. student Nelson Vadori for his proactive attitude to the research and continuous curiosity.

Many thanks go to my daughter, Julianna Svishchuk, who helped me a lot with editing some chapters and especially the Preface. I would also like to thank my daughter, Julianna, for capturing and refining the cover photo for this book. She really helped me bring my idea to life.

Thanks to our department and to our Chair, Tony Ware, for providing a very friendly and healthy environment for writing the book. I wrote and prepared it during my R&D leave (a.k.a. sabbatical leave) without actually leaving, but staying in Calgary just to be close to my family at home, and to be very close to the coffee machine at work: Tony installed it in our lounge right after becoming our Chair! The more coffee you drink, the more theorems you produce! Probably the spirit of Paul Erdős (recalling his quote: 'A mathematician is a device for turning coffee into theorems'), who visited our department so many times, is still here!

Finally, many thanks go to my dear family, including my adorable granddaughter, Ivanka Maria, for continuing support and inspiration.

Introduction

"The beauty in mathematical structures cannot be appreciated without understanding of a group of numerical formulae that express laws of logic. Only mathematicians can read 'musical score' containing many numerical formulae, and play that 'music' in their hearts."

Kiyosi Itô

In brief, random evolutions are dynamic systems with operators depending on some stochastic process (Markov, semi-Markov, Levy process, etc.). Verbally, a random evolution is a model for a dynamic system in a random environment whose equation of state is subject to random variation; for example, a stock/asset that switches between different volatilities. Mathematically, a random evolution is an operator integro-differential equation with a generator depending on a parameter, and this parameter is a stochastic process. The stochastic processes define the name for the random evolutions. Also, depending on the structure of the operator equation, we have continuous, discontinuous/jump, discrete, homogeneous, inhomogeneous random evolutions, etc. Markov random evolutions in Euclidian spaces are usually termed in the literature as 'hidden Markov' or 'regime-switching' models. Random evolutions began to be studied in the 1970s, because of their potential applications in finance, insurance, biology, queuing or risk theories, to name a few.

In this book, we introduce a new class of random evolutions, namely, inhomogeneous random evolutions (IHRE), including semi-Markov random evolutions, and consider their many applications, including finance and insurance (risk theory). The novelty of the book is fourfold: i) studying of new inhomogeneous random evolutions (IHREs), constructed by a family of inhomogeneous semi-group of operators, which describe the evolution of our dynamic system; ii) the study, in particular, of new inhomogeneous semi-Markov random evolutions, based on inhomogeneous semi-Markov process, which switch regimes of evolution; iii) considerations of many applications, including finance and insurance (risk theory); iv) present new results on the Law of Large Numbers (LLN) and Functional Central Limit Theorems (FCLT) for IHREs and their applications. To our best knowledge, only homogeneous random evolutions (based on homogeneous Markov or semi-Markov processes, and on homogeneous semi-group of operators) and their applications have been studied.

There are three parts in the book: 1) Stochastic Calculus in Banach Spaces, 2) Homogeneous and Inhomogeneous Random Evolutions (IHREs), and 3)

Applications of IHREs. The first part consists of Chapters 1-3, and contains some preliminaries and basics, stochastic calculus in Banach spaces, and convergence of random bounded linear operators in the Skorokhod space. The second part consists of Chapters 4-5, and deals with general theory for homogeneous and inhomogeneous random evolutions, their martingale characterization and main theorems for them, namely Law of Large Numbers (LLN) and Functional Central Limit Theorems (FCLT). The third part consists of Chapters 6-8 and is devoted to the applications of IHREs in finance and insurance. In Chapter 6 we present financial applications to illiquidity modeling using regime-switching time-inhomogeneous Lévy price dynamics and regime-switching Lévy driven diffusion based price dynamics. We also present a generalized version of the multi-asset model of price impact from distressed selling, for which we retrieve (and generalize) their diffusion limit result for the price process. Chapter 7 introduces a semi-Markovian modeling of the limit order book, and presents the main probabilistic results, such as duration until the next price change, probability of price increase and characterization of the Markov renewal process driving the stock price process. Here, we also present LLN and FCLT for this semi-Markovian model, The diffusive limits for the latter asset models will be computed, connecting the parameters driving the high-frequency activities to the daily volatilities, and thus making the explicit link between the microscopic activity and the macroscopic one. In Chapter 8, we introduce a new model for the risk process based on general compound Hawkes process (GCHP) for the arrival of claims. We call it risk model based on general compound Hawkes process (RMGCHP). The LLN and the FCLT are proved. We use the diffusive limit for the risk process to calculate the ruin probabilities for the RMGCHP.

There are many books that consider random evolutions, including [1]–[9]; however, again, they are all devoted to the homogeneous random evolutions.

This book is intended for graduate students and academics working in stochastic processes and their applications in finance and insurance. The practitioners may also find it handy as an instrument that can be used in applied problems associated with finance and insurance.

Bibliography

[1] Korolyuk, V. S. Swishchuk, A. *Semi-Markov Random Evolutions.* Naukova Dumka Publ., Kyiv. 1992, 256 pages. (In Russian).

[2] Korolyuk V. S. and Swishchuk A. V. *Evolution of Systems in Random Media.* CRC Press, Boca Raton, USA. 1995, 356 pages.

[3] Korolyuk V. S. and Swishchuk A. V. (1995) *Semi-Markov Random Evolutions.* Kluwer Academic Publishers, Dordrecht, The Netherlands. 1995, 310 pages. (English Translation of [1]).

[4] Pinsky, M. *Lectures in Random Evolutions.* World Sci. Publ., 1991.

[5] Swishchuk, A. *Limit Theorems for Semi-Markov Random Evolutions and their Applications.* D.Sc. Thesis, Institute of Mathematics, Nation. Acad. Sci. Ukraine, Kyiv. 1992, 356 pages.

[6] Swishchuk, A. *Random Evolutions and their Applications.* Kluwer AP, Dordrecht, The Netherlands, 1997.

[7] Swishchuk A. V. and Islam S. *Random Dynamical Systems in Finance.* Chapman & Hall/CRC Press, Boca Raton, FL, USA, 2013.

[8] Swishchuk A. V. and Wu J. *Evolutions of Biological Systems in Random Media. Limit Theorems and Stability.* Kluwer AP, Dordrecht, The Netherlands, 218p. Series: Mathematical Modelling: Theory and Applications, 2003, v. 18.

[9] Swishchuk A. V. *Random Evolutions and Their Applications. New Trends.* Kluwer Academic Publ., Dordrecht, The Netherlands, 2000, v.504, 315p.

Part I

Stochastic Calculus in Banach Spaces

1

Basics in Banach Spaces

This chapter contains some necessary preliminary notions and facts associated with separable Banach spaces, including random (or stochastic) processes, stochastic integration, martingales and martingale problems (see 1.1), weak convergence (see 1.2) and semigroup theory (see 1.3). All references ([1]–[11]) may be found in the end of this chapter.

1.1 Random Elements, Processes and Integrals in Banach Spaces

Some Notations and Spaces. Suppose that $(\Omega, \mathcal{F}, \mathcal{F}_t, P)$ is a probability space with filtration $\mathcal{F}_t$, $t \in R_+ := [0, +\infty)$, and $(B, \mathcal{B}, || \cdot ||)$ is a *separable Banach space* with σ-algebra of Borel sets $\mathcal{B}$ and with norm $|| \cdot ||$. We recall that B is a normed linear space that is a complete metric space with respect to the metric derived from its norm. And a space is called *separable* if it contains a countable, dense subset. Let B^* be a *dual space* to B (it contains all linear functionals on B, together with the vector space structure of pointwise addition and scalar multiplication by constants) separating the point of B. We recall that collections of functionals l from B^* *separates points*, if for every $x, y \in B$ with $x \neq y$, $l(x) \neq l(y)$. A couple of examples of Banach spaces and their dual spaces include l^p, L^p and l^q, L^q, respectively, where $p > 1$ and $1/p + 1/q = 1$. Let also $C_B[0, T]$ be a *space of continuous functions* on $[0, T]$ with values in B, and $D_B[0, T]$ be a *space of right-continuous and having left limits functions* on $[0, T]$ with values in B.

    Random Elements in B. Let $V(\omega)$ maps the probability space Ω into the Banach space $B : V(\omega) : \Omega \to B$. If $V(\omega)$ is measurable, i.e., $V^{-1}\mathcal{B} \in \mathcal{F}$, (we also call it strongly measurable to make a difference with weakly measurable, i.e., $l^{-1}(V(\omega))\mathcal{R} \in \mathcal{F}$, $\forall l \in B^*$, where $\mathcal{R}$ is a Borel sets in R) then $V(\omega)$ is called a *random element*. If $B = R^1 = R := (-\infty, +\infty)$, then V is called a *random variable*; if $B = R^n, n > 1$, then V is called a *random vector*. *Distribution of random element* V is a probability measure $\mu := PV^{-1}$ on $(B, \mathcal{B})$. It means that $(B, \mathcal{B}, \mu)$ is a probability space. We note that the distribution μ is always taken on a metric space in contrast to the probability measure P defined on a space of any nature.

Random Processes in B. For any $\omega \in \Omega$, we denote by $V(t, \omega)$, $t \in R_+$, the element of either $C_B[0, T]$ or $D_B[0, T]$: $V(t, \omega)$ maps the probability space Ω and R_+ into B in a such way that $V(t, \omega)$ is either continuous or right-continuous and having left limits w.r.t. $t \in R_+$ for any ω, respectively. The random element $V(t, \omega)$ is said to be *adapted* to a filtration $\mathcal{F}_t$ if, for each t, $V(t, \cdot)$ is $\mathcal{F}_t$-measurable, namely, $\{V(t, \cdot) \in A\} \in \mathcal{F}_t$, $\forall A \in \mathcal{B}$. Thus, we consider the maps $V(t, \omega) : R_+ \times \Omega \to C_B[0, T]$ (or $V(t, \omega) : R_+ \times \Omega \to D_B[0, T]$) to be progressively measurable w.r.t. the σ-algebras of Borel sets $\mathcal{C}$ (or $\mathcal{D}$) in $C_B[0, T]$ (or $D_B[0, T]$). Let $(t_1, ..., t_n)$ be arbitrary points from $[0, T]$, and $\pi_{t_1, ..., t_n}$ be the map from $C_B[0, T]$ ($D_B[0, T]$) into $B^n := B \times ... \times B$: $C_B[0, T] \to (V(t_1, \omega), ..., V(t_n, \omega))) \in B^n$ ($D_B[0, T] \to (V(t_1, \omega), ..., V(t_n, \omega))) \in B^n$). The set $\pi_{t_1, ..., t_n}^{-1} \mathcal{R}$, $\forall \mathcal{R} \in B^n$, is called a finite-dimensional set in $C_B[0, T]$ (or in $D_B[0, T]$). Distributions of the form $\mu \pi_{t_1, ..., t_n}^{-1}$ are called *finite-dimensional distributions* corresponding to probability measure μ on $(B, \mathcal{B})$. We note that finite-dimensional sets generate the σ-algebra of Borel sets $\mathcal{C}$ (or $\mathcal{D}$) on $(B, \mathcal{B})$. A random function $V(t, \omega)$ is a *random process* in $C_B[0, T]$ (or in $D_B[0, T]$) if and only if for any $t \in [0, T]$ $V(t, \omega)$ is the random element, i.e., $V^{-1}\mathcal{B} \in \mathcal{C}$ (or $V^{-1}\mathcal{B} \in \mathcal{D}$). We will frequently use the following short notation $V(t)$ for a random process $V(t) \in B$ instead of $V(t, \omega)$.

Weak Random Elements. A random element $V(\omega)$ is called a *weak random element* if the map $V(\omega) : \Omega \to B$ is a weakly measurable, i.e., $l(V(\omega)) : \Omega \to R$ is measurable for any $l \in B^*$: $l^{-1}(V(\omega))\mathcal{R} \in \mathcal{F}$, where $\mathcal{R}$ is a Borel set in R. To make a difference, the random element $V(\omega)$ defined above is in fact a *strong random element*. We use the notion of random element for a strong random element.

Integrability and Expectations in B. A random element $V(\omega)$ is *integrable in the Bochner sense* if there exists a sequence $V_n(\omega)$ of simple functionals in B which converges to $V(\omega)$ with probability 1, and

$$\lim_{m,n \to +\infty} \int_\Omega \|V_m(\omega) - V_n(\omega)\| dP = 0.$$

Then the limit $\lim_{n \to +\infty} \int_\Omega V_n(\omega) dP$ exists (B is a Banach space) and, by definition

$$\int_\Omega V(\omega) dP := \lim_{n \to +\infty} \int_\Omega V_n(\omega) dP.$$

The limit defines the *strong expectation EV* of V :

$$EV(\omega) = \int_\Omega V(\omega) dP.$$

A *weak expectation* may be defined as

$$El(V(\omega)) = \int_\Omega l(V(\omega)) dP, \quad \forall l \in B^*.$$

Let V be integrable in the Bochner sense and $\mathcal{F}$-measurable (i.e., $V^{-1}\mathcal{B} \in \mathcal{F}$). Let $\mathcal{G} \subset \mathcal{F}$ be a sub-σ-algebra of $\mathcal{F}$. Then there exists a function $E[V(\omega)/\mathcal{G}]$: $\Omega \to B$, $\mathcal{G}$-measurable, which is integrable in the Bochner sense, strongly measurable, unique a.s., and such that

$$\int_A V(\omega)dP = \int_A E[V(\omega)/\mathcal{G}]dP, \quad \forall A \in \mathcal{G}.$$

The function $E[V(\omega)/\mathcal{G}]$ is called a *strong conditional expectation* of V w.r.t. $\mathcal{G}$. A *weak conditional expectation* is defined as $E[l(V(\omega))/\mathcal{F}]$, $\forall l \in B^*$. We can also define the *strong conditional variance* of random element V : $Var(V/\mathcal{G}) = E((V - E(V/\mathcal{G}))^2/\mathcal{G})$. We note that the strong and weak conditional expectations satisfy all the main properties and laws of standard conditional expectations for random variables, e.g., *linearity, positivity, monotonicity, monotone and dominated convergences, laws of total expectations, tower property*, etc.

By analogy, we can define the *strong and weak expectations*, as well as *strong and weak conditional expectations* for random process $V(t,\omega)$ in B w.r.t. filtration $\mathcal{F}_t$. For example, the strong expectation is defined as

$$EV(t,\omega) = \int_\Omega V(t,\omega)dP, \quad \forall t \in R_+,$$

and the strong conditional expectation is defined as a function $E[V(t,\omega)/\mathcal{G}]$: $\Omega \times R_+ \to B$, $\mathcal{G}$-measurable, which is integrable in the Bochner sense, strongly measurable, unique a.s., and such that

$$\int_A V(t,\omega)dP = \int_A E[V(t,\omega)/\mathcal{G}]dP, \quad \forall A \in \mathcal{G}, \quad \mathcal{G} \subset \mathcal{F}, \quad \forall t \in R_+.$$

Martingales in B. Let $M(t,\omega)$ be a map from Ω to B which is integrable in the Bochner sense and strongly measurable w.r.t. $\mathcal{F}_t$, $t \in R_+$. A random process $M(t,\omega)$ is called a *strong $\mathcal{F}_t$-martingale in Banach space B*, if 1) $E\|M(t,\omega)\| < +\infty, t \in R_+$, and 2) $E[M(t,\omega)/\mathcal{F}_s] = M(s,\omega), s < t$, with $P = 1$. A random process $M(t,\omega)$ is called a *weak $\mathcal{F}_t$-martingale* in B, if 1) is satisfied and 2') $E[l(M(t,\omega))/\mathcal{F}_s] = l(M(s,\omega))$, $s < t$, with $P = 1$, $\forall l \in B^*$. Thus, $l(M(t,\omega))$ is a real $\mathcal{F}_t$-martingale, $\forall l \in B^*$. *Strong and weak semi-martingales (sub- and super-martingales)* are defined quite similarly. We note that if $l(M(t,\omega))$ is a real $\mathcal{F}_t$-martingale $\forall l \in B^*$, and $M(t,\omega)$ is integrable in the Bochner sense, then $M(t,\omega)$ is a martingale in B.

Let τ be a *stopping time* w.r.t. $\mathcal{F}_t$, namely, an $\mathcal{F}_t$-measurable function $\tau : \Omega \to R_+$ such that $\{\tau < t\} \in \mathcal{F}_t$. $M(t,\omega)$ is called a *local strong martingale* in B if and only if there exists a sequence of stopping times $(\tau_k; k \in N)$ such that $\tau_k \to +\infty$ a.s., and for any k $M(t \wedge \tau_k, \omega)$ is a strong martingale, where $t \wedge \tau_k = min(t, \tau_k)$. Here, $N = \{1, 2, ..., n, ...\}$.

A martingale $M(t,\omega)$ in B is called a *square integrable martingale* if and only if $\sup_{t \in [0,T]} E\|M(t,\omega)\|^2 < +\infty$. The process $v(t,\omega)$ is a *quadratic* *variation of martingale* $l(M(t,\omega))$ if the process $l^2(M(t,\omega) - v(t,\omega))$ is an $\mathcal{F}_t$-martingale in R. We note that $v(t,\omega) \in R$ and we denote $v(t,\omega) := \langle l(M(t,\omega)) \rangle$. We also note that if there exists a process $V(t,\omega)$ in B such that $l(V(t,\omega)) = v(t,\omega)$, $\forall l \in B^*$, then $V(t,\omega)$ is a Banach-valued analogue of the quadratic variation for the martingale $M(t,\omega)$ in B. For the calculation of the quadratic variation of the martingale $M(t,\omega)$ in B, the following definition is also used:

$$\langle l(M(t,\omega)) \rangle = \lim_{\Delta \to 0} \sum_{k=1}^{n} E[l^2(M(t_k,\omega) - M(t_{k-1},\omega))/\mathcal{F}_{t_{k-1}}],$$

where $0 = t_0 < t_1 < ... < t_n$, $\Delta := \max_i[t_i, t_{i+1})$, $\mathcal{F}_{t_k} := \sigma\{M(s,\omega); 0 \le s \le t_k\}$, $k = 0, 1, ..., n$, $\forall l \in B^*$.

$M(t,\omega)$ is called a *discrete martingale (strong or weak)* in B, if $t \in N$. It is denoted as $M_n(\omega)$. For discrete martingale we have the following definition of quadratic variation:

$$\langle l(M_n(\omega)) \rangle := \sum_{k=1}^{n} E[l^2(M_k(\omega) - M_{k-1}(\omega))/\mathcal{F}_{k-1}],$$

where $\mathcal{F}_k := \sigma\{l(M_i(\omega)); 0 \le i \le k\}$, $\forall l \in B^*$.

By definition, the quadratic variation $\langle l(M(t,\omega)) \rangle$ satisfies the relation:

$$E[l^2(M(t,\omega) - M(s,\omega))/\mathcal{F}_s] = E[\langle l(M(t,\omega)) \rangle - \langle l(M(s,\omega)) \rangle/\mathcal{F}_s], \quad s < t.$$

We note the important inequality for martingales:

$$E \sup_{t \in [0,T]} |l(M(t,\omega))| \le 3E\sqrt{\langle l(M(T,\omega)) \rangle}, \quad \forall l \in B^*.$$

From here we have that if $\langle l(M(t,\omega)) \rangle = 0$, $\forall t \in [0,T]$, then $l(M(t,\omega)) = 0$, and $M(t,\omega) = 0$ because the space B^* separates points of B.

Stochastic Integrals in B. Let $V(t,\omega)$ be a $\mathcal{F}_t$-measurable random process in B, and $\forall l \in B^*$: $\int_0^T l^2(V(t,\omega))dt < +\infty$ a.s. Let also $m(t)$ be a right-continuous local $\mathcal{F}_t$-martingale in R. If $m(t)$ has the locally bounded variation and $V(t,\omega)$ is a continuous adapted process, then $\int_0^t l(V(s,\omega))dm(s)$ is defined as the *Riemann-Stieltjes integral* for any t and $\forall l \in B^*$. Thus, we can define the *integral* $\int_0^t V(s,\omega)dm(s)$ in B w.r.t. martingale $m(t)$. One of the examples can be obtained if we take $m(t) = N(t) - \lambda t$, where $N(t)$ is a standard Poisson process with intensity $\lambda > 0$. In this way, the integral w.r.t. $m(t)$ is realized by paths, because of the regularity of the Poisson process $N(t)$, which has a finite number of jumps on any finite time interval. However, if we take $m(t) = W(t)$, a standard Wiener process, then the integral $\int_0^t V(s,\omega)dW(s)$ cannot be defined by paths because the Wiener process is not a function with locally bounded variation. This integral is defined further as an isometry between the space of processes $V(t,\omega)$ and the space of square integrable stochastic integrals $\int_0^t V(s,\omega)dW(s)$. We will use more general construction for this integral w.r.t. orthogonal martingale measure in Chapter 2.

1.2 Weak Convergence in Banach Spaces

Let $(B, \mathcal{B}, || \cdot ||)$ be a separable Banach space with σ-algebra of Borel sets $\mathcal{B}$ and with norm $|| \cdot ||$. Let also $\nu(\cdot)$ be a non-negative σ-additive set function on $\mathcal{B}$ such that $\nu(B) = 1$, i.e., $\nu(\cdot)$ is a probability measure on $\mathcal{B}$. Probability measures ν_n *converges weakly* to a probability measure ν ($\nu_n \Rightarrow \nu$) if and only if

$$\int_B f d\nu_n \to \int_B f d\nu, \quad n \to +\infty,$$

where f is any bounded real-valued continuous function on B, i.e., $f \in C_B(R)$. A probability measure ν on $(B, \mathcal{B})$ is called *tight*, if for any $\epsilon > 0$ there exists a *compact set* K_ϵ such that $\nu(K_\epsilon) > 1 - \epsilon$. Ulam's theorem states that any probability measure on a *Polish space* (i.e., separable completely metrizable topological space) is tight. A sequence V_n of random elements *converges in distribution* to a random element V ($V_n \xrightarrow{D} V$), $n \to +\infty$, if distributions μ_n of elements V_n converges weakly to the distribution μ of element $V : \mu_n \Rightarrow \mu$. A sequence V_n *converges in probability* to $a \in B$, $V_n \xrightarrow{P} a$, $n \to +\infty$, if for any $\epsilon > 0$, $P\{||V_n - a|| \geq \epsilon\} \to 0$, $n \to +\infty$. The following theorem holds:

Theorem 1.2.1. *If $V_n \xrightarrow{D} V$ and $||V_n - W_n|| \xrightarrow{P} 0$, then $W_n \xrightarrow{D} V$, $n \to +\infty$.*

A family of probability measures ν_n on $(B, \mathcal{B})$ is called weakly compact (or, equivalently, relatively compact), if any sequence of elements ν_n contains a weakly converging subsequence. The relationship between tightness and weak compactness is established in the following theorem:

Theorem 1.2.2. *The next two statements are true:*
1) if ν_n is tight, then it is weakly compact for any metric space;
2) if metric space is separable and complete, then weak compactness implies tightness.

Thus, in the case of the Banach space B, the notions of tightness and weak compactness coincide. Because of the continuity of projections $\pi_{t_1, \ldots, t_n}$ on $(C_B[0, T], \mathcal{C})$ we have that weak convergence for probability measures on $(C_B[0, T], \mathcal{C})$ implies weak convergence of the corresponding finite-dimensional distributions. Inverse is not true. However, the following result holds:

Theorem 1.2.3. *Let ν_n and ν be probability measures on $(C_B[0, T], \mathcal{C})$. If finite-dimensional distributions of measure ν and the family ν_n are tight, then $\nu_n \Rightarrow \nu$, $n \to +\infty$.*

The Arzelà-Ascoli's theorem gives the characterization of compact sets in $C_B[0, T]$. However, the space $C_B[0, T]$ is not appropriate, for example, for the description of processes containing jumps. In this case we use the space $D_B[0, T]$.

A sequence V_n of random elements of the space $D_B[0,T]$ is tight, if the sequence of the corresponding distributions is tight. Weak compactness of the family of probability measures on $(D_B[0,T], \mathcal{D})$ is not analogous to the weak compactness in $(C_B[0,T], \mathcal{C})$, because the projections $\pi_{t_1,...,t_n}$ are not continuous. Let T_ν be a set containing the points of $[0,T]$ such that projections $\pi_{t_1,...,t_n}$ are continuous everywhere, except the set of points with ν-measure 0. The analogue of Theorem 1.2.3 in $(D_B[0,T], \mathcal{D})$ is the following one:

Theorem 1.2.4. *If the family ν_n is tight and if $\nu_n \pi_{t_1,...,t_k}^{-1} \Rightarrow \nu \pi_{t_1,...,t_k}^{-1}$ for any $(t_1, ..., t_k) \in T_\nu$, then $\nu_n \Rightarrow \nu$, $n \to +\infty$.*

Tightness of probability measures on $(D_B[0,T], \mathcal{D})$ is established by the modulus of continuity for elements $V \in D_B[0,T]$. Modulus of continuity in $D_B[0,T]$ is defined by the following function:

$$w'(V, \delta, T) := \inf_{t_i} \max_i w(V, t_{i-1}, t_i),$$

where $w(V, t_{i-1}, t_i) := \sup\{||V(s) - V(s')||; t \le s < s' < t'\}$, $0 = t_0 < t_1 < ... < t_n = T$, $t_i - t_{i-1} > \delta$, $\forall i = 1, 2, ..., n$.

We note that the weak convergence in $D_B[0, +\infty)$ is the weak convergence in $D_B[0, T_k]$ for any sequence $T_k : \lim_{k \to +\infty} T_k = +\infty$.

Theorem 1.2.5. *The set $A \subset D_B[0, +\infty)$ has a compact closure if and only if the following conditions hold:*
1) $\forall T > 0$ there exists a compact set $K_T \subseteq B : \forall V(t) \in A$ it follows that $V(t) \in K_T$;
2) $\lim_{\delta \to 0} \sup_{V \in A} w'(V, \delta, t) = 0$, $\forall t \in [0, +\infty)$.

We formulate compact conditions for random processes in B in the form of sequences w.r.t. $\epsilon \to 0$ instead of $n \to +\infty$, which is only a convention but not a restriction.

Theorem 1.2.6. *A sequence of processes $V_\epsilon(t) \in B$, $\epsilon > 0$, is weakly compact if and only if the following conditions hold:*
1) $\forall \Delta > 0$ and $\forall T > 0$ there exists a compact set $K_T^\Delta \subseteq B :$
$\lim_{\epsilon \to 0} \inf P\{V_\epsilon(t) \in K_T^\Delta; 0 \le t \le T\} \ge 1 - \Delta$;
2) $\forall \Delta > 0$ and $\forall t \in [0, T]$, there exists $\delta > 0 :$
$\lim_{\epsilon \to 0} \inf P\{w'(V_\epsilon(t, \delta, t) \ge \Delta\} \le \Delta$.

We formulate a more convenient version of Theorem 1.2.6 which follows from properties of modulus w' and some estimations of variations of the elements of $D_B[0, +\infty) :$

Theorem 1.2.7. *A set of processes $V_\epsilon(t) \in B$, $\epsilon > 0$, is a weakly compact if and only if the condition 1) of Theorem 1.2.6 holds, and also there exists a random function $C_\epsilon(\eta)$, $0 < \eta < 1$, such that $\lim_{\eta \to 0} \lim_{\epsilon \to 0} EC_\epsilon(\eta) = 0$, and the following estimation takes place as $t \in [0, T] :$*
$E[||V_\epsilon(t + h_1) - V_\epsilon(t)||/F_t^\epsilon] \times ||V_\epsilon(t - h_2) - V_\epsilon(t)|| \wedge 1 \le E[C_\epsilon(\eta)/F_t^\epsilon]$.

This result implies the following criteria of weak compactness of real-valued processes $v_\epsilon(t) \in D_R[0, +\infty)$, $\epsilon > 0$:

Theorem 1.2.8. *Let $v_\epsilon(t) \in D_R[0, +\infty)$, $\epsilon > 0$, and there exists a compact set $K_T^\Delta \in R : \lim_{\epsilon \to 0} \inf P\{v_\epsilon(t) \in K_T^\Delta; 0 \leq t \leq T\} \geq 1 - \Delta$.*
Then a weak compactness of $v_\epsilon(t) \in D_R[0, +\infty)$, $\epsilon > 0$, takes place if and only if there exists a random function $C_\epsilon(\eta)$, $0 < \eta < 1$, such that $\lim_{\epsilon \to 0} \lim_{\eta \to 0} EC_\epsilon(\eta) = 0$, and the followeing inequality holds for some $r > 0$:
$E[|v_\epsilon(t+h) - v_\epsilon(t)|^r \wedge 1/F_t^\epsilon] \leq E[C_\epsilon(\eta)/F_t^\epsilon]$,
where $F_t^\epsilon := \sigma\{v_\epsilon(s); 0 \leq s \leq t\}$, $\quad 0 \leq h \leq \eta$, $\quad t \in [0, T]$.

We can state now the criteria of weak compactness for processes in B with the help of Theorem 1.2.8 and functionals from B^*. Let B^* be a dual space to B (see sec. 1.1) which separate points of B, and B_0^* be a dense set in B^*.

Theorem 1.2.9. (CCC) *A set of processes $V_\epsilon(t) \in D_B[0, +\infty)$, $\epsilon > 0$, is a weakly compact with the limit points in $C_B[0, +\infty)$ if and only if the following conditions hold:*
1) $\forall \Delta > 0$ and $\forall T > 0$ there exists a compact set $K_T^\Delta \subseteq B$:
$\lim_{\epsilon \to 0} \inf P\{V_\epsilon(t) \in K_T^\Delta; 0 \leq t \leq T\} \geq 1 - \Delta$;
2) $\forall l \in B_0^*$ the family $l(V_\epsilon(t))$ is weakly compact with the limit points in $C_R[0, +\infty)$.

The condition 1) in Theorems 1.2.6 and 1.2.9 is called the compact containment criterion (CCC). This criterion is both the most stringent requirement and the most difficult requirement to verify. But it is also the most important in the limit theorems for operator-valued random processes and random evolutions.

Theorem 1.2.9 requires the limit points to be continuous processes. Consider then necessary and sufficient conditions for a limiting process to have sample paths in $C_B[0, +\infty)$. For $V \in D_B[0, +\infty)$ define

$$J(V) := \int_0^{+\infty} e^{-u}[J(V, u) \wedge 1]du,$$

where $J(V, u) := \sup_{0 \leq t \leq u} ||V(t) - V(t-)||$. Since the map $V \to J(V, \cdot)$ from $D_B[0, +\infty)$ is continuous, it follows that J is continuous on $D_B[0, +\infty)$. For each $V \in D_B[0, +\infty)$, $J(V, \cdot)$ is nondecreasing, thus $J(V) = 0$ if and only if $V \in C_B[0, +\infty)$. The next two Theorems gives some results about the relationship between the limiting random processes in $D_B[0, +\infty)$, function $J(V)$, and CCC:

Theorem 1.2.10. *Let $V_\epsilon(t)$ and $V(t)$ be a random processes with sample paths in $D_B[0, +\infty)$, and suppose that $V_\epsilon(t) \Rightarrow V(t)$ when $\epsilon \to 0$. Then:*
1) $V(t)$ is a.s. continuous if and only if $J(V_\epsilon) \to 0$;
2) if $V(t)$ is a.s. continuous, then $f(V_\epsilon(t) \Rightarrow f(V(t))$, where f is a measurable continuous function on $D_B[0, +\infty)$.

Theorem 1.2.11. *Let $V_\epsilon(t)$ and $V(t)$ be a random processes with sample paths in $D_B[0, +\infty)$, and suppose that the CCC holds (see Theorem 1.2.9). Let M be a dense subset of $C_B[0, +\infty)$ in the topology of uniform convergence (w.r.t. the norm $||f|| := \sup_{x \in B} |f(x)|, f \in C_B[0, +\infty))$ on a compact set. Then $V_\epsilon(t)$ is weakly compact if and only if $f(V_\epsilon(t))$ is weakly compact $\forall f \in M$.*

We say that $M \subseteq B^*$ strongly separates points if for every $x \in S$ and $\delta > 0$, $\inf_{y \in S} \max_{1 \le i \le k} |h_i(x) - h_i(y)| > 0$, where $S := \{y : ||x - y|| \ge \delta\}$, $h_i(x) \in M$, $i = 1, 2, ..., k$.

The following result explains the relationship between weakly compactness and strongly separates points condition:

Theorem 1.2.12. *Let $V_\epsilon(t) \in D_B[0, +\infty)$ and M be a subspace of $C_B[0, +\infty)$ which strongly separates points. If the finite-dimensional distributions of $V_\epsilon(t)$ converges to those of the process $V(t)$ with sample paths in $D_B[0, +\infty)$, and if $f(V_\epsilon(t))$ is weakly compact in $D_B[0, +\infty)$, $\forall f \in M$, then $V_\epsilon(t) \Rightarrow V(t)$, when $\epsilon \to 0$.*

1.3 Semigroups of Operators and Their Generators

Let $(B, \mathcal{B}, || \cdot ||)$ be a real separable Banach space B with σ-algebra of Borel sets $\mathcal{B}$ and the norm $|| \cdot ||$.

A one-parameter family $(\Gamma(t))_{t \in R^+}$ of bounded linear operators on B is called a *semigroup of operators*, if

(i)
$$\Gamma(0) = I,$$

is the identity operator;

(ii)
$$\Gamma(t + s) = \Gamma(t) \cdot \Gamma(s),$$

for all $s, t \ge 0$.

The semigroup $(\Gamma(t))_{t \in R^+}$, is said to be a *contraction* semigroup of $|| \Gamma(t) || \le 1, \forall t \ge 0$; and *strongly continuous* semigroup if

$$\lim_{t \to 0} || (\Gamma(t) - I)f || = 0, \forall f \in B.$$

The *generator (or infinitesimal operator)* of a semigroup $(\Gamma(t))_{t \in R^+}$ is the linear operator A defined by

$$Af := \lim_{t \to 0} t^{-1}[(\Gamma(t) - I)f],$$

with the *domain* $D(A)$ given by

$$D(A) := \{f \in B : \lim_{t \to 0} t^{-1}[(\Gamma(t) - I)f] \quad \text{exists}\}.$$

It is known that the generator A of a strongly continuous semigroup $(\Gamma(t))_{t \in R^+}$ on B is a densely defined closed operator, i.e., $\overline{D(A)} = B$. Moreover, the set $\bigcap_{n=1}^{+\infty} D(A^n)$ is dense in B.

Also, for a strongly continuous semigroup $(\Gamma(t))_{t \in R^+}$ on $\mathbf{B}$ with the generator A, we have

$$\begin{cases} \frac{d\Gamma(t)}{dt} = \Gamma(t) \cdot A = A \cdot \Gamma(t), \\ \Gamma(0) = I. \end{cases}$$

We now give a few examples of semigroups of operators and generators.

(E1). *Exponential semigroup.* Let A be a bounded linear operator on B. Define the following one-parameter family

$$\Gamma(t) = e^{tA} := \sum_{k=0}^{+\infty} \frac{t^k}{k!} A^k, t \geq 0.$$

It is easy to verify that $(\Gamma(t))_{t \in R^+}$ defined above is a strongly continuous semigroup with generator A. Here $D(A) = B$.

(E2). *Uniform motion on the real line.* Let $B = C(R)$ be a Banach space of bounded continuous functions on R equipped with sup-norm, and let $\Gamma(t)f(x) := f(x + vt)$, where $v > 0$ is a constant velocity, $f \epsilon C(R)$. $(\Gamma(t))_{t \in R^+}$ is a semigroup with $Af(x) = v \cdot f'(x)$ and $D(A) = C^1(R)$ is the space of differentiable functions on R with continuous derivatives.

(E3). *Motion with velocity depending on the state.* Let $q(t, z)$ solves the Cauchy problem:

$$\frac{dq(t, z)}{dt} = v(q(t, z)), q(0, z) = z.$$

Then

$$\Gamma(t)f(z) := f(q(t, z)), f \epsilon C(R),$$

gives a strongly continuous contraction semigroup and

$$\Gamma f(z) = v(z)f'(z), \quad \forall f \epsilon C'(R).$$

The semigroup property follows from the equality

$$q(t + s, z) = q(s, q(t, z)), \forall z \ \epsilon, \forall s, t \ \epsilon R^+.$$

(E4). *Continuous-time Markov chain and its infinitesimal matrix.* Here, $\Gamma(t) = \mathbf{P}(t)$ is the infinite dimentional matrix with

$$\mathbf{P}(t) = (p_{ij}(t); i, j = 1, 2, \ldots), t \geq 0,$$

and

$$A = Q = (q_{ij}; i, j = 1, 2, \ldots).$$

(E5). *Bellman-Harris branching process.* Let $\xi(t)$ be a Bellman-Harris process (a homogeneous Markov branching process) with generating function

$$\Phi(t, u) := Eu^{\xi(t)}, |u| \leq 1,$$

and $b(a) := a[p(u) - u]$, where $a > 0$ is the intensity of the exponential distribution of the lifetimes of particles of $\xi(t)$ and

$$p(u) := \sum_{k=0}^{\infty} p_k u^k$$

is the generating function of the number of direct descendants of one particle. Then

$$\begin{cases} \frac{d\Phi(t,u)}{dt} = b(\Phi(t, u)) \\ \Phi(0, u) = u. \end{cases}$$

Define $\Gamma(t)f(u) := f(\Phi(t, u)), f \epsilon C(R) = B$. Then we obtain a semigroup $(\Gamma(t))_{t \in R_+}$ with the generator given by

$$\Gamma f(u) = b(u)\frac{df(u)}{du}, \quad f \in C^1(R) = D(\Gamma).$$

(E6). *Diffusion processes.* Let $y(t)$ be a diffusion process with drift $a(t, y) \equiv a(y)$ and diffusion $\sigma(t, y) \equiv \sigma(y)$. As these are independent of t, we obtain the so-called *homogeneous diffusion process* with transition probabilities $P(t, y, A), t \in R^+, y \in R, A \in \mathcal{R}$. The associated semigroup and its generator are

$$\Gamma(t)f(z) := \int_Y f(y)P(t, z, dy), f(y)\epsilon C(R),$$

and

$$Af(z) = a(y)\frac{df(z)}{dz} + 1/2\ \sigma^2(y)\frac{d^2 f(z)}{dz^2}, f\epsilon C^2(R).$$

In the special case of a Wiener process, we will have

$$P(t, y, A) = \frac{1}{\sqrt{2\pi t}} \int_A \exp^{-\frac{(z-y)^2}{2t}} dz$$

and

$$Af(z) = 1/2\frac{d^2 f(z)}{dz^2}.$$

(E7). *Jump Markov process* For a regular homogeneous jump Markov process, the semigroup defined by

$$\Gamma(t)f(y) = \int_Y P(t, y, dz)f(z) = \int_Y P(y, dz)f(z) \cdot (1 - e^{-\lambda(y)t}),$$

where $f \in C'(R)$, is a strongly continuous contraction semigroup, with the generator

$$Af(y) = \lambda(y) \cdot \int_Y [P(y, dz)f(z) - f(y)], \forall f(y) \epsilon C(R) := D(A).$$

(E8). *Semi-Markov process.* Let $y(t) := y_{\nu(t)}$ be the semi-Markov process. Each of the auxiliary processes $\theta(t), \gamma(t)$ and $\gamma^+(t)$ compliments $y(t)$ to a Markov process. In particular, for $\gamma(t) := t - \tau_{\nu(t)}$, $(y(t), \gamma(t))$ is a Markov process on $Y \times R^+$ with the generator

$$Af(y, t) = \frac{df(y, t)}{dt} + \frac{g_y(t)}{\bar{G}_y(t)}[Pf(y, 0) - f(y, t)],$$

where

$$g_y(t) := \frac{dG_y(t)}{dt}, \bar{G}_y(t) := 1 - G_y(t), \qquad f \in C(Y) \times C^1(R^+).$$

We should mention that a semi-Markov process $y(t)$ does not generate a semigroup due to the arbitrary distribution function for the sojourn times, rather than the exponential one in the case of a Markov process.

Bibliography

[1] Adams, R. (1979). *Sobolev Spaces.* Academic Press, NY.

[2] Billingsley, P. (1999). *Convergence of Probability Measures.* Wiley.

[3] Doob, J. (1953) *Stochastic Processes.* Wiley.

[4] Eithier, R. and Kurtz, T. (1986). *Markov Processes: Characterization and Convergence.* Wiley.

[5] Hille, E. and Fillips, R. (1957). *Functional Analysis and Semigroups.* American Mathematical Society.

[6] Korolyuk, V. and Swishchuk, A. (1995). *Evolution of Systems in Random Media.* CRC Press, NY.

[7] Lindvall, T. (1973). Weak convergence of probability measures and random functional space $D[0, +\infty]$. *J. Appl. Prob.*, 10, 109-121.

[8] Liptser, R. and Shiryaev, A. (1987). *Theory of Martingales.* Nauka, Moscow.

[9] Skorokhod, A. (1956). Limit theorems for random processes. *Theor. Prob. Appl.*, 1(3), 289-319.

[10] Skorokhod, A. (1978). *Linear Random Operators*, Naukova Dumka, Kiev.

[11] Varadarajan, V. (1958). Weak convergence of measures on separable metric spaces. *Sankhya*, 19, 15-22.

2

Stochastic Calculus in Separable Banach Spaces

This chapter is devoted to the development of stochastic calculus in separable Banach spaces, including construction of integrals over martingale measures (see Section 2.1), such as Itô, Stratonovich and Skorokhod integrals. We also study multiplicative operator functionals (MOF) in Banach spaces (see 2.2), which are a generalization of random evolutions (RE) [8]. One of the results includes Dynkin's formula for MOF (see 2.2.5). Boundary values problems for MOF in Banach spaces are investigated as well . Applications are given to the evolutionary stochastic systems (see 2.2.6).

We introduce first three classes of stochastic integrals over martingale measures in a way similar to that of Itô [3], Stratonovich [6] and Skorokhod [5] integrals, and study some their properties in 2.1.2. We also study some stochastic evolutionary operator equations driven by a space–time white noise. Examples of those equations arise from the limiting RE in diffusion approximation (see Chapter 4). We can obtain them from the solution of the martingale problem over the martingale measure. It is a way to investigate the evolutionary operator equations driven by Wiener martingale measure, introduced and studied in 2.1.1.

We also investigate the characteristic operator and equations for resolvent and potential for the multiplicative operator functionals (MOF) of Markov processes in 2.2. In 2.2.1 we consider the definitions of MOF of Markov processes and equations for expectations. Section 2.2.2 deals with properties of infinitesimal and characteristic operators of MOFs and some equations for them. In Section 2.2.3 we find the resolvent and potential for MOF of Markov processes. Equations for resolvent and potential of MOF of Markov processes are derived in Section 2.2.4. An analogue of Dynkin's formula for MOF of Markov processes are considered in Section 2.2.5. Applications of these formula are given to traffic, storage and diffusion processes in random media are discussed (Section 2.2.6).

2.1   Stochastic Calculus for Integrals over Martingale Measures

2.1.1   The Existence of Wiener Measure and Related Stochastic Equations

Let $(\Omega, \mathcal{F}, \mathsf{P})$ be a probability space and let $(X, \mathcal{X}, \rho)$ be a measurable space with probability measure ρ and countable generated σ-algebra $\mathcal{X}$. Let also $W(t, A)$ be a Wiener martingale measure with quadratic variation $t \cdot \rho(A)$, $A \in \mathcal{X}, t \in R_+$.

Theorem (The existence of Wiener measure $W(t,A)$). *Let* $\xi_0(\omega, A)$, $\xi_1(\omega, A), \ldots, \xi_n(\omega, A), \ldots$ *be a sequence of independent Gaussian random measures such that*

$$\forall\, A \in \mathcal{X}\ \mathsf{E}\xi_k(\omega, A) = 0, \quad \mathsf{E}\xi_k^2(\omega, A) = \rho(A), \qquad k = 0, 1, 2, \ldots. \tag{2.1}$$

Then for some sequence $N(k) \to \infty$, $k \to \infty$, *the measures*

$$W^k(t, A) := \frac{1}{\sqrt{\pi}} t \xi_0(\omega, A) + \sqrt{\frac{2}{\pi}} \sum_{n=1}^{N(k)} \xi_i(\omega, A) \frac{1}{n} \sin nt \tag{2.2}$$

converge uniformly on $t \in [0, \pi]$ *and their continuous limit is a Wiener measure* $W(t, A)$ *on* $[0, \pi]$, $\forall\, A \in \mathcal{X}$.

Proof. *For the Parseval equality for the expansions in Fourier series on* $[-\pi, \pi]$ *of functions* $I(|x| < t)$ *and* $I(|x| < s)$ *we obtain that if* $t, s \in [0, \pi]$ *then*

$$\begin{aligned}
|t - s| &= \frac{1}{2} \int_{-\pi}^{\pi} [I(|x| < t) - I(|x| < s)]^2 \, dx \\
&= \frac{1}{\pi}(t - s)^2 + \frac{2}{\pi} \sum_{n=1}^{\infty} \frac{1}{n^2} (\sin nt - \cos ns)^2.
\end{aligned} \tag{2.3}$$

It is known [[10], p. 279] that for given $\alpha < 1/p$, $p \in [1, +\infty)$, *there is a constant* c_0 *such that for any measurable on* $[0, \pi]$ *function* $f(t)$ *for almost all* $t, s \in [0, \pi]$ *we have the inequality:*

$$|f(t) - f(s)| \leq c_0 \cdot |t - s|^{\alpha - \frac{1}{p}} \cdot \left(\int_0^{\pi} \int_0^{\pi} \frac{|f(x) - f(y)|^p}{|x - y|^{1 + \alpha p}} \, dx \, dy \right)^{1/p}. \tag{2.4}$$

Set $\delta^k(t, A) := W^{k+1}(t, A) - W^k(t, A)$. Then from (2.4) as $p = 4$ and $s = 0$ we obtain:

$$\mathsf{E}\sup_{t \in [0, \pi]} |\delta^k(t, A)|^4$$
$$\leq c_0^4 \cdot t^{4\alpha - 1} \cdot \int_0^\pi \int_0^\pi \frac{\mathsf{E}|\delta^k(x, A) - \delta_y^k(A)|^4}{|x - y|^{1+4\alpha}} \, dx \, dy$$
$$\leq c_0^4 \pi^{4\alpha - 1} \int_0^\pi \int_0^\pi$$
$$\frac{\mathsf{E}|\frac{1}{\sqrt{\pi}}(x - y)\xi_0(A) + \sqrt{\frac{2}{\pi}} \sum_{n=N(k)+1}^{N(k+1)} \frac{1}{n}\xi_n(A)(\sin nx - \cos ny)|^4}{|x - y|^{1+4\alpha}} \, dx \, dy.$$

$$(2.5)$$

It follows from (2.1)–(2.3) that

$$\mathsf{E}|\delta^k(x, A) - \delta^k(y, A)|^4 \leq \rho^2(A)\left[\left(\frac{1}{\pi}(x - y)^2\right.\right.$$
$$\left.\left. + \frac{2}{\pi} \sum_{N(k)+1}^{N(k+1)} \frac{1}{n^2}(\sin nx - \cos ny)\right)^2\right]^2$$
$$\leq \rho^2(A)(x - y)^2.$$

$$(2.6)$$

We note that

$$\int_0^\pi \int_0^\pi \frac{|\frac{1}{\pi}(x - y)^2 + \frac{2}{\pi} \sum_1^\infty \frac{1}{n^2}(\sin nx - \cos ny)^2|^2}{(x - y)^{1+4\alpha}} \, dx \, dy$$
$$\leq \int_0^\pi \int_0^\pi \frac{1}{(x - y)^{4\alpha - 1}} \, dx \, dy \quad < +\infty, \quad \forall \, \alpha \in \left(\frac{1}{4}, \frac{1}{2}\right) \quad (\text{see (2.3)}),$$

$$(2.7)$$

and

$$b(n) := \int_0^\pi \int_0^\pi \frac{(\frac{1}{\pi}(x - y)^2 + \frac{2}{\pi} \sum_{k=n}^\infty \frac{1}{k^2}(\sin kx - \cos ky)^2)^2}{(x - y)^{4\alpha + 1}} \, dx \, dy \to 0,$$

$$(2.8)$$

$n \to \infty$, then we can define $N(k)$ such that $b(N(k)) \leq 2$.
From (2.5)-(2.8) we obtain:

$$\mathsf{E}\sup_{t \in [0, \pi]} |\delta^k(t, A)|^4 \leq c_0^4 \cdot \pi^{4\alpha - 1} \rho^2(A) b(N(k)) \leq c_0^4 \pi^{4\alpha - 1} \cdot 2^{-k}, \qquad (2.9)$$

since $\rho(A) \leq 1$, $\forall \, A \in \mathcal{X}$.
From (2.9) and Hölder inequality it follows that the measure

$$\sum_{k=1}^\infty \sup_{t \in [0, \pi]} |\delta^k(t, A)|$$

has a finite mean and hence it is a finite, $\forall \, A \in \mathcal{X}$. Hence, the sequence (2.2) converges uniformly on t to some continuous measure that we indicate by $W(t, A)$. Obviously, that $W(t, A)$ is a Gaussian process, $\mathsf{E}W(t, A) = 0$, $\forall \, A \in \mathcal{X}$. Finally, from (2.2)–(2.3) we obtain $\mathsf{E}W^2(t, A) = t \cdot \rho(A)$.

Let's consider connection $W(t, A)$ with martingale measures [16].

Let $N(t, A)$ be an $\mathcal{F}_t$-measurable martingale measure, $\forall A \in \mathcal{X}$, $t \in R_+$ such that:

(i) $EN^2(t, A) < +\infty$, $\forall A \in \mathcal{X}$, $\forall t \in R_+$;

(ii) $N(t, A)$ is a continuous function on t, $\forall A \in \mathcal{X}$;

(iii) there exists a measurable nonnegative function $f(t, x, \omega)$ with respect to the measure $dt\rho(dx)\,d\mathsf{P}$ such that $f(t, x, \omega)$ is a measurable on ω with respect to $\mathcal{F}_t$ under fixed t and A, and such that

$$\langle N(t, A)\rangle = \int_0^t \int_A f^2(s, x)\, ds\rho(dx). \qquad (2.10)$$

Then if the function $f(t, x, \omega)$ is not equal to zero a.s. on (t, x, ω), then there exists a Wiener martingale measure $W(t, A)$ such that $\forall t \in R_+$ with probability 1 and $\forall A \in \mathcal{X}$:

$$N(t, A) = \int_0^t \int_A f(s, x, \omega) W(ds, dx). \qquad (2.11)$$

Let's define the process:

$$W(t, A) := \int_0^t \int_A \frac{N(ds, dx)}{f(s, x, \omega)}, \qquad (2.12)$$

(if $f \equiv 0$ then we set $1/f = 0$). The process $W(t, A)$ is an $\mathcal{F}_t$-martingale since $N(t, A)$ is an $\mathcal{F}_t$-martingale and

$$\langle W(t, A)\rangle = \int_0^t \int_A \frac{\langle N(ds, dx)\rangle}{f^2(s, x, \omega)} = \int_0^t \int_A \frac{f^2(s, x, \omega)\, ds\rho(dx)}{f^2(s, x, \omega)} = t \cdot \rho(A)$$

Measure $W(t, A)$ is a continuous on t, $\forall A \in \mathcal{X}$, and $W(t, A)$ is a Wiener martingale measure, thus we obtain:

$$N(t, A) = \int_0^t \int_A f(s, x, \omega) W(ds, dx).$$

Let us consider the analogue of Girsanov's theorem for a Wiener measure $W(t, A)$ [3].

Proposition (Analogue of Girsanov's Theorem for Wiener Measure).
Let $W(t, A)$ be a Wiener martingale measure on the space $(\Omega, \mathcal{F}, \mathcal{F}_t, \mathsf{P})$ and let $f(t, x, \omega) : R_+ \times X \times \Omega \mapsto R$ be an anticipating process such that:

$$\int_0^T \int_X f^2(t, x)\, dt\rho(dx) < +\infty \quad a.s.$$

Set

$$\eta_t^A(f) := \exp\left\{\int_0^t \int_A f(s, x) W(ds, dx) - \frac{1}{2} \int_0^t \int_A f^2(s, x)\, ds\rho(dx)\right\},$$

and suppose that

$$\mathsf{E}\eta_T^X(f) = 1. \tag{2.13}$$

If Q *is a probability measure on* $(\Omega, \mathcal{F})$ *such that*

$$\frac{d\mathsf{Q}}{d\mathsf{P}} = \eta_T^X(f),$$

then

$$\tilde{W}(t, A) := W(t, A) - \int_0^t \int_A f(s, x)\, ds\rho(dx) \tag{2.14}$$

is a Wiener martingale measure on $(\Omega, \mathcal{F}, \mathcal{F}_t, \mathsf{Q})$.

Proof. *Let* $N(t, \cdot) = W(t, \cdot)$ *and* $X(t, \cdot) = \int_0^t \int_A f(s, x) W(ds, dx)$. *Then we obtain*

$$\tilde{N}(t, A) := N(t, A) - \langle N, X \rangle$$

$$= W(t, A) - \int_0^t \int_A f(s, x)\, ds\rho(dx)$$

$$= \widetilde{W}(t, A)$$

is a continuous martingale measure with respect to Q *and*

$$\langle \widetilde{W}, \widetilde{W} \rangle^\mathsf{Q} = \langle W, W \rangle^\mathsf{P} = t \cdot \rho(A).$$

Hence, $\widetilde{W}(t, A)$ *is a continuous Wiener martingale measure.*

Remark. *It follows from (2.14) that* $W(t, A)$ *is a solution of the integral stochastic equation*

$$W(t, A) = \int_0^t \int_A f(s, x)\, ds\rho(dx) + \widetilde{W}(t, A).$$

Remark. *Since* $W(t) := W(t, X)$ *is a Wiener process, then under* $f(s, x) \equiv f(s), \forall x \in X$, *and* $A \equiv X$, *we have that the above Theorem is an analogue of Girsanov theorem.*

Remark. *The condition (2.13) is fulfilled under analogue of Novikov's condition, namely, the following result holds.*

Theorem (Analogue of Novikov's Condition). *Let* $W(t, A)$ *be a Wiener martingale measure and let*

$$f(s, x, \omega)\colon R_+ \times X \times \Omega \mapsto R$$

be a function such that

$$\int_0^T \int_X f^2(s,x)\,ds\rho(dx) < +\infty \quad a.s.$$

Then, if

$$\mathsf{E}\left[\exp\left(\frac{1}{2}\int_0^T \int_X f^2(s,x)\,ds\rho(dx)\right)\right] < +\infty$$

the $\mathcal{F}_t$-semimartingale $\eta_t^A(f)$ in Analogue of Girsanov's Theorem is then a martingale $\forall\, A \in \mathcal{X}$ and $\mathsf{E}\eta_t^A(f) = 1$, $\forall\, t \in [0,T]$, $\forall\, A \in \mathcal{X}$.

2.1.2   Stochastic Integrals over Martingale Measures

2.1.2.1   Orthogonal Martingale Measures

Let $(\Omega, \mathcal{F}, \mathcal{F}_t, \mathsf{P})$ be a probability space, $t \in R_+$, and let $(X, \mathcal{X}, \rho)$ be a measurable space with probability measure ρ. Let's remember a definition of orthogonal martingale measure.

A family of (locally) square integrated martingales $N(A,t)$ indexed by $A \in \mathcal{X}$, σ-algebra on X, and adapted to the filtration $\mathcal{F}_t$, is an orthogonal (local) martingale measure if the following conditions are satisfied for all A, A_1, A_2 in $\mathcal{X}$ and all $t \in R_+$:

(i) additivity: $N(A_1,t) + N(A_2,t) = N(A_1 \cup A_2, t)$ a. s. for $A_1 \cap A_2 = \varnothing$;

(ii) orthogonality: $N(A_1,t) \cdot N(A_2,t)$ is a (local) martingale for $A_1 \cap A_2 = \varnothing$;

(iii) $\langle N(A,t), N(A,t) \rangle = \pi(A,t)$, where $\pi(A,t)$ is a random function which for fixed t is a measure on $\mathcal{X}$ with probability one, and for fixed A, is a continuous monotonically increasing function of t.

Remark. *$\pi(A,t)$ is called the quadratic variation of the martingale measure $N(A,t)$. If $\pi(A,0) = 0$, $\forall\, A \in \mathcal{X}$, then $\pi(A,t)$ is unique.*

Remark. *For $A_1, A_2 \in \mathcal{X}$*

$$\langle N(A_1,t), N(A_2,t) \rangle = \pi(A_1 \cap A_2, t).$$

Remark. *If $A_1 \cap A_2 = \varnothing$, then*

$$\pi(A_1,t) + \pi(A_2,t) = \pi(A_1 \cup A_2, t),$$

i. e., π is additive.

If $\mathcal{X}$ has atoms $\{A_1, A_2, \ldots, A_n\}$, then

$$\{N(A_1,t), N(A_2,t), \ldots, N(A_n,t)\}$$

form a finite family of orthogonal martingales.

Let X be Gaussian white noise measure on the positive quadrant in R^2. Define $N(A,t) := X(A \times [0,t])$, where A is a Borel set in $[0,T]$. If $\{\mathcal{F}_s^t; 0 \leq s \leq T; t \in R_+\}$ is the filtration for X, then N is a martingale measure with respect to the filtration $\mathcal{F}_t := \sigma\{\mathcal{F}_s^t; 0 \leq s \leq T\}$. The quadratic variation process π is: $\langle \pi(A,t) \rangle = m(A) \cdot t$, where m is Lebesque measure.

Let $\pi(A,t) := \rho(A) \cdot t$, where ρ is a probability measure on $\mathcal{X}$. Upon appealing to Levy's characterization of Brownian motion the associated martingale measure $N(A,t)$ having continuous sample paths is Brownian motion with variance $\rho(A)$. We shall denote this martingale measure by $W_\rho(A,t)$.

Space-time white noise. Consider the generalized Gaussian zero mean random field $\{W(B); B \in \mathcal{B}(R_+ \times D), D \subset R^n\}$ ($\mathcal{B}(R_+ \times D)$ denotes the Borel field of subsets of $R_+ \times D$), defined on a probability space $(\Omega, \mathcal{F}, \mathsf{P})$, whose covariance function is given by

$$\mathsf{E}[W(B)W(C)] = m(B \cap C),$$

where m denote Lebesque measure on $R_+ \times D$.

In the case $n = 1$, $D = [0,1]$, for instance, the continuous random field $\{W_{t,x} := W([0,t] \times [0,x]); (t,x) \in \mathbb{R}_+ \times [0,1]\}$ is the so-called Brownian sheet.

2.1.2.2 Ito's Integrals over Martingale Measures

The construction of this integral follows three major stages of development.

Let N be a continuous square integrated martingale measure with quadratic variation π, and let $\mathcal{Z}_1$ denote the class of functions of the form $v \times I_{A \times \Delta}$, where $A \in \mathcal{X}$, $\Delta = (s,t]$, and v is a bounded $\mathcal{F}_s$-measurable random variable, where $I_{A \times \Delta}$ is an indicator function for the set $A \times \Delta$.

Set

$$\iint v \cdot I_{A \times \Delta} N(dx, ds) := v \times (N(A,t) - N(A,s)), \qquad s \leq t, \quad A \in \mathcal{X}.$$

The simple functions $\mathcal{Z}_2$ are finite linear combinations of elements $\mathcal{Z}_1$: $\forall V \in \mathcal{Z}_2$:

$$V = \sum_{k=1}^{n} \sum_{j=1}^{m} v_{jk} I_{A_{jk} \times \Delta_k},$$

where $0 \leq t_1 < t_2 < \ldots < t_n$, $\Delta_k = (t_{k-1}, t_k]$, $A_{jk} \in \mathcal{X}$, v_{jk} is a bounded $\mathcal{F}_{t_{k-1}}$-measurable random variable. Let us introduce the notation:

$$\int_A \int_s^t V \, dN := \int_A \int_0^t V(x,s) N(dx, ds)$$

defines a continuous square integrated martingale measure. In addition:

1) linearity: $\forall\, V_1, V_2 \in \mathcal{Z}_2,\; c_1, c_2 \in R$:

$$\iint (c_1 V_1 + c_2 V_2)\, dN = c_1 \iint V_1\, dN + c_2 \iint V_2\, dN;$$

2) $\mathsf{E}[(\iint V_1\, dN)(\iint V_2\, dN)] = \mathsf{E}[\iint V_1 \cdot V_2\, d\pi];$

3) $\mathsf{E}[(\iint V\, dN)^2] = \mathsf{E}[\iint V^2\, d\pi]$ (L^2-isometry);

4) $\mathsf{E}[(\iint V_1\, dN - \iint V_2\, dN)^2] = \mathsf{E}[\iint (V_1 - V_2)^2\, d\pi];$

5) $\forall\, A_1, A_2 \in \mathcal{X}$, let

$$M_i(A_i, t) := \int_A \int_0^t V_i\, dN, \qquad i = 1, 2,$$

      then

$$\langle M_1(A_1, t), M_2(A_2, t) \rangle = \int_{A_1 \cap A_2} \int_0^t V_1 \cdot V_2\, d\pi;$$

6) for an $\mathcal{F}_t$-stopping time τ,

$$\int_A \int_0^{t \wedge \tau} V\, dN = \int_A \int_0^t V\, dN^\tau,$$

where $N^\tau(A, t) := N(A, t \wedge \tau)$.

Let $\mathcal{Z}_3 := \{\, V = (V(x, t)\colon t \geq 0)\colon V$ is progressively measurable and

$$\mathsf{E}[\int_X \int_0^t V^2(x, s)\pi(dx, ds)] < +\infty$$

for all $t \geq 0\,\}$. The proposition above implies that $\mathcal{Z}_2 \subset \mathcal{Z}_3$. Since $\mathcal{Z}_3$ is a closed subspace then $\overline{\mathcal{Z}_2} \subset \mathcal{Z}_3$, where $\overline{\mathcal{Z}_2}$ denotes the completion of $\mathcal{Z}_2$ in the isometry stated in 3), but we would like $\overline{\mathcal{Z}_2} = \mathcal{Z}_3$. It follows from the following

Lemma. *Let $V \in \mathcal{Z}_3$, then there exist a sequence $\{V_n\} \subset \mathcal{Z}_2$ such that for all $t \geq 0$:*

$$\lim_{n \to \infty} \mathsf{E}[\iint (V_n(x, s) - V(x, s))^2 \pi(dx, ds)] = 0.$$

Let $V \in \mathcal{Z}_3$, and choose a sequence $\{V_n\} \subset \mathcal{Z}_2$, then by equation 4),

Proposition. $\int_X \int_0^t V_n(x,s) N(dx,ds)$ *is a Cauchy sequence in* $L^2(\mathsf{P})$ *for each* t.

Thus we specify the limiting process, denoted

$$\int_X \int_0^t V(x,s) N(dx,ds), \qquad \forall\, t > 0,$$

as an equivalence class in $L^2(\mathsf{P})$.

In the future, the term stochastic integral will mean a continuous version of this process, since $\int_A \int_0^t V\,dN$ has a continuous version.

Further, if N is a local martingale measure, then we are able to integrate over N the processes

$$\mathcal{Z}_4 := \Big\{ V(x,s) \colon V \text{ is progressively measurable and}$$

$$\int_X \int_0^t V^2(x,s) \pi(dx,ds) < +\infty \Big\},$$

a. s. $\forall\, t > 0$ *using the standard arguments.*

In the summary we have:

Proposition. *Let* N *be a local martingale measure with quadratic variation* π, *and* $V \in \mathcal{Z}_4$. *Then*

1) $M(A,t) := \int_A \int_0^t V(x,s) N(dx,ds)$ *is a continuous locally square integrated martingale measure with continuous quadratic variation*

$$\langle M(A,t), M(A,t) \rangle = \int_A \int_0^t V^2(x,s) \pi(dx,ds);$$

2) *if* $V_1, V_2 \in \mathcal{Z}_4$, $c_1, c_2 \in R$, *then*

$$\iint (c_1 V_1 + c_2 V_2)\,dN = c_1 \iint V_1\,dN + c_2 \iint V_2\,dN;$$

3) $\int_A \int_0^{t \wedge \tau} V_1\,dN = \int_A \int_0^t V_1\,dN^\tau$, *where*

$$N^\tau := N(A, t \wedge \tau)$$

and τ *is an* $\mathcal{F}_t$-*stopping time.*

Let N_1 and N_2 be two local martingale measures. Then the process

$$\langle N_1, N_2 \rangle := \frac{1}{2}\{ \langle N_1 + N_2, N_1 + N_2 \rangle - \langle N_1, N_1 \rangle - \langle N_2, N_2 \rangle \}$$

is called the covariation of N_1 *and* N_2.

This notation is owed to $N_1 \cdot N_2 - \langle N_1, N_2 \rangle$ being a martingale.

Proposition. *Let N_1 and N_2 be a local martingale measure with quadratic variation π_1 and π_2 respectively. Let $V_i \in \mathcal{Z}_4$ with π_i, $i = 1, 2$. Then*

$$\left(\int_A \int_0^t |V_1 \cdot V_2| \langle N_1, N_2 \rangle \right)^2 \leq \left(\int_A \int_0^t V_1^2 \, d\pi_1 \right) \left(\int_A \int_0^t V_2^2 \, d\pi_2 \right)$$

and

$$\left\langle \int_A \int_0^t V_1 \, dN_1, \int_A \int_0^t V_2 \, dN_2 \right\rangle = \int_A \int_0^t V_1 \cdot V_2 \langle N_1, N_2 \rangle.$$

Proposition. *Let $\pi(dx, ds)$ be a positive measure on $X \times R_+$ and denote by $\pi(A, t) := \int_A \int_0^t \pi(dx, ds)$. Then there exists a process W, unique in distribution, such that for each $A \in \mathcal{X}$:*

1) *$W(\cdot, t)$ has sample paths in $C_{\mathbb{R}}[0, +\infty)$;*

2) *$W(\cdot, t)$ is a martingale;*

3) *$W^2(A, t) - \pi(A, t)$ is a martingale $\forall A \in \mathcal{X}$;*

4) *$W(A, t)$ is a martingale measure.*

A white noise based on π satisfies 1)–4) and therefore such a process W exists.

Remark. *Let $(B, \mathcal{B}, \| \cdot \|)$ be a separable Banach space, and $a(x, s, f) \colon X \times R_+ \times B \mapsto B$ be a measurable bounded function. Then we can define the stochastic integral $\int_A \int_0^t a(x, s, V(s)) N(dx, ds)$ as integral in a weak sense:*

$$\int_A \int_0^t l(a(x, s, V(s))) N(dx, ds)$$

for all $l \in B^*$, where B^* is a dual space to B. Since $l(a(x, s, V(s)))$ is a measurable random variable then we can use the mentioned above definition of stochastic integral, $V(s) \in B$, $\forall s \in R_+$.

Remark. *One can define an Ito's integral with respect to the space-time white noise as follows.*

Let $\mathcal{F}_t := \sigma\{W(B); B \in \mathcal{B}([0, t] \times D); D \subset R^n\}$, and ζ denotes the σ-field of $\mathcal{F}_t$-progressively measurable subsets of $\Omega \times R_+$. If $\varphi \in L^2(\Omega \times R_+ \times D, \zeta \oplus \mathcal{B}(D), P(dw) \, dt \, dx)$, then one can define the process

$$\int_D \int_0^t \varphi(x, s) W(dx, ds)$$

as a continuous martingale whose associated increasing process is given by

$$\int_D \int_0^t \varphi^2(x, s) \, dx \, ds, \qquad t \in R_+, \quad n = 1.$$

It's considered as a particular case of an integral with respect to a martingale measure.

Remark. *We may also define an Ito's integral with respect to the colored noise [16].*

2.1.2.3 Symmetric (Stratonovich) Integral over Martingale Measure

In previous section we have developed the stochastic integral similar to the Ito's integral. In a similar fashion, we may begin with a symmetric approximating sum and follow the Stratonovich development of the integral.

Let N be an orthogonal martingale measure on X. If V is a progressively measurable function and $V(x, \cdot)$ is a fixed function on the disjoint sets $A_1, \ldots, A_m$, then we may define the following integral

$$
\begin{aligned}
&\int_0^t V(x_j, s) \circ N(A_j, ds) \\
&:= \lim_{|\Delta| \to 0} \sum_{k=0}^{n-1} \frac{1}{2} (V(x_j, t \wedge t_{k+1}) + V(x_j, t \wedge t_k)) \\
&(N(A_j, t \wedge t_{k+1}) - N(A_j, t \wedge t_k)),
\end{aligned}
\tag{2.15}
$$

where $0 \le t_0 < t_1 < t_2 < \ldots < t_n = t$, $\Delta := \max_k (t_{k+1} - t_k)$, $A_i \cap A_j = \varnothing$, $x_j \in A_j$, $i, j = \overline{1, m}$.

$$
\int_X \int_0^t V(x, s) \circ N(dx, ds) := \sum_{j=1}^m \int_0^t V(x_j, s) \circ N(A_j, ds),
\tag{2.16}
$$

where integral in the righthand side of (2.16) is defined in (2.15), $x_j \in A_j$, $j = \overline{1, m}$.

In this way, the raised small circle on the righthand side of (2.16) denotes the usual Stratonovich integral with respect to martingales.

Further, we can use this definition as a basis for an approximation scheme for more general V. We can pass to $V(x, \cdot)$ is a fixed function on the disjoint sets $A_{1,k}, \ldots, A_{n_k, k}$ for $t_k \le s \le t_{k+1}$, and then on to more general V. And finally we have.

Proposition. *Let N be a local martingale measure with quadratic variation π and $V \in \mathcal{Z}_4$. Then:*

1) *If $V_1, V_2 \in \mathcal{Z}_4$ and $c_1, c_2 \in R$, then*

$$
\iint (c_1 V_1 + c_2 V_2) \, dN = c_1 \iint V_1 \circ dN + c_2 \iint V_2 \circ dN;
$$

2) *$\int_A \int_0^{t \wedge \tau} V \circ dN = \int_A \int_0^t V \circ dN^\tau$, where $N^\tau := N(A, t \wedge \tau)$ and τ is an $\mathcal{F}_t$-stopping time.*

Let us give a formula which shows the relationship of the two types of integral.

Proposition. *Let $V \in \mathcal{Z}_4$ and N be a local martingale measure with quadratic variation π. Then*

$$
\begin{aligned}
&\int_X \int_0^t V(x,s) \circ N(dx, ds) \\
&= \int_X \int_0^t V(x,s) N(dx, ds) + \frac{1}{2} \int_X \{V(x,t), N(dx,t)\},
\end{aligned}
\tag{2.17}
$$

where

$$
\{V, N\} := \lim_{|\Delta| \to 0} \sum_{k=0}^{n-1} (V(x, t \wedge t_{k+1}) - V(x, t \wedge t_k))(N(A, t \wedge t_{k+1}) - N(A, t \wedge t_k)).
$$

Proof. *Let's consider the integral*

$$
\int_0^t V(x_j, s) \circ N(A_j, ds).
\tag{2.18}
$$

Observe the relation:

$$
\begin{aligned}
&\sum_{k=0}^{n-1} \frac{1}{2}(V(x_j, t \wedge t_{k+1}) + V(x_j, t \wedge t_k))(N(A_j, t \wedge t_{k+1}) - N(A_j, t \wedge t_k)) \\
&= \sum_{k=0}^{n-1} V(x_j, t \wedge t_k)(N(A_j, t \wedge t_{k+1}) - N(A_j, t \wedge t_k)) \\
&+ \frac{1}{2} \sum_{k=0}^{n-1}(V(x_j, t \wedge t_{k+1}) - V(x_j, t \wedge t_k))(N(A_j, t \wedge t_{k+1}) - N(A_j, t \wedge t_k)).
\end{aligned}
\tag{2.19}
$$

Take the limit in (2.15) as $\Delta := \max_k(t_{k+1} - t_k)$ in the sense of the convergence in probability and taking into account (2.15) and (2.18) we obtain:

$$
\int_0^t V(x_j, s) \circ N(A_j, ds) = \int_0^t V(x_j, s) N(A_j, ds) + \frac{1}{2}\{V(x_j, t), N(A_j, t)\}.
\tag{2.20}
$$

From (2.16) and (2.20) we obtain:

$$
\begin{aligned}
&\sum_{j=1}^{m} \int_0^t V(x_j, s) \circ N(A_j, ds) = \sum_{j=1}^{m} \int_0^t V(x_j, s) N(A_j, ds) \\
&+ \frac{1}{2} \sum_{j=1}^{m} \{V(x_j, t), N(A_j, t)\},
\end{aligned}
\tag{2.21}
$$

for the fixed function $V(x_j, \cdot)$ on the disjoint sets A_j, $j = \overline{1, m}$, $x_j \in A_j$. Using the standard arguments for more general V and from (2.21) we have (2.17).

Remark. *We can define the same integral as in (2.16) for the Banach-valued function $a(x, s, f)$ as the integral in weak sense: $\forall l \in B^*$ we define the integral $\int_A \int_0^t a(x, s, V(s)) \circ N(dx, ds)$ by*

$$
\int_A \int_0^t l(a(x, s, V(s))) \circ N(dx, ds).
$$

Remark. *The analogical definition of the integral in (2.16) has the following form:*

$$\int_0^t V(x_j, s) \circ N(A_j, ds) = \lim_{\Delta \to 0} \sum_{k=0}^{n-1} \frac{1}{t \wedge t_{k+1} - t \wedge t_k}$$
$$\times \left(\int_{t \wedge t_k}^{t \wedge t_{k+1}} V(x_j, s) ds \right) (N(A_j, t \wedge t_{k+1}) - N(A_j, t \wedge t_k))$$

where the limit is taken in the sense of the convergence in probability, $x_j \in A_j$, $j = \overline{1, m}$.

2.1.2.4 Anticipating (Skorokhod) Integral over Martingale Measure

Let $\mathcal{F}_{[t_k, t_{k+1}]^c}$ be a sigma-algebra generated by the increments of the martingale measure $N(A, t)$, $\forall A \in \mathcal{X}$, on the complement of the interval $[t_k, t_{k+1}]$. The anticipating integral over martingale measure can be approximated in L^2 by Riemann sums defined in terms of the conditional expectation of the values of the process V in each small interval $[t_k, t_{k+1}]$ given the σ-algebra $\mathcal{F}_{[t_k, t_{k+1}]^c}$. In such a way, let's define the following integral:

$$\int_0^t V(x_j, s) * N(A_j, ds) := \lim_{|\Delta| \to 0} \sum_{k=0}^{n-1} \frac{1}{t \wedge t_{k+1} - t \wedge t_k}$$
$$\times \mathsf{E} \left(\int_{t \wedge t_k}^{t \wedge t_{k+1}} V(x_j, s) \, ds / \mathcal{F}_{[t_k, t_{k+1}]^c} \right) (N(A_j, t \wedge t_{k+1}) - N(A_j, t \wedge t_k)),$$
(2.22)

$A_i \cap A_j = \varnothing$, $x_j \in A_j$, $i, j = \overline{1, m}$, $i \neq j$, where the limit is taken in the sense of the convergence in probability.

$$\int_X \int_0^t V(x, s) * N(dx, ds) := \sum_{j=1}^m \int_0^t V(x_j, s) * N(A_j, ds), \qquad (2.23)$$

where integral in the righthand side of (2.23) is defined in (2.23).

In the same manner, as in previous section, we use this definition as a basis for an approximation scheme for more general V. The approximation procedure in (2.22) works in the L^2 norm, if the process belongs to the space

$$L^{1,2}(D) := \{ V(A, s) \colon dV(A, s)/ds \in L^2(D), \forall A \in \mathcal{X}, \forall s \in D,$$

with the norm

$$\|V\|_{1,2} := [\mathsf{E} \int_X \int_D V^2(x, s) \pi(dx, ds) + \mathsf{E} \int_X \int_D (dV(x, s)/ds)^2 \pi(dx, ds)]^{1/2},$$

where D is an open set of R_+.

Let's give a formula that relates integrals in (2.16) and (2.23).

Proposition. *Let* $V \in L^{1,2}(D)$ *and* N *be a local martingale measure with quadratic variation* π. *Then*

$$\int_X \int_0^t V(x,s) * N(dx,ds) = \int_X \int_0^t V(x,s) \circ N(dx,ds)$$
$$- \frac{1}{2} \int_X \int_0^t \left(\frac{dV(x,s^+)}{ds} + \frac{dV(x,s^-)}{ds} \right) \pi(dx,ds).$$

The following result concerns the formula that relates integrals in (2.23) and in previous section.

Proposition. *Let* $V \in L^{1,2}(D)$ *and* N *be a local martingale measure with quadratic variation* π. *Then*

$$\int_X \int_0^t V(x,s) N(dx,ds) = \int_X \int_0^t V(x,s) * N(dx,ds)$$
$$+ \int_X \int_0^t \frac{dV(x,s^-)}{ds} \pi(dx,ds).$$

2.1.2.5   Multiple Ito's Integral over Martingale Measure

Let's define the measurable functions

$$a_m(x_1,s_1;x_2,s_2;\ldots;x_m,s_m), \qquad \forall\, x_i \in X, \quad s_i \in R_+, \quad i = \overline{1,m} :$$

$a_m : X \times R_+ \times \ldots \times X \times R_+ \mapsto R$ and let N be a local martingale measure with nonrandom quadratic variation π.

Multiple Ito integrals of a_m over N are defined by the expression:

$$\int_X \int_0^t \int_X \int_0^{t_1} \ldots \int_X \int_0^{t_{m-1}} a_m(x_1,t_1;x_2,t_2;\ldots;x_m,t_m) \qquad (2.24)$$
$$\times N(dx_m,dt_m)\ldots N(dx_1,dt_1).$$

This multiple integral has sense for the functions a_m such that:

$$\int_X \int_0^t \int_X \int_0^{t_1} \ldots \int_X \int_0^{t_{m-1}} a_m^2(x_1,t_1;\ldots;x_m,t_m)$$
$$\times \pi(dx_m,dt_m)\ldots\pi(dx_1,dt_1) < +\infty,$$
$$\forall\, t \in R_+.$$

Let $\mathcal{H} := L^2(X \times R_+)$ and let denote by $\mathcal{H}^n$ the n-times tensor product of $\mathcal{H}$:

$$\mathcal{H}^n := \mathcal{H} \otimes \mathcal{H} \otimes \ldots \otimes \mathcal{H}.$$

We set $\mathcal{H}^0 := R$ and $\mathcal{F}(\mathcal{H}) := \bigoplus_{n=0}^\infty \mathcal{H}^n$. Element of $\mathcal{F}(\mathcal{H})$ is the sequence of functions such that $\forall\, \psi \in \mathcal{F}(\mathcal{H})$:

$$\psi = \{a_0, a_1(x_1,s_1), a_2(x_1,s_1;x_2,s_2), \ldots, a_m(x_1,s_1;\ldots;x_m,s_m), \ldots\}$$

and

$$|a_0|^2 + \sum_{n=0}^\infty \int_{X^n} \int_{R_+^n} |a_m(x_1,s_1;\ldots;x_n,s_n)|^2 \pi(dx_n,ds_n)\ldots\pi(dx_1,ds_1) < +\infty.$$

Remark. *We can define the same integral as in (2.24) for the Banach-valued function* $a_m : X \times R_+ \times \ldots \times X \times R_+ \mapsto B$ *in a weak sense:*

$$\int_X \int_0^t \int_X \int_0^{t_1} \cdots \int_X \int_0^{t_{m-1}} l(a_m(x_1, t_1; \ldots; x_m, t_m))$$
$$\times N(dx_m, dt_m) \ldots N(dx_1, dt_1),$$

and this multiple integral has sense if

$$\int_X \int_0^t \int_X \int_0^{t_1} \cdots \int_X \int_0^{t_{m-1}} l^2(a_m(x_1, t_1; \ldots; x_m, t_m)) \times \pi(dx_m, dt_m) \ldots$$

$$\pi(dx_1, dt_1) < +\infty,$$

$$\forall\, t \in R_+, \qquad \forall\, l \in B^*.$$

Remark. *The space* $\mathcal{F}(\mathcal{H})$ *is the analogue of Fock space over* $\mathcal{H} = L^2(X \times R_+)$.

Let $W(A, t)$ *be a Wiener martingale measure with quadratic variation* $\rho(A) \cdot t$, *i.e., structure equation has the following form:*

$$d[W(A, t)] = \rho(A)\, dt.$$

2.1.3 Stochastic Integral Equations over Martingale Measures

Let $\pi(A, t)$ be a positive measure on $(X \times R_+, \mathcal{X} \times \mathbf{B}(R_+))$, and let N be the continuous martingale measure with $\langle N(A, t) \rangle = \pi(A, t)$. Let $a, b : X \times R_+ \times B \mapsto B$ be measurable bounded functions, where space B is a separable Banach space.

Stochastic integral equation for the process $V(t)$ in B over martingale measure is defined by the equation:

$$V(t) = V(0) + \int_X \int_0^t a(x, s, V(s))\pi(dx, ds) + \int_X \int_0^t b(x, s, V(s))N(dx, ds).$$
$$(2.25)$$

This equation is read after pairing both sides of the equation with an element from B^*: $\forall\, l \in B^*$ we have from (2.25):

$$\begin{aligned} l(V(t)) &= l(V(0) + \int_X \int_0^t l(a(x, s, V(s)))\pi(dx, ds) \\ &+ \int_X \int_0^t l(b(x, s, V(s)))N(dx, ds). \end{aligned} \quad (2.26)$$

Proposition (analogue of Ito's formula). *Let* $F \in C_b^2(R)$. *Then we have:*

$$\begin{aligned} &F(l(V(t))) - F(l(V(0))) \\ &= \int_X \int_0^t \frac{dF(l(V(s)))}{dz} l(a(x, s, V(s)))\pi(dx, ds) \\ &+ \frac{1}{2} \int_X \int_0^t \frac{d^2 F(l(V(s)))}{dz^2} \cdot l^2(b(x, s, V(s)))\pi(dx, ds) \\ &+ \int_X \int_0^t \frac{dF(l(V(s)))}{dz} l(b(x, s, V(s)))N(dx, ds). \end{aligned} \quad (2.27)$$

Remark. *If $a, b : X \times R_+ \times R \mapsto R$, then we have from (2.27): $\forall\ F \in C_b^2(R)$*

$$
\begin{aligned}
F(V(t)) - F(V(0)) \\
= \int_X \int_0^t F'(V(s))a(x, s, V(s))\pi(dx, ds) \\
+ \frac{1}{2} \int_X \int_0^t F''(V(s))b^2(x, s, V(s))\pi(dx, ds) \\
+ \int_X \int_0^t F'(V(s))b(x, s, V(s))N(dx, ds).
\end{aligned}
\tag{2.28}
$$

Let us define another forms of stochastic equations.

The symmetric stochastic integral equation for the process $V(t)$ in **B** over martingale measure is defined by the equation:

$$
V(t) = V(0) + \int_X \int_0^t a(x, s, V(s))\pi(dx, ds) + \int_X \int_0^t b(x, s, V(s)) \circ N(dx, ds),
\tag{2.29}
$$

where the third term in the righthand side of (2.29) is a symmetric integral (see 2.1.2) such that is defined in remark.

Anticipating stochastic integral equation for the process $V(t)$ in B over martingale measure is defined by the equation:

$$
V(t) = V(0) + \int_X \int_0^t a(x, s, V(s))\pi(dx, ds) + \int_X \int_0^t b(x, s, V(s)) * N(dx, ds),
\tag{2.30}
$$

where the third term in the righthand side of (2.30) is anticipating integral (see previous section) in weak sense.

Let $W(A, t)$ be a Wiener martingale measure with quadratic variation $\pi(A, t) := \rho(A) \cdot t$ and let we have following equation:

$$
\xi_t = 1 + \int_X \int_0^t \xi_s \cdot \sigma(x, s)W(dx, ds).
\tag{2.31}
$$

With the formula (2.28) it is an elementary check that the solution of (2.31) has the form:

$$
\xi_t = \exp\left\{ \int_X \int_0^t \sigma(x, s)W(dx, ds) - \frac{1}{2} \int_X \int_0^t \sigma^2(x, s)\rho(dx)\, ds \right\}.
$$

W and π are defined by Example 5.6. Let's define the equation:

$$
\xi_t = G + \int_X \int_0^t (\xi_s \cdot \sigma(x, s)) * W(dx, ds),
\tag{2.32}
$$

where σ is a deterministic and square integrated function, and in the righthand side of (2.32) stands the anticipated integral. If $G \in L^p(\Omega)$ for some $p > 2$, there exists a unique solution of the equation (2.32) which is given by

$$
\xi_t = (G \circ A_t)M_t,
\tag{2.33}
$$

where

$$M_t := \exp\left\{\int_X \int_0^t \sigma(x,s)W(dx,ds) - \frac{1}{2}\int_X \int_0^t \sigma^2(x,s)\rho(dx)\,ds\right\},$$

and

$$A_t(\omega)_s = \omega_s - \int_X \int_0^{t\wedge s} \sigma(x,s)\rho(dx)\,ds.$$

The solution of the equation (2.32) can be represented by the Wick product [16]:

$$\xi_t = G\Diamond M_t,$$

where M_t is defined in (2.33).

2.1.4 Martingale Problems Associated with Stochastic Equations over Martingale Measures

Let $\pi(A,t)$ be a positive measure on $(X\times R_+, \mathcal{X}\times \mathbf{B}(R_+))$ and let N be the continuous process for which $\langle N(A,t)\rangle = \pi(A,t)$. Let $a,b: X\times R_+ \times B \mapsto B$ be measurable bounded functions.

A progressively measurable process $V(A,t) \in C_B(R_+)$ is said to be a solution to the (a,b,π)-martingale problem if the following conditions hold with respect to the measure P and the filtration $\mathcal{F}_t^V := \sigma\{V(A,t); A\in \mathcal{X}, t\in R_+\}$:

1) V is additive in A: $\forall\, A_1, A_2: A_1\cap A_2 = \emptyset$

$$V(A_1,t) + V(A_2,t) = V(A_1\cup A_2,t) \quad \text{a. s. P};$$

2) $V(t) := V(X,t)$ and $\forall\, l\in B^*$:

$$m^l(A,t) := l(V(A,t) - V(A,0) - \int_A \int_0^t a(x,s,V(s))\pi(dx,ds))$$

is a continuous orthogonal martingale measure;

3) the quadratic variation of $m^l(A,t)$ is

$$v^l(A,t) = \int_A \int_0^t l^2(b(x,s,V(s)))\pi(dx,ds).$$

An (a,b,π)-martingale problem is said to be well posed if there exists a solution and every solution has the same finite dimensional distribution.

Assume the existence of the process in equation (2.25) and define

$$V(A,0) := \rho(A)\cdot V(0),$$

where ρ is some probability measure, and

$$\begin{aligned} V(A,t) &= V(A,0) + \int_A \int_0^t a(x,s,V(s))\pi(dx,ds) \\ &+ \int_A \int_0^t b(x,s,V(s))N(dx,ds). \end{aligned} \tag{2.34}$$

Thus one easily sees that the existence of a solution of the stochastic equation in (2.34) gives a solution of the (a,b,π)-martingale problem. In addition, if the (a,b,π)-martingale problem is well posed, then the solution of the equation (2.33) is unique. This gives us one direction in each of the following two statements [16].

Remarks. *1. The stochastic integral equation has a solution if and only if the martingale problem has a solution. 2. The solution to the stochastic integral equation is unique if and only if the (a,b,π)-martingale problem is well posed.*

The interest of proof comes in the converse. The main idea is to construct a process that behaves as N, and we must build it from the processes m and v in 2), 3) respectively, by definition. First, a process $z(A,t)$ is defined as a stochastic integral over m. By the definition z will be a martingale measure. Second, we can show that $\langle z \rangle = \pi(A,t)$. By proposition this guarantees us that z and N have the same distribution. The finishing touch is to show that $V(t)$ solves the stochastic integral equation.

2.1.5 Evolutionary Operator Equations Driven by Wiener Martingale Measures

We want to study the following class of equations:

$$\begin{cases} \dfrac{du(t,x,z)}{dt} &= \Gamma(x)u(t,x,z) + Qu(t,x,z) + f(t,x;u) \\ & \quad + g(t,x;u)W'(t,x) \\ u(0,x,z) &= u_0(x,z), \qquad z \in B, \end{cases} \tag{2.35}$$

where $f,g : R_+ \times X \times B \mapsto B$ are some functions, Q is an infinitesimal operator on $C(X)$, $\Gamma(x)$ are defined in Section 2.1, $W'(t,x)$ is a formal expression to denote a derivative of a Wiener martingale measure (for example, space-time white noise).

The formulation which we have given above in (2.35) is formal since $W'(t,x)$ does not make sense. There is one way of giving a rigorous meaning to the equation. We note that the operators $\Gamma(x)$ and Q act by different variables: $\Gamma(x)$-by z, and Q-by x.

It is known [16] that the solution of the equation

$$\begin{cases} \dfrac{dg(t,x,z)}{dt} &= \Gamma(x)g(t,x,z) + Qg(t,x,z) \\ g(0,x,z) &= U_0(x,z) \end{cases} \tag{2.36}$$

has the following form:

$$g(t,x,z) = \mathsf{E}_x[V(t)U_0(x(t),z)],$$

where $x(0) = x$, and $V(t)$ acts by variable z.

It is also known that the operator $\Gamma(x) + Q$ generates the semigroup $T(t)$ and

$$T(t)U_0(x, z) = \mathsf{E}_x[V(t)U_0(x(t), z)] = g(t, x, z). \qquad (2.37)$$

Its semigroup is strong, continuous and contractive.

Let's write the equation (2.35) in integral form using the semigroup (2.37). In that formulation $U(t, x, z)$ is a solution if at any point $(t, x, z) \in R_+ \times X \times B$:

$$U(t, x, z) = T(t)U_0(x, z) + \int_0^t T(t - s)f(s, x; u)\, ds \\ + \int_X \int_0^t T(t)(t - s)g(s, x; u)W(dx, ds), \qquad (2.38)$$

where the last term in the expression (2.38) is defined in the Remark in Section 2.1.2.2, and makes sense if and only if for any $l \in B^*$:

$$\int_X \int_0^t l^2(T(t - s)g(s, x; u))\pi(dx, ds) < +\infty,$$

where $\pi(A, t) := \langle W(A, t) \rangle$.

Let's consider the equation of the form:

$$\begin{cases} \dfrac{du(t, x, z)}{dt} & = \ \Gamma(x)u(t, x, z) + Qu(t, x, z) + f(t, x)W'(t, x) \\ u(0, x, z) & = \ u_0(x, z), \end{cases} \qquad (2.39)$$

where $f : R_+ \times X \mapsto B$ is a some bounded function.

It follows from (2.38) that its solution is given explicitly by the following formula:

$$U(t, x, z) = T(t)U_0(x, z) + \int_X \int_0^t T(t - s)f(s, x)W(dx, ds).$$

We also want to study the following class of equations:

$$\begin{cases} \dfrac{du(t, x, z)}{dt} & = \ \Gamma(x)u(t, x, z) + \int_X P(x, dy)[\mathcal{D}(x, y) - I] \cdot u(t, z, x) \\ & \quad + Qu(t, z) + f(t, x; u) + g(t, x; u) \cdot W'(t, x) \\ u(0, x, z) & = \ u_0(x, z), \qquad \forall\, x \in X; \quad \forall\, z \in B, \end{cases}$$
$$(2.40)$$

where $\Gamma(x)$, Q, f, g are defined in (2.35) and operators $\{\mathcal{D}(x, y); x, y \in X\}$ are defined in Section 2.1, with $\mathcal{D}^\varepsilon(x, y) \equiv \mathcal{D}(x, y)$, $\forall\, x, y \in X$, $\forall\, \varepsilon > 0$ and $P(x, A)$ is a stochastic kernel, $\forall\, x \in X$, $\forall\, A \in \mathcal{X}$, $W'(t, x)$ is defined in (2.35). We note that operators $\mathcal{D}(x, y)$ act by variable.

It is known [16] that the solution of the equation

$$\begin{cases} \dfrac{dG(t, x, z)}{dt} & = \ \Gamma(x)G(t, x, z) + \int_X P(x, dy)[\mathcal{D}(x, y) - I]G(t, x, z) \\ & \quad + QG(t, x, z) \\ G(0, x, z) & = \ u_0(x, z) \end{cases}$$
$$(2.41)$$

has the following form:

$$G(t, x, z) = \mathsf{E}_x[V(t)u_0(x(t), z)],$$

where $x(0) = x$, $V(t)$ acts by variable z.

Operator $\Gamma(x) + P[\mathcal{D}(x, \cdot) - I] + Q$ in the righthand side of (2.41) generates the semigroup $S(t)$ and

$$S(t)u_0(x, z) = \mathsf{E}_x[V(t)u_0(x(t), z)] = G(t, x, z). \qquad (2.42)$$

Let's write the equation (2.40) in an integrated form using the semi-group (2.42). In that formulation $u(t, x, z)$ in (2.40) is a solution if at any point $(t, x, z) \in R_+ \times X \times B$:

$$\begin{aligned} u(t, x, z) &= S(t)u_0(x, z) + \int_0^t S(t - s)f(s, x; u)\, ds \\ &+ \int_X \int_0^t S(t - s)g(s, x; u)W(dx, ds), \end{aligned} \qquad (2.43)$$

where the last term in the expression (2.43) is defined as well as the last term in (2.39).

Let's consider, for example, the equation of the form:

$$\begin{cases} \dfrac{du(t, x, z)}{dt} &= \Gamma(x)u(t, x, z) + \int_X P(x, dy)[\mathcal{D}(x, y) - I]u(t, x, z) \\ &\quad + Qu(t, x, z) + f(t, x)W'(t, x) \\ u(0, x, z) &= u_0(x, z), \end{cases}$$

$$(2.44)$$

where function f is defined in (2.39).

It follows from (2.43) that the solution of the equation (2.44) is given explicitly by the following formula:

$$U(t, x, z) = S(t)u_0(x, z) + \int_X \int_0^t S(t - s)f(s, x)W(dx, ds). \qquad (2.45)$$

2.2 Stochastic Calculus for Multiplicative Operator Functionals (MOF)

We investigate here the characteristic operator and equations for resolvent and potential for the multiplicative operator functionals (MOF) of Markov processes.

In 2.2.1 we consider the definitions of MOF of Markov processes and equations for expectations. Section 2.2.2 deals with properties of infinitesimal and characteristic operators of MOFs and some equations for them. In Section 2.2.3 we find the resolvent and potential for MOF of Markov processes. Equations for resolvent and potential of MOF of Markov processes are derived in Section 2.2.4. An analogue of Dynkin's formula for MOF of Markov processes are considered in Section 2.2.5. Applications of these formula are given to traffic, storage and diffusion processes in random media (Section 2.2.6).

2.2.1 Definition of MOF

Let

$$(\Omega, \mathcal{F}, \mathcal{F}_t, P_x) \tag{2.46}$$

be a probability space, let

$$X := (x_t, \zeta, \mathcal{F}_t, P_x) \tag{2.47}$$

be a right continuous Markov process on a phase space $(\mathbf{X}, \mathcal{X})$, let

$$(B, \mathcal{B}, \| \cdot \|) \tag{2.48}$$

be a separable Banach space, and let $L(B)$ be the the space of bounded linear operators on B.

MOF of a Markov process X is a map $t \mapsto V(t, \omega) \in L(\mathbf{B})$, which satisfies the following properties:

a) $\omega \mapsto V(t, \omega)$ is measurable with respect to the *sigma*-algebra generated by $x_s, 0 \le s \le t$;

b) the map $t \mapsto V(t, \omega)$ is strongly right continuous, a. s.;

c) $V(t + s, \omega) = V(t, \omega) \vartheta_t V(s, \omega)$ a. s., $\forall\, 0 \le s \le t$;

d) $V(0, \omega) = I$ — identity operator on B.

The symbol ϑ_t in c) is the shift operator along the trajectories of X.

Suppose that

$$\widetilde{B} := \prod_{x \in X} B \tag{2.49}$$

and

$$\widetilde{B}_\infty = \{\tilde{f} \in \widetilde{B} : \ x \mapsto f(x) \text{ is } \mathcal{X}/\mathcal{B}\text{-measurable and } sup_{x \in X} \|f(x)\| < +\infty\}. \tag{2.50}$$

In what follows we set

$$(\tilde{f})_x := f(x) \in B, \ \forall\, x \in X. \tag{2.51}$$

The expectation semigroup of MOF of a Markov process is defined by the operator

$$(\widetilde{T}(t)f)(x) := \mathbf{E}_x[V(t, \omega)f(x_t)], \ \forall\, \tilde{f} \in \widetilde{B}_\infty. \tag{2.52}$$

If MOF $V(t, \omega)$ is contractive, the semigroup $\widetilde{T}(t)$ is also contractive. We will suppose that this condition will be satisfied.

The infinitesimal operator of semigroup $\widetilde{T}(t)$ is defined by the formula

$$\widetilde{A}\tilde{f} := s - \lim_{t \downarrow 0} \frac{\widetilde{T}(t)\tilde{f} - \tilde{f}}{t}. \tag{2.53}$$

The domain $\mathcal{D}_{\widetilde{A}}$ of operator $\widetilde{A}$ consists of those $\tilde{f} \in \widetilde{B}_\infty$ for which there exists the limit in (2.53).

Let τ_U be the first exit moment from the set $U \in \mathcal{X}$:

$$\tau_U := \inf_t \{t : x_t \notin U\}. \tag{2.54}$$

We call $\tilde{U}$ a characteristic operator of MOF $V(t, \omega)$ if

$$\tilde{U}\tilde{f} := \lim_{U \downarrow x} \frac{E[V(\tau_U)f(x_{\tau_U})] - \tilde{f}}{E\tau_U} \tag{2.55}$$

as $\tau_U \downarrow 0$.

The weak infinitesimal operator $\tilde{\tilde{A}}$ of the semigroup $\tilde{T}(t)$ is defined by formula

$$\tilde{\tilde{A}}\tilde{f} := w - \lim_{t \downarrow 0} \frac{\tilde{T}(t)\tilde{f} - \tilde{f}}{t}, \ \forall \ \tilde{f} \in \tilde{\mathcal{B}}_\infty, \tag{2.56}$$

where the symbol w means weak convergence in $\tilde{\mathcal{B}}_\infty$.

Evidently, operator $\tilde{\tilde{A}}$ in (2.56) is an extension of operator $\tilde{A}$ in (2.53).

Lemma 2.2.1. *Let $\tilde{\tilde{A}}$ be a weak infinitesimal operator of strong Markov process X in the phase space $(\mathbf{X}, \mathcal{X})$. Let $x \in \mathbf{X}$, and let for all $U \in \mathcal{X}, x \in U$ there exists a stopping time $\tau : S_\tau \subseteq U$ and*

$$0 < E_x \tau < +\infty, \tag{2.57}$$

where

$$S_\tau := \{x_t(\omega) : \omega \in \Omega, \ 0 \le t < \tau(\omega)\}. \tag{2.58}$$

Then $\forall \ \tilde{f} \in \mathcal{D}_{\tilde{\tilde{A}}}$ such that $\tilde{\tilde{A}}f(x)$ is continuous in x the following formula is valid:

$$(\tilde{\tilde{A}}\tilde{f})_x = \lim_{s_\tau \downarrow x} \frac{E_x[V(\tau)f(x_\tau)] - f(x)}{E_x \tau} = \lim_{s_\tau \downarrow x} \frac{\tilde{T}(\tau)f(x_\tau) - f(x)}{E_x \tau}. \tag{2.59}$$

Proof of Lemma 2.2.1 follws from the analogue of Dynkin's formula (see Subsection 2.2.5).

Lemma 2.2.2. *Let X be a right continuous strong Markov process on semicompact $(\mathbf{X}, \mathcal{X})$, let $V(t, \omega)$ be a MOF of X, let $\tilde{U}$ be the characteristic operator of MOF $V(t, \omega)$, and let $\tilde{\tilde{A}}$ be a weak infinitesimal operator. Suppose that there exists in the neighbourhood U of the point $x \in \mathbf{X}$ such that*

$$E_x \tau_U < +\infty. \tag{2.60}$$

If $\tilde{f} \in \mathcal{D}_{\tilde{\tilde{A}}}$ and function $\tilde{\tilde{A}}f(x)$ is continuous in x, then $f \in \mathcal{D}_{\tilde{U}}$ and

$$\tilde{U}\tilde{f} = \tilde{\tilde{A}}\tilde{f} \ \forall \ \tilde{f} \in \tilde{\mathcal{B}}_\infty. \tag{2.61}$$

Proof of Lemma 2.2.2 follows from (2.59).

Theorem 2.2.3. *Let $C(\mathbf{X})$ be the space of continuous bounded functions on the semicompact $(\mathbf{X}, \mathcal{X})$ containing all finite functions. If X is right continuous process on $(\mathbf{X}, \mathcal{X})$, then*

$$\tilde{A} \subseteq \tilde{\tilde{A}} \subseteq \tilde{U}.$$

Proof of this theorem follows from previous lemmas.

2.2.2 Properties of the Characteristic Operator of MOF

The characteristic operator $\tilde{U}$ is defined on many functions that do not belong to domain $\mathcal{D}_{\tilde{A}}$ of the operator A. Some of the functions are described in the following theorem.

Theorem 2.2.4. *Let X be a right continuous strong Markov process on semi-compact $(\mathbf{X}, \mathcal{X})$. Let $G \in \mathcal{X}$ and $\tau_G := \inf\{t : x_t \notin G\}$, let $V(t)$ be a MOF of X. Further, let $h(x)$ be a measurable function, and let $h(x_{\tau_G})$ be a P_x-integrable function $\forall\, x \in \mathbf{X}$.*
Then $\forall\, x \in G$ function

$$\tilde{b} := E[V(\tau_G)h(x_{\tau_G})], \tag{2.62}$$

i.e. $b(x) := E_x[V(\tau_G)h(x_{\tau_G})]$ belongs to $\mathcal{D}_{\tilde{U}}$, and

$$\tilde{U}\tilde{b} = 0. \tag{2.63}$$

Furthermore, let

$$H(x) := E_x \int_0^{\tau_G} V(s)g(x_s)\, ds, \tag{2.64}$$

where $g(x)$ is some measurable function on G.
Let function $g(x)$ be continuous, let function $H(x)$ be finite, and let there exist a neighbourhood U_0 of x such that $E_x\tau_{U_0} < +\infty$. Then $H(x) \in \mathcal{D}_{\tilde{U}}$ and

$$\tilde{U}\tilde{H} = -\tilde{g}, \tag{2.65}$$

i.e., $\tilde{U}H(x) = -g(x)$, $\forall\, x \in \mathbf{X}$.

Proof. *We notice that $\vartheta_{\tau_U} x_{\tau_G} = x_{\tau_G}$ and we get from (2.62):*

$$E_x b(x_{\tau_U}) = E_x E_{x_{\tau_U}} V(x_G)h(x_{\tau_G}) = E_x E_x \vartheta_{\tau_U} V(\tau_G)h(x_{\tau_G})$$
$$= E_x E_x V(\tau_G)h(x_{\tau_G}) = E_x V(\tau_G)h(x_{\tau_G}) = b(x)$$

which proves (2.63), where $U \in \mathcal{X}$ and $U \subseteq G$. Futhermore, we obtain from expression $\vartheta_{\tau_U}\tau_G = \tau_G - \tau_U$:

$$\vartheta_{\tau_U} \int_0^{\tau_G} V(s)g(x_s)\,ds = \vartheta_{\tau_U} \int_0^{\infty} V(s)g(x_s)\chi\{\tau_G > s\}\,ds$$

$$= \int_0^{\infty} \vartheta_{\tau_U} V(s)g(x_s)\chi\{\tau_G > s\}\,ds$$

$$= \int_0^{\infty} V(s+\tau_U)g(s+\tau_U)\chi\{\tau_G - \tau_U > s\}\,ds = \int_{\tau_U}^{\tau_G} V(s)g(x_s)\,ds$$

$$= \int_0^{\tau_G} V(s)g(x_s)\,ds - \int_0^{\tau_U} V(s)g(x_S)\,ds.$$

From this expression and from (2.64) we have

$$E_x H(x_{\tau_U}) = E_x E_{x_{\tau_U}} \int_0^{\tau_G} V(s)g(x_s)\,ds = E_x E_x \vartheta_{\tau_U} \int_0^{\tau_G} V(s)g(x_s)\,ds$$

$$= E_x \int_0^{\tau_G} V(s)g(x_s)\,ds - E_x \int_0^{\tau_U} V(s)g(x_s)\,ds$$

$$= H(x) - E_x \int_0^{\tau_U} V(s)g(x_s)\,ds.$$

Hence we obtain:

$$\frac{E_x H(x_{\tau_U}) - H(x)}{E_x \tau_u} = \frac{-E_x \displaystyle\int_0^{\tau_U} V(s)g(x_s)\,ds}{E_x \tau_U}. \tag{2.66}$$

Since $V(s)$ is contractive we have:

$$\left\| E_x \int_0^{\tau_U} V(s)g(x_s)\,ds \right\| \le \sup_{x \in \mathbf{X}} \|g(x)\| E_x \tau_U < +\infty. \tag{2.67}$$

Finally, (2.66), (2.67) and continuity of $V(s)$ and x_s imply (2.65). Theorem 2.2.4 is completely proved.

Let X be a continuous Markov process on $(\mathbf{X}, \mathcal{X})$. By a Markov random evolution here we mean the solution of the following random operator equation:

$$\begin{cases} \dfrac{dV(t)}{dt} &= V(t)\Gamma(x_t) \\ V(0) &= I, \end{cases} \tag{2.68}$$

where $\{\Gamma(x); \ x \in \mathbf{X}\}$ is a family of closed operators with common domain $\mathcal{B}_0 \subset \mathcal{B}$ which is dense in $\mathcal{B}$.

Let us define the following function:

$$U(t,x) := E_x[V(t)f(x_t)] = (\tilde{T}(t)\tilde{f})_x, \ \forall \, x \in \mathbf{X}, \tag{2.69}$$

where $V(t)$ is defined in (2.68).

It is known, that this function satisfies the following equation:

$$\begin{cases} \dfrac{dU(t,x)}{dt} & = & QU(t,x) + \Gamma(x)U(t,x) \\ U(0,x) & = & f(x) \in \mathcal{B}_0, \end{cases} \tag{2.70}$$

where Q is an infinitesimal operator of X.

From (2.70) and (2.52) we conclude that the infinitesimal operator of semigroup $\tilde{T}(t)$ (see sec. 3.1) has the form:

$$\tilde{A} = Q + \tilde{\Gamma} \ (\text{i. e.}, (\tilde{A})_x = Q + \Gamma(x), \ \forall \, x \in \mathbf{X}). \tag{2.71}$$

2.2.3 Resolvent and Potential for MOF

Let $X = (x_t, \zeta, \mathcal{F}_t, P_x)$ be a measurable Markov process, and let $V(t)$ be a MOF of X.

Let $\tilde{R}_\lambda$ and $\tilde{R}$ be a resolvent and potential of semigroup $\tilde{T}(t)$ respectively.

Lemma 2.2.5. *Resolvent $\tilde{R}_\lambda$ and potential $\tilde{R}$ of semigroup $\tilde{T}(t)$ are defined by the following formulas:*

$$(\tilde{R}_\lambda \tilde{f})_x = E_x \int_0^\zeta e^{-\lambda t} V(t) f(x_t)\, dt, \tag{2.72}$$

$$(\tilde{R}\tilde{f})_x = E_x \int_0^\zeta V(t) f(x_t)\, dt. \tag{2.73}$$

Proof. *The resolvent $\tilde{R}_\lambda$ is the solution of the equation:*

$$\lambda h(x) - A(x)h(x) = f(x) \ (\text{i. e.}, \lambda \tilde{h} - \tilde{A}\tilde{h} = \tilde{f}). \tag{2.74}$$

This equation has the solution:

$$\tilde{h} = \tilde{R}_\lambda \tilde{f} = \int_0^\zeta e^{-\lambda t} \tilde{T}(t) f(x_t)\, dt. \tag{2.75}$$

From (2.69) and (2.75) we obtain:

$$(\tilde{R}_\lambda \tilde{f})_x = \int_0^\zeta \int_\Omega e^{-\lambda t} V(t,\omega) f(x_t(\omega)) P_x(d\omega)\, dt. \tag{2.76}$$

Because of measurability of both $V(t)$ and x_t the integrand in (2.76) is measurable.

Changing the order of integration, by Fubini's theorem we get from (2.75), (2.76):

$$(\tilde{R}\tilde{f})_x = E_x \int_0^\zeta e^{-\lambda t} V(t) f(x_t)\, dt. \tag{2.77}$$

Formula (2.73) follows from (2.77) when $\lambda = 0$.

2.2.4 Equations for Resolvent and Potential for MOF

Let $X = (x_t, \zeta, \mathcal{F}_t, P_x)$ be a strongly measurable strong Markov process on phase space $(\mathbf{X}, \mathcal{X})$ and τ is a stopping time for X.

Theorem 2.2.6. *If $\tilde{h} = \tilde{R}_\lambda \tilde{f}$, then*

$$E[e^{-\lambda t} V(\tau) h(x_\tau)] - \tilde{h} = -E \int_0^\tau e^{-\lambda t} V(t) f(x_t)\, dt. \tag{2.78}$$

If $\tilde{h} = \tilde{R}\tilde{f}$, then

$$E[V(\tau) h(x_\tau)] - \tilde{h} = -E \int_0^\tau V(t) f(x_t)\, dt. \tag{2.79}$$

Proof. *From formula (2.77) we get:*

$$h(x) = E_x \int_0^\tau e^{-\lambda t} V(t) f(x_t)\, dt + \int_\tau^\zeta e^{-\lambda t} V(t) f(x_t)\, dt. \tag{2.80}$$

For the second term in the righthand side of (2.80) we have:

$$
\begin{aligned}
E_x \int_\tau^\zeta e^{-\lambda t} V(t) f(x_t)\, dt &\, E_x \int_0^{\zeta-\tau} e^{-\lambda(t+\tau)} V(t+\tau) f(x_{t+\tau})\, dt \\
&= E_x e^{-\lambda \tau} V(\tau) \int_0^{\zeta-\tau} e^{-\lambda t} \vartheta_\tau V(t) f(x_t)\, dt \\
&= E_x e^{-\lambda \tau} V(\tau) \vartheta_\tau \int_0^{\zeta-\tau} e^{-\lambda t} V(t) f(x_t)\, dt \\
&= E_x e^{-\lambda \tau} V(\tau) E_x \int_0^\zeta e^{-\lambda t} V(t) f(x_t)\, dt = E_x e^{-\lambda \tau} V(\tau) h(x_\tau).
\end{aligned} \tag{2.81}
$$

In (2.81) we have used the property (Section 2.2.1) for MOF and the following equality:

$$E_x(\xi \vartheta_\tau \eta) = E_x(\xi E_{x_\tau} \eta),$$

for $\mathcal{F}_t$-measurable ξ and P_x-integrable ξ and $\xi \vartheta_\tau \eta$. Finally, we get from (2.80) and (2.81):

$$E_x[e^{-\lambda \tau} V(\tau) h(x_\tau)] - h(x) = -E_x \int_0^\tau e^{-\lambda t} V(t) f(x_t) \, dt. \qquad (2.82)$$

Equation (2.82) is equivalent to that of (2.78). Formula (2.79) is being obtained from (2.82) as $\lambda = 0$.

2.2.5   Analogue of Dynkin's Formulas (ADF) for MOF

Let X be a strongly measurable strong Markov process, let $V(t)$ be a MOF of X, let $\tilde{A}$ be the infinitesimal operator of semigroup $\tilde{T}(t)$ in (2.50), and let τ is a stopping time for X.

Theorem 2.2.7. *If $\tilde{A}\tilde{h} = \tilde{g}$ and $E_x \tau < +\infty$, then*

$$E[V(\tau) h(x_\tau)] - \tilde{h} = E \int_0^\tau V(t) A h(x_t) \, dt. \qquad (2.83)$$

Proof. *Let $\tilde{f}_\lambda := \lambda \tilde{h} - \tilde{g}$. Then*

$$\tilde{R}_\lambda \tilde{f}_\lambda = [\lambda I - A]^{-1}[\lambda \tilde{h} - \tilde{A}\tilde{h}] = \tilde{h}.$$

Equation (2.72) is satisfied for functions $\tilde{h}$ and $\tilde{f} = \tilde{f}_\lambda$, i. e.,

$$E[e^{-\lambda \tau} V(\tau) h(x_\tau)] - \tilde{h} = -E \int_0^\tau e^{-\lambda t} V(t)[\lambda \tilde{h} - g(x_t)] \, dt. \qquad (2.84)$$

Passing to the limit $\lambda \to 0$ in (2.84), we obtain formula (2.83).

Remark. *Formula (2.83) is an analogue of Dynkin's formula for Markov processes.*

In fact, if we set $V(t) \equiv I$, then from (2.83) we obtain:

$$E_x[h(x_\tau)] - h(x) = E_x \int_0^\tau Q h(x_t) \, dt, \qquad (2.85)$$

where Q is the infinitesimal operator of x_t.

Formula (2.85) coincides with Dynkin's formula.

Corollary. *Let $V(t)$ be a Markov random evolution (2.68), and let the conditions of Theorem 2.2.7 be satisfied. Then we have the following formula:*

$$E_x[V(\tau) h(x_\tau)] - h(x) = E_x \int_0^\tau V(t)[Q + \Gamma(x)] h(x_t) \, dt. \qquad (2.86)$$

Proof. *Follows from (2.69)–(2.71) and formula (2.83),*

$$\tau_G := \inf_t \{t > 0 : \ w(t) \notin G\}. \tag{2.87}$$

2.2.6 Analogue of Dynkin's Formulae (ADF) for SES

We consider applications of the ADF for such SES as traffic, storage and diffusion processes in random media. Random media here is described by Markov or semi-Markov processes.

2.2.6.1 ADF for Traffic Processes in Random Media

Let $v(z, x)$ be a smooth function on z and continuous bounded function on x, $z \in$ R, $x \in$ **X**, let x_t be a Markov (or semi-Markov) process with infinitesimal operator Q (see [7]: Basics in Probability Theory and Stochastic Processes).

Traffic process z_t in a Markov (or semi-Markov) random media x_t is defined by the Cauchy problem:

$$\begin{cases} \dfrac{dz_t}{dt} &= v(z_t, x_t) \\ z_0 &= z. \end{cases} \tag{2.88}$$

We note that the process (z_t, x_t) is also Markov process on R$_+ \times$ **X** with infinitesimal operator

$$v(z, x)\frac{d}{dz} + Q. \tag{2.89}$$

Here $\Gamma(x) = v(z, x)d/dz$.

If we have semi-Markov process x_t, then the process $(z_t, x_t, \gamma(t))$ is a Markov process with an infinitesimal operator

$$v(z, x)\frac{d}{dz} + Q_1, \tag{2.90}$$

where $\gamma(t)$ and Q_1 are defined in (2.95) and (2.97) respectively.

A Markov random evolution $V(t)$ of x_t generated by z_t is defined by the equality

$$V(t)f(z, x) = f(z_t, x), \ f(z, x) \in \mathrm{C}_b^1(\mathrm{R} \times \mathbf{X}). \tag{2.91}$$

Corollary. *Let τ be a stopping time for Markov process x_t (see Appendix: Basics in Probability Theory and Stochastic Processes). The analogue of Dynkin's formula for traffic process z_t in Markov random media x_t follows from (2.100), (2.101), and formula (2.86):*

$$E_{z,x}[f(z_\tau, x_\tau)] - f(z, x) = E_{z,x} \int_0^\tau \left[Q + v(z_s, x_s)\frac{d}{dz} \right] f(z_s, x_s)\, ds, \tag{2.92}$$

where $f(z, x) \in \mathrm{C}_b^1(\mathrm{R} \times \mathbf{X})$.

Corollary. *Let x_t be a semi-Markov process and let z_t be a traffic process in (2.100), let τ be a stopping time for $y_t = (x_t, \gamma(t))$ (see (2.95)). The analogue of Dynkin's formula for traffic process z_t in semi-Markov random media x_t follows from (2.100), (2.102), and formula (2.99) with $D(x, y) \equiv I$, $\forall x, y \in \mathbf{X}$:*

$$E_{z,y}[f(z_\tau, y_\tau)] - f(z, y) = E_{z,y} \int_0^\tau \left[Q_1 + v(z_s, x_s) \frac{d}{dz} \right] f(z_s, y_s) \, ds, \quad (2.93)$$

where $f(z, y) \in \mathrm{C}_b^1(\mathrm{R} \times \mathbf{X} \times \mathrm{R}_+)$.

2.2.6.2 ADF for Storage Processes in Random Media

Let $a(x, y)$ be a bounded function on $\mathbf{X} \times \mathbf{X}$.

Storage process z_t in a Markov (or semi-Markov) random media x_t is represented by the solution of the following equation:

$$z_t = z + \sum_{k=1}^{\nu(t)} a(x_{k-1}, x_k) - \int_0^t v(z_s, x_s) \, ds, \quad (2.94)$$

where x_t is a jump Markov (or semi-Markov) process.

We note that the process (z_t, x_t) is also a Markov process on $\mathrm{R} \times \mathbf{X}$ with an infinitesimal operator

$$-v(z, x) \frac{d}{dz} f(z, x) + \int_{\mathbf{X}} P(x, dy)[f(z + a(x, y), x) - f(z, x)] + Q, \quad (2.95)$$

where Q is an infinitesimal operator of x_t.

We note that

$$\Gamma(x) = -v(z, x) \frac{d}{dz}, \quad D(x, y) f(z, x) = f(z + a(x, y), x). \quad (2.96)$$

For semi-Markov process x_t we obtain that process $(z_t, x_t, \gamma(t))$ is a Markov process on $\mathrm{R} \times \mathbf{X} \times \mathrm{R}_+$ with an infinitesimal operator

$$-v(z, x) \frac{d}{dz} f(z, x) + \int_{\mathbf{X}} P(x, dy)[f(z + a(x, y), x) - f(z, x)] + Q_1, \quad (2.97)$$

where Q_1 is defined in (2.97).

A Markov random evolution $V(t)$ of x_t generated by storage process z_t in (2.94) is also defined by (2.91).

Corollary. *The analogue of Dynkin's formula for storage process z_t (see (2.94)) in Markov random media x_t follows from (2.95), (2.96), (2.91) and formula (2.93):*

$$E_{z,x}[f(z_\tau, x_\tau)] - f(z, x) = E_{z,x} \int_0^\tau \{[Q - v(z_s, x_s)\frac{d}{dz}]f(z_s, x_s)$$

$$+ \int_X P(x, , dy)[f(z_{s-} + a(x_{s-}, y), x_s) - f(z_{s-}, x_s)]\} \, ds,$$

(2.98)

where τ is a stopping time for x_t, $f(z, x) \in C_b^1(R \times X)$.

Corollary. *The analogue of Dynkin's formula for storage process z_t in semi-Markov random media x_t follows from (2.96), (2.97) and formula (2.99):*

$$E_{z,y}[f(z_\tau, y_\tau)] - f(z, y) = E_{z,y} \int_0^\tau \{[Q_1 - v(z_s, \gamma_s)\frac{d}{dz}]f(z_s, y_s)$$

$$+ \int_X P(x, dy)[f(z_{s-} + a(x_{s-}, y), y_s) - f(z_{s-}, y_s)]\} \, ds,$$

(2.99)

where τ is a stopping time for $y_t = (x_t, \gamma(t))$ (see (2.94)), Q_1 is defined in (2.96), $f(z, y) \in C_b^1(R \times X \times R_+)$.

2.2.6.3 ADF for Diffusion Process in Random Media

Let $v(z, x)$ and $\sigma(z, x)$ be bounded and continuous functions on $R \times X$.

Diffusion process z_t in a Markov (or semi-Markov) random media x_t is represented by the solution of the following stochastic differential equation:

$$dz_t = v(z_t, x_t)dt + \sigma(z_t, x_t)dw_t,$$

(2.100)

where w_t is a standard Wiener process which is independent on x_t. We remark that process $(z_t, x_t, \gamma(t))$ is a Markov process on $R \times X \times R_+$ with an infinitesimal operator

$$v(z, x)\frac{d}{dz} + 2^{-1}\sigma^2(z, x)\frac{d^2}{dz^2} + Q_1,$$

(2.101)

where Q_1 is defined in (2.97). We also note that

$$\Gamma(x) = v(z, x)\frac{d}{dz} + 2^{-1}\sigma^2(z, x)\frac{d^2}{dz^2}; \mathcal{D}(x, y) = I,$$

(2.102)

for all $x, y \in X$.

Corollary. *ADF for diffusion process z_t in semi-Markov random media x_t follows from (2.101), (2.102) and formula (2.99) and has the following form:*

$$E_{z,y}[f(z_\tau, y_\tau] - f(z, y) =$$

$$E_{z,y} \int_0^\tau \left[Q_1 + v(z,x)d/dz + 2^{-1}\sigma^2(z,x)d^2/dz^2 \right] f(z_s, y_s)ds,$$

where τ is a stopping time for process $y_t = (x_t, \gamma(t))$, Q_1 is defined in (2.96), $f(z,y) \in C_b^2(R \times X \times R_+)$.

Bibliography

[1] Dynkin E. *Theory of Markov Processes*. Prentice-Hall. 1961.

[2] Elliott R. *Stochastic Calculus and its Applications*. Springer-Verlag, Berlin, 1982.

[3] Itô K. Stochastic integral. *Imperial Academy, Tokyo, Proceedings*, v. 20, 1944, 519-524.

[4] Nikolsky S.M. *Approximation of the Functions of Many Variables and Embedding Theorems*, Nauka, Moscow, 1969. (In Russian).

[5] Skorokhod A. *Studies in the Theory of Random Processes*. Dover, 1982.

[6] Stratonovich R. *Conditional Markov Processes and their Application to the Theory of Optimal Control*. Elsevier, 1968.

[7] Swishchuk A. *Random Evolutions and their Applications*. Kluwer AP, Dordrecht, 1997.

[8] Swishchuk A. *The analogue of Dynkins formula and boundary values problems for multiplicative operator functionals of Markov processes*. Preprint 93.44, Institute of Mathematics, Kiev, Ukraine, 1993, 19 pages.

3

Convergence of Random Bounded Linear Operators in the Skorokhod Space

In order to prove our main results, weak law of large numbers (WLLN) and central limit theorem (CLT) for inhomogeneous RE in Chapter 5, because of the backward (in time) nature of our setting (cf. the martingale representation), we will need a suitable topology to study the convergence of operator-valued random variables in the Skorokhod space [12]: this is the main purpose of this chapter.

We introduce in this chapter the space of random bounded linear operators on a separable Banach space such that their range belongs to the Skorokhod space of right-continuous with lefthand limit functions. We call these random operators D-valued random variables. Almost sure and weak convergence results for the sequences of such random variables are proved by martingale methods.

We also mention that the main contribution of this chapter is in the so-called *compact containment criterion* (CCC): we need it in order to prove our limit theorems. In all the aforementioned literature on random evolutions (e.g. [13]), it is mentioned that there exists a compact embedding of a Hilbert space into $C_0(R^d)$. Unfortunately, this is not correct, and we show in this chapter how to overcome this problem in practice, which allows us to deal rigorously with various applications. The results of this chapter are based on [11], [12].

3.1 Introduction

Let $(\Omega, \mathcal{F}, \mathbb{P})$ a probability space, $(Y, ||\cdot||)$ a separable Banach space, and $\mathcal{B}(Y)$ the space of bounded linear operators $Y \to Y$. Introduce the space:

$$\mathbb{D} := \{S \in \mathcal{B}(Y)^{\mathbb{R}_+} : Sf \in D(\mathbb{R}_+, Y) \quad \forall f \in Y\}, \tag{3.1}$$

where $D(\mathbb{R}_+, Y)$ is the Skorokhod space of right-continuous with lefthand limit functions $\mathbb{R}_+ \to Y$, $\mathbb{R}_+ := [0, \infty)$ and $\mathcal{B}(Y)^{\mathbb{R}_+}$ the space of maps $\mathbb{R}_+ \to \mathcal{B}(Y)$. Say we are given a sequence $\{S_n\}_{n \in \mathbb{N}}$ of maps $\Omega \to \mathbb{D}$, $\mathbb{N}$ being the set of positive integers. This is, for example, the setting of [13]–[15] which involves random evolutions in Banach spaces. Using well-known techniques (see e.g. [3], Chapter 3), one might be able to prove that for each $f \in Y$, the family $\{S_n f\}$ is relatively compact in $D(\mathbb{R}_+, Y)$, and study the limit points. The question we want to answer is the following: assume that we understand well the convergence of the family $\{S_n f\}$ for each $f \in Y$, what can we say about the convergence of the family $\{S_n\}$? In other words: are the limit points of $\{S_n f\}$ of the form Sf for some $\mathbb{D}$–valued random variable S? And if yes, which kind of random variable?

One of the problems is the following: $\mathcal{B}(Y)$, topologized with the operator norm $\| \cdot \|_{op}$ is not separable in the general case, and, therefore, the usual approach of [3] (Chapter 3) fails. Further, this operator norm generates a topology which is too strong and not well suited for our purpose. Therefore, we cannot simply equip $\mathbb{D}$ with the usual Skorokhod metric on $D(\mathbb{R}_+, (\mathcal{B}(Y), \| \cdot \|_{op}))$ (see [3], Chapter 3, equation 5.2). The heart of this book consists in finding a suitable topology for $\mathbb{D}$, from which we will be able to prove our main almost certain and weak convergence results. Indeed, there are at least two other metrics we can think of when considering elements of $\mathbb{D}$:

1. the first possibility is to see $\mathbb{D}$ as $D(\mathbb{R}_+, Y^\infty)$. Indeed, let $\mathbf{f} := \{f_n\}_{n \in \mathbb{N}}$ a countable dense subset of Y (fixed throughout the rest of the book) and define the following metric r on $\mathcal{B}(Y)$:

$$r(A_1, A_2) = \sum_{n=1}^{\infty} 2^{-n}(\|A_1 f_n - A_2 f_n\| \wedge 1), \qquad A_1, A_2 \in \mathcal{B}(Y).$$

$$(3.2)$$

We notice that:

$$r(A_1, A_2) = d_{Y^\infty}(\{A_1 f_n\}_{n \in \mathbb{N}}, \{A_2 f_n\}_{n \in \mathbb{N}}), \qquad (3.3)$$

where d_{Y^∞} is the usual metric on Y^∞:

$$d_{Y^\infty}(\{h_n\}_{n \in \mathbb{N}}, \{g_n\}_{n \in \mathbb{N}}) = \sum_{n=1}^{\infty} 2^{-n}(\|h_n - g_n\| \wedge 1),$$

$$\{h_n\}_{n \in \mathbb{N}}, \{g_n\}_{n \in \mathbb{N}} \in Y^\infty. \qquad (3.4)$$

For convenience we will denote throughout the rest of the book, sequences $\{g_n\}_{n \in \mathbb{N}}$ by $\mathbf{g}$. In the continuity of this approach, we equip $\mathbb{D}$ with the usual Skorokhod metric d on $D(\mathbb{R}_+, (\mathcal{B}(Y), r))$ (see [3], Chapter 3, equation 5.2). The latter is possible because $\mathbb{D}$ is included in $D(\mathbb{R}_+, (\mathcal{B}(Y), r))$, by definition of $\mathbb{D}$ and of the product topology. The latter inclusion is strict, cf. Proposition 3.2.1. In the following, d (without any subscript) will always represent the Skorokhod metric on $D(\mathbb{R}_+, E)$, possibly for different metric spaces E. We observe, using (3.3) and the definition of the Skorokhod metric that:

$$d(S_1, S_2) = d(S_1\mathbf{f}, S_2\mathbf{f}), \qquad S_1, S_2 \in \mathbb{D}, \tag{3.5}$$

where we recall that $d(S_1\mathbf{f}, S_2\mathbf{f})$ is the Skorokhod metric on $D(\mathbb{R}_+, Y^\infty)$, whereas $d(S_1, S_2)$ is the Skorokhod metric on $D(\mathbb{R}_+, (\mathcal{B}(Y), r))$.

2. The second possibility is to see $\mathbb{D}$ as $D(\mathbb{R}_+, Y)^\infty$ by defining the following metric d' on $\mathbb{D}$:

$$d'(S_1, S_2) = \sum_{n=0}^{\infty} 2^{-n}(d(S_1 f_n, S_2 f_n) \wedge 1), \qquad S_1, S_2 \in \mathbb{D}. \tag{3.6}$$

Again, recall that each $d(S_1 f_n, S_2 f_n)$ represents the Skorokhod metric in $D(\mathbb{R}_+, Y)$. We observe that:

$$d'(S_1, S_2) = d_{D(\mathbb{R}_+,Y)^\infty}(S_1\mathbf{f}, S_2\mathbf{f}), \qquad S_1, S_2 \in \mathbb{D}, \tag{3.7}$$

with $d_{D(\mathbb{R}_+,Y)^\infty}$ the usual metric on $D(\mathbb{R}_+, Y)^\infty$:

$$d_{D(\mathbb{R}_+,Y)^\infty}(\mathbf{s}^{(1)}, \mathbf{s}^{(2)}) = \sum_{n=1}^{\infty} 2^{-n}(d(s_n^{(1)}, s_n^{(2)}) \wedge 1), \quad \mathbf{s}^{(1)},$$
$$\mathbf{s}^{(2)} \in D(\mathbb{R}_+, Y)^\infty. \tag{3.8}$$

To the best of our knowledge, this problem has not been treated before. For various results on random linear operators, see for example [5]–[10]. In fact, it will be proved that the two metrics d and d' on $\mathbb{D}$ above are equivalent (Proposition 3.2.2). A case of interest is the following martingale problem: assume we are given a sequence of $\mathbb{D}$–valued martingales $\{M_n\}$, together with the generator $A(t)$ of a backward propagator $U(s, t)$ ($s \leq t$) on Y (see e.g. [7], Definition 2.2.1, Chapter 2), and that we have for all f in some dense subset of Y and $t \in \mathbb{R}_+$:

$$M_n(t)f = S_n(t)f - S_n(0)f - \int_0^t S_n(u)A(u)f du. \tag{3.9}$$

In this case, the convergence of the family of $D(\mathbb{R}_+, Y)$ random variables $\{S_n f\}$ for each f is not enough: we need the convergence of $\{S_n\}$ to some $\mathbb{D}$–valued random variable S to be able to conclude - modulo some technicalities - that the limit (in some sense) of the righthand side in the previous equation is:

$$S(t)f - S(0)f - \int_0^t S(u)A(u)f du, \tag{3.10}$$

and - if M is the limit of $\{M_n\}$ (in some sense) - study the unicity of solutions S to the martingale problem:

$$M(t)f = S(t)f - S(0)f - \int_0^t S(u)A(u)f\,du. \qquad (3.11)$$

For instance, if $M = 0$ almost surely, it can be proved under some conditions related to the well-posedness of the problem (in particular, $t \to S(t)f$ is continuous in the $Y-$norm for each $f \in Y$) that a solution S to the above problem is unique and equal to the backward propagator U whose generator is A. More generally, the main corollary of Theorem 3.4.1 is the following: assume that you have proved that the family $\{S_n f\}$ is relatively compact in $D(\mathbb{R}_+, Y)$ for each $f \in Y$, using for example the well-known techniques of e.g. [3], Chapter 3. In this case, because marginal relative compactness implies joint relative compactness, you get that $\{S_n f_1, S_n f_2, S_n f_3, ...\}$ is relatively compact in $D(\mathbb{R}_+, Y)^\infty$. Take a weakly converging subsequence k_n (denoting weak convergence by $\Rightarrow$):

$$(S_{k_n} f_1, S_{k_n} f_2, S_{k_n} f_3, ...) \Rightarrow (\alpha_1, \alpha_2, \alpha_3, ...) \text{ in } D(\mathbb{R}_+, Y)^\infty. \qquad (3.12)$$

Theorem 3.4.1 then allows you to conclude that there exists a $(\mathbb{D}, d)-$valued random variable S such that:

$$S_{k_n} \Rightarrow S \text{ in } (\mathbb{D}, d). \qquad (3.13)$$

From there, one is able to study the unicity in distribution of the operator S - as in the setting of the martingale problem discussed above - and finally conclude that $S_n \Rightarrow S$ in $(\mathbb{D}, d)$.

The book is organized as follows: $\mathbb{D}-$valued random variables are defined in Proposition 3.2.6, and the main results of the book are Theorems 3.3.1 and 3.4.1 about almost sure and weak convergence of $\mathbb{D}-$valued random variables, respectively. Section 3.2 is mostly technical and gives various properties about elements of $\mathbb{D}$ which will be used in the proofs of Theorems 3.3.1 and 3.4.1.

3.2 $\mathbb{D}-$Valued Random Variables and Various Properties on Elements of $\mathbb{D}$

Let $(\Omega, \mathcal{F}, \mathbb{P})$ a probability space on which we consider all our random variables (by which we mean as it is often done so: measurable maps), unless mentioned otherwise. In the following we use the notations introduced in the introduction, in particular we fix throughout the book a countable dense subset of Y, $\mathbf{f} := \{f_n\}_{n \in \mathbb{N}}$, assumed without loss of generality to be closed under finite rational linear combinations. Every metric space considered in this book will be equipped with its metric topology. Further, we will denote $||\cdot||_{op}$ the operator norm on $\mathcal{B}(Y)$.

In the definition of $\mathbb{D}$ (3.1), we required $Sf \in D(\mathbb{R}_+, Y)$ for every $f \in Y$. In fact we could have required only $Sf_j \in D(\mathbb{R}_+, Y)$ for every j under local boundedness in the operator norm, as the following simple proposition shows it:

Proposition 3.2.1. *Let $S \in \mathcal{B}(Y)^{\mathbb{R}_+}$. The following propositions are equivalent:*

    1. $S \in \mathbb{D}$.

    2. $Sf_j \in D(\mathbb{R}_+, Y)$, $\forall j$ and:

$$\sup_{t \in [0,T]} ||S(t)||_{op} < \infty, \quad \forall T \in \mathbb{R}_+. \tag{3.14}$$

Proof. *Assume 1). Then for every $g \in Y$ we have $Sg \in D(\mathbb{R}_+, Y)$, and therefore $\sup_{t \in [0,T]} ||S(t)g|| < \infty$. By the principle of uniform boundedness we have $\sup_{t \in [0,T]} ||S(t)||_{op} < \infty$.*

Now assume 2). Let $h \in [0,1]$ and $g \in Y$. Take a converging sequence $\mathbf{g} \subseteq \mathbf{f}$ such that $g_n \to g$. We have:

$$||S(t+h)g - S(t)g|| \leq ||S(t+h)g - S(t+h)g_n|| + ||S(t+h)g_n - S(t)g_n||$$
$$+ ||S(t)g_n - S(t)g||$$
$$\leq 2 \sup_{u \in [0,t+1]} ||S(u)||_{op}||g - g_n|| + ||S(t+h)g_n - S(t)g_n||. \tag{3.15}$$

First choose n such that $||g - g_n||$ is small, then choose h such that $||S(t + h)g_n - S(t)g_n||$ is small: this shows the right-continuity of Sg. To show the existence of the left-limit we observe that $S(t^-)g = \lim_{n \to \infty} S(t^-)g_n$. Indeed we have:

$$||S(t^-)g_n - S(t^-)g_m|| \leq ||S(t^-)g_n - S(t-h)g_n|| + ||S(t-h)g_n - S(t-h)g_m||$$
$$+ ||S(t-h)g_m - S(t^-)g_m||. \tag{3.16}$$

Let $\epsilon > 0$. Since the second term is bounded by $\sup_{u \in [0,t]} ||S(u)||_{op}||g_m - g_n||$, we can choose N_ϵ such that for $m, n \geq N_\epsilon$, the second term is less than ϵ. Then, for a given pair n, m, choose $h_{n,m}^\epsilon$ such that $||S(t^-)g_n - S(t - h_{n,m}^\epsilon)g_n|| < \epsilon$ and $||S(t^-)g_m - S(t - h_{n,m}^\epsilon)g_m|| < \epsilon$. This shows that the sequence $S(t^-)g_n$

is Cauchy in Y and, therefore, has a limit s^*. To see that $s^* = S(t^-)g$, just observe that:

$$||S(t-h)g - s^*|| \leq ||S(t-h)g - S(t-h)g_n|| + ||S(t-h)g_n - S(t^-)g_n||$$
$$+ ||S(t^-)g_n - s^*||.$$

$$(3.17)$$

The first term is bounded by $\sup_{u \in [0,t]} ||S(u)||_{op}||g - g_n||$. So choose n such that the first and third terms are small, then choose h such that the second term is small.

The following result shows that d and d' are in fact equivalent.

Proposition 3.2.2. *Let $\{S_n\}_{n \in \mathbb{N}}$, S elements of $\mathbb{D}$. Then $d(S_n, S) \to 0$ if and only if $d'(S_n, S) \to 0$ (which is equivalent to $d(S_n f_j, S f_j) \to 0$ for every j).*

Proof. *By (3.5) and (3.7), we get that $d(S_n, S) \to 0$ if and only if $S_n\mathbf{f} \to S\mathbf{f}$ in $D(\mathbb{R}_+, Y^\infty)$, and $d'(S_n, S) \to 0$ if and only if $S_n\mathbf{f} \to S\mathbf{f}$ in $D(\mathbb{R}_+, Y)^\infty$. Therefore $d(S_n, S) \to 0$ implies $d'(S_n, S) \to 0$. Now to see the converse, assume $d'(S_n, S) \to 0$. Because $\mathbf{f}$ is closed under finite rational linear combinations, we get that $S_n f_i + S_n f_j = S_n(f_i + f_j)$ is relatively compact in $D(\mathbb{R}_+, Y)$ for every i, j. By problems 22, 23 of [3], Chapter 3, we get that $S_n\mathbf{f}$ is relatively compact in $D(\mathbb{R}_+, Y^\infty)$. This implies that $S_n\mathbf{f} \to S\mathbf{f}$ in $D(\mathbb{R}_+, Y^\infty)$. Note that Problems 22, 23 of [3], Chapter 3, deal with the case $Y = \mathbb{R}$. Nevertheless, the results are still true for any separable Banach space Y.*

We would like our metrics d and d' to be independent of the choice of the family $\mathbf{f}$ in (3.5) and (3.7), because we would like limit points S of sequences $\mathbf{S}$ of elements of $\mathbb{D}$ to be independent of this choice. In the next proposition, we show that this choice has no particular importance under the condition that $\sup_{\substack{n \in \mathbb{N} \\ t \in [0,T]}} ||S_n(t)||_{op} < \infty$ for all $T \in \mathbb{R}_+$. Also, we show that this condition is necessary and sufficient in the following sense: assume we have $d(S_n, S) \to 0$ (resp. $d'(S_n, S) \to 0$) for some sequence $\mathbf{S}$ of elements of $\mathbb{D}$ and $S \in \mathbb{D}$. If we would also like that $d_{\mathbf{g}}(S_n, S) \to 0$ (resp. $d'_{\mathbf{g}}(S_n, S) \to 0$) for every $\mathbf{g}$ countable dense subset of Y, where $d_{\mathbf{g}}$ (resp. $d'_{\mathbf{g}}$) is the metric of (3.5) (resp. (3.7)) associated to the family $\mathbf{g}$, then we must have $\sup_{\substack{n \in \mathbb{N} \\ t \in [0,T]}} ||S_n(t)||_{op} < \infty$ for all $T \in \mathbb{R}_+$.

Proposition 3.2.3. *Let $\mathbf{S}$ be a sequence of elements of $\mathbb{D}$ and $S \in \mathbb{D}$. The following propositions are equivalent:*

1. *$d'(S_n, S) \to 0$ and:*

$$\sup_{\substack{n \in \mathbb{N} \\ t \in [0,T]}} ||S_n(t)||_{op} < \infty, \quad \forall T \in \mathbb{R}_+. \tag{3.18}$$

2. $d_{\mathbf{g}}(S_n, S) \to 0$ *for every* $\mathbf{g}$ *countable dense subset of* Y, *where* $d_{\mathbf{g}}$ *is the metric of (3.5) associated to the family* $\mathbf{g}$.

Further, if one of the above is true, in fact we have $d(S_n\mathbf{g}, S\mathbf{g}) \to 0$ *for every* $\mathbf{g} \in Y^\infty$.

Proof. a) *Assume 1). Let* $g \in Y$. *Let's first show that* $S_n g \to S g$ *in* $D(\mathbb{R}_+, Y)$. *There exists a sequence* $\mathbf{g} \subseteq \mathbf{f}$ *such that* $g_m \to g$ *in* Y *as* $m \to \infty$. *We have:*

$$d(S_n g, S g) \le d(S_n g, S_n g_m) + d(S_n g_m, S g_m) + d(S g_m, S g). \tag{3.19}$$

By definition of the Skorokhod topology:

$$d(S_n g, S_n g_m) \le \sup_{\substack{n \in \mathbb{N} \\ t \in [0,T]}} ||S_n(t)||_{op} ||g - g_m|| + e^{-T}, \tag{3.20}$$

$$d(S g_m, S g) \le \sup_{t \in [0,T]} ||S(t)||_{op} ||g - g_m|| + e^{-T}. \tag{3.21}$$

In the previous inequality, we have used the fact that $\sup_{t \in [0,T]} ||S(t)||_{op} < \infty$. *Indeed, since* $S \in \mathbb{D}$, *we get:*

$$\sup_{t \in [0,T]} ||S(t)h|| < \infty, \quad \forall T \in \mathbb{R}_+, \forall h \in Y. \tag{3.22}$$

Therefore by the principle of uniform boundedness we have $\sup_{t \in [0,T]} ||S(t)||_{op} < \infty \ \forall T \in \mathbb{R}_+$. *So first choose* m, T *such that* $||g - g_m||$ *and* e^{-T} *are small. Then choose* n *such that* $d(S_n g_m, S g_m)$ *is small. This shows that* $d(S_n g, S g) \to 0$.

By problems 22, 23 of [3], Chapter 3, we get that $S_n\mathbf{g}$ *is relatively compact in* $D(\mathbb{R}_+, Y^\infty)$ *for every* $\mathbf{g} \in Y^\infty$. *This implies that* $S_n\mathbf{g} \to S\mathbf{g}$ *in* $D(\mathbb{R}_+, Y^\infty)$. *Note that Problems 22, 23 of [3], Chapter 3, deal with the case* $Y = \mathbb{R}$. *Nevertheless, the results are still true for any separable Banach space* Y.

b) *Assume 2). We get* $d_{\mathbf{g}}(S_n, S) \to 0$ *for every countable dense subset* $\mathbf{g}$ *of* Y. *In particular, by Proposition 3.2.2, for each* $h \in Y$ *we have* $d(S_n h, S h) \to 0$. *Therefore by [3] (Proposition 5.3, Chapter 3) there exists a sequence of Lipschitz continuous functions* λ_n^h *such that:*

$$\sup_{t \in [0,T]} ||S_n(t)h - S(\lambda_n^h(t))h|| \to 0, \quad \forall T \in \mathbb{R}_+, \tag{3.23}$$

$$\lim_{n \to \infty} \sup_{s > t \ge 0} \left| \ln \left(\frac{\lambda_n^h(s) - \lambda_n^h(t)}{s - t} \right) \right| = 0. \tag{3.24}$$

But we have:

$$\sup_{t \in [0,T]} ||S_n(t)h|| \le \sup_{t \in [0,T]} ||S_n(t)h - S(\lambda_n^h(t))h|| + \sup_{t \in [0,T]} ||S(\lambda_n^h(t))h||. \tag{3.25}$$

There exists a N such that $\lambda_n(t) < 2t$, $\forall t \in [0,T]$, $\forall n \geq N$. For $n \geq N$ we have:

$$\sup_{t\in[0,T]} ||S_n(t)h|| \leq \sup_{t\in[0,T]} ||S_n(t)h - S(\lambda_n^h(t))h|| + \sup_{t\in[0,2T]} ||S(t)h||. \quad (3.26)$$

and because $\sup_{t\in[0,T]} ||S_n(t)h - S(\lambda_n^h(t))h|| \to 0$ we get:

$$\sup_{\substack{n\in\mathbb{N} \\ t\in[0,T]}} ||S_n(t)h|| < \infty. \quad (3.27)$$

By the principle of uniform boundedness, we get $\sup_{\substack{n\in\mathbb{N} \\ t\in[0,T]}} ||S_n(t)||_{op} < \infty$.

For convenience, let's introduce the subset $\mathbb{C}$ of $\mathbb{D}$ consisting of continuous functions:

$$\mathbb{C} := \{S \in \mathbb{D} : Sf \in C(\mathbb{R}_+, Y) \;\; \forall f \in Y\}. \quad (3.28)$$

We have a result similar to Proposition 3.2.1:

Proposition 3.2.4. *Let $S \in \mathcal{B}(Y)^{\mathbb{R}_+}$. The following propositions are equivalent:*

1. $S \in \mathbb{C}$.

2. $Sf_j \in C(\mathbb{R}_+, Y)$, $\forall j$ and:

$$\sup_{t\in[0,T]} ||S(t)||_{op} < \infty, \quad \forall T \in \mathbb{R}_+. \quad (3.29)$$

Proof. *Same proof as Proposition 3.2.1.*

The following result shows that under the setting of Proposition 3.2.3, if $d'(S_n, S) \to 0$ (equivalently, $d(S_n, S) \to 0$), then the limit S is bounded in the operator norm and we can quantify this bound.

Proposition 3.2.5. *Let $\mathbf{S}$ a sequence of elements of $\mathbb{D}$ and $S \in \mathbb{D}$. Assume as in Proposition 3.2.3 that $d'(S_n, S) \to 0$ and:*

$$\sup_{\substack{n\in\mathbb{N} \\ t\in[0,T]}} ||S_n(t)||_{op} < \infty, \quad \forall T \in \mathbb{R}_+. \quad (3.30)$$

Then we have:

$$||S(t)||_{op} \leq \liminf_{n\to\infty} \sup_{u\in[t-\epsilon t, t+\epsilon t]} ||S_n(u)||_{op}, \quad \forall t \in \mathbb{R}_+, \forall \epsilon \in (0,1]. \quad (3.31)$$

In addition, if $Sf_j \in C(\mathbb{R}_+, Y)$ for every j, then $S \in \mathbb{C}$ and:

$$||S(t)||_{op} \leq \liminf_{n\to\infty} ||S_n(t)||_{op}, \quad \forall t \in \mathbb{R}_+. \quad (3.32)$$

Proof. Let $h \in Y$ such that $||h|| = 1$. By Proposition 3.2.3 we have $d(S_n h, Sh) \to 0$. By [3] (Proposition 5.3, Chapter 3) there exists a sequence of Lipschitz continuous functions λ_n^h such that:

$$\sup_{t \in [0,T]} ||S_n(t)h - S(\lambda_n^h(t))h|| \to 0, \quad \forall T \in \mathbb{R}_+, \tag{3.33}$$

$$\lim_{n \to \infty} \sup_{s>t\geq 0} \left| \ln\left(\frac{\lambda_n^h(s) - \lambda_n^h(t)}{s-t} \right) \right| = 0. \tag{3.34}$$

Let $t \in \mathbb{R}_+$. We have:

$$||S(t)h|| \leq ||S(t)h - S_n(\lambda_n^h(t))h|| + ||S_n(\lambda_n^h(t))h||. \tag{3.35}$$

Fix $\epsilon \in (0,1]$. There exists a N_ϵ such that $-\epsilon t \leq \lambda_n^h(t) - t \leq \epsilon t, \forall n \geq N_\epsilon$. Therefore we get:

$$\liminf_{n \to \infty} ||S_n(\lambda_n^h(t))h|| = \lim_{n \to \infty} \inf_{k \geq n} ||S_k(\lambda_k^h(t))h||$$

$$= \lim_{n \to \infty} \inf_{k \geq (n \vee N_\epsilon)} ||S_k(\lambda_k^h(t))h||$$

$$\leq \lim_{n \to \infty} \inf_{k \geq (n \vee N_\epsilon)} \sup_{u \in [t-\epsilon t, t+\epsilon t]} ||S_k(u)h|| \tag{3.36}$$

$$\leq \liminf_{n \to \infty} \sup_{u \in [t-\epsilon t, t+\epsilon t]} ||S_n(u)||_{op}.$$

Therefore, because $\lim_{n \to \infty} ||S(t)h - S_n(\lambda_n^h(t))h|| = 0$:

$$||S(t)h|| \leq \liminf_{n \to \infty} \sup_{u \in [t-\epsilon t, t+\epsilon t]} ||S_n(u)||_{op} \tag{3.37}$$

$$\Rightarrow ||S(t)||_{op} \leq \liminf_{n \to \infty} \sup_{u \in [t-\epsilon t, t+\epsilon t]} ||S_n(u)||_{op}. \tag{3.38}$$

On the other hand, if $Sf_j \in C(\mathbb{R}_+, Y)$ for every j, then $S \in \mathbb{C}$ by Propositions 3.2.1 and 3.2.4 (since $S \in \mathbb{D}$). Therefore we have for every $h \in Y$:

$$\lim_{n \to \infty} \sup_{t \in [0,T]} ||S_n(t)h - S(t)h|| = 0. \tag{3.39}$$

And therefore:

$$||S(t)h|| \leq \liminf_{n \to \infty} ||S_n(t)h|| \leq \liminf_{n \to \infty} ||S_n(t)||_{op}, \tag{3.40}$$

$$\Rightarrow ||S(t)||_{op} \leq \liminf_{n \to \infty} ||S_n(t)||_{op}. \tag{3.41}$$

The following proposition introduces the notion of $\mathbb{D}$-valued random variables, which will be useful in the remaining of the book.

Proposition 3.2.6. *Assume that $S : \Omega \to \mathbb{D}$. Then S is a $(\mathbb{D}, d)$ (resp. $(\mathbb{D}, d')$) valued random variable if and only if Sf_n is a $D(\mathbb{R}_+, Y)$ valued random* *variable for every n. Further, in this case, for every countable dense subset $\mathbf{g}$ of Y, S is a $(\mathbb{D}, d_{\mathbf{g}})$ valued random variable, where $d_{\mathbf{g}}$ is the metric of (3.5) associated with the family $\mathbf{g}$. In the following we will use the terminology "$\mathbb{D}-$valued random variable" for such random variables.*

Further, if S_1, S_2 are $\mathbb{D}-$valued random variables, the following propositions are equivalent (denoting equality in distribution by $\overset{d}{=}$):

$S_1 \overset{d}{=} S_2$ in $(\mathbb{D}, d)$.

$S_1 \overset{d}{=} S_2$ in $(\mathbb{D}, d')$.

$S_1 \mathbf{f} \overset{d}{=} S_2 \mathbf{f}$ in $D(\mathbb{R}_+, Y^\infty)$.

$S_1 \mathbf{f} \overset{d}{=} S_2 \mathbf{f}$ in $D(\mathbb{R}_+, Y)^\infty$.

Proof. *$S f_n$ being a $D(\mathbb{R}_+, Y)$ valued random variable for every n is equivalent to $S\mathbf{f}$ being a $D(\mathbb{R}_+, Y)^\infty$ valued random variable. By (3.7), the latter is equivalent to $S\mathbf{f}$ being a $(\mathbb{D}, d')$ valued random variable. By Proposition 3.2.2, the metrics d and d' are equivalent and, therefore, the latter is equivalent to $S\mathbf{f}$ being a $(\mathbb{D}, d)$ valued random variable, which is equivalent by (3.5) to $S\mathbf{f}$ being a $D(\mathbb{R}_+, Y^\infty)$ valued random variable.*

Now, let $\omega \in \Omega$, and a sequence $\mathbf{f^i} \subseteq \mathbf{f}$ such that $f_n^i \to g_i$ as $n \to \infty$. We have for each $T \in \mathbb{R}_+$:

$$d(S(\omega)g_i, S(\omega)f_n^i) \leq \sup_{t \in [0,T]} ||S(\omega)(t)||_{op} ||g_i - f_n^i|| + e^{-T}, \qquad (3.42)$$

which shows that $S(\omega)f_n^i \to S(\omega)g_i$ in $D(\mathbb{R}_+, Y)$. Therefore Sg_i is a $D(\mathbb{R}_+, Y)$ valued random variable as the sure limit of a sequence of $D(\mathbb{R}_+, Y)$ valued random variables. By what we showed above, we get that S is a $(\mathbb{D}, d_{\mathbf{g}})$ (resp. $(\mathbb{D}, d'_{\mathbf{g}})$) valued random variable.

Further, (3.5) (resp. (3.7)) gives us that i) equivalent to iii) (resp. ii) equivalent to iv)). By Proposition 3.2.2, the metrics d and d' are equivalent, and therefore generate the same topology. Therefore i) equivalent to ii).

3.3　Almost Sure Convergence of $\mathbb{D}-$Valued Random Variables

Here is one of the two main results of the book on almost sure convergence of $\mathbb{D}-$valued random variables, in which the boundedness assumption is motivated by Proposition 3.2.3:

Theorem 3.3.1. *Let $\{S_n\}$ be a sequence of $\mathbb{D}-$valued random variables defined on $(\Omega, \mathcal{F}, \mathbb{P})$, uniformly bounded in the operator norm on compact inter-*

vals almost everywhere, namely for each $T \in \mathbb{R}_+$:

$$\sup_{\substack{t \in [0,T] \\ n \in \mathbb{N}}} ||S_n(\omega)(t)||_{op} < \infty, \quad \text{for a.e. } \omega \in \Omega. \tag{3.43}$$

If $\{S_n \mathbf{f}\}$ converges almost everywhere in $D(\mathbb{R}_+, Y)^\infty$ as $n \to \infty$, then $\{S_n\}$ converges almost everywhere to some $\mathbb{D}$–valued random variable S as $n \to \infty$ in $(\mathbb{D}, d)$. In this case, S doesn't depend on the sequence $\mathbf{f}$ in the sense that for every countable dense subset $\mathbf{g}$ of Y, $\{S_n\}$ are $(\mathbb{D}, d_\mathbf{g})$–valued random variables by Proposition 3.2.6 and $\{S_n\}$ converges almost everywhere to S as $n \to \infty$ in $(\mathbb{D}, d_\mathbf{g})$, where $d_\mathbf{g}$ is the metric of (3.5) associated to the family $\mathbf{g}$. Furthermore:

$$\mathbb{P}\left[||S(t)||_{op} \le \liminf_{n \to \infty} \sup_{u \in [t-\epsilon t, t+\epsilon t]} ||S_n(u)||_{op}, \quad \forall t \in \mathbb{R}_+, \forall \epsilon \in (0,1] \right] = 1. \tag{3.44}$$

In addition, if $\mathbb{P}\left[S f_j \in C(\mathbb{R}_+, Y) \right] = 1$ for every j we have $\mathbb{P}[S \in \mathbb{C}] = 1$ and:

$$\mathbb{P}\left[||S(t)||_{op} \le \liminf_{n \to \infty} ||S_n(t)||_{op}, \quad \forall t \in \mathbb{R}_+ \right] = 1. \tag{3.45}$$

Proof. *Assume that $\{S_n \mathbf{f}\}$ converges almost everywhere in $D(\mathbb{R}_+, Y)^\infty$ to some $\{S(\mathbf{f})\}$, say on Ω^*. Denote the random variable:*

$$M_T(\omega) := \sup_{\substack{t \in [0,T] \\ n \in \mathbb{N}}} ||S_n(\omega)(t)||_{op}. \tag{3.46}$$

If $\Omega_0 := \bigcap_{T \in \mathbb{Q}_+} \{M_T < \infty\}$, then we consider without loss of generality Ω^* to be $\Omega^* \cap \Omega_0$.

Let $g \in Y$. There exists a sequence $\mathbf{g} \subseteq \mathbf{f}$ such that $g_m \to g$ as $m \to \infty$. On Ω^* we have for every i, j, n, T:

$$d(S_n g_i, S_n g_j) \le M_T ||g_i - g_j|| + e^{-T}. \tag{3.47}$$

By continuity of d, we may take the limit as $n \to \infty$ in the previous inequality and get for every i, j, T:

$$d(S(g_i), S(g_j)) \le M_T ||g_i - g_j|| + e^{-T}, \tag{3.48}$$

which shows that the sequence $S(\mathbf{g})$ is Cauchy in $D(\mathbb{R}_+, Y)$, and therefore converges to some $S(g) \in D(\mathbb{R}_+, Y)$. Now to see that the latter is the pointwise limit of $S_n g$ on Ω^*, we observe that:

$$d(S(g), S_n g) \le d(S(g), S(g_i)) + d(S(g_i), S_n g_i) + d(S_n g_i, S_n g). \tag{3.49}$$

We have on Ω^* that $d(S_n g_i, S_n g) \leq M_T \|g_i - g\| + e^{-T}$ for every T. There-fore first choose i, T such that $d(S(g), S(g_i))$ and $d(S_n g_i, S_n g)$ are small, then choose n such that $d(S(g_i), S_n g_i)$ is small.

Now that $S(g)$ is well defined on Ω^* for every $g \in Y$, let's show the linearity of S on Ω^*. Let $\lambda \in \mathbb{R}$, $g, h \in Y$. We have just shown that $S_n g + \lambda S_n h = S_n(g + \lambda h) \to S(g + \lambda h)$. Since the only possible limit for $S_n g + \lambda S_n h$ is $S(g) + \lambda S(h)$, then $S(g) + \lambda S(h) = S(g + \lambda h)$. Usually, the addition is not continuous in the Skorokhod Space (the limit of the sum is not necessarily the sum of the limits). Nevertheless, in our setting, we can bypass this problem using the linearity and boundedness of S_n.

To show that $S \in \mathbb{D}$ on Ω^*, it remains to show the boundedness of S. Let $g \in Y$. By [3] (Proposition 5.3, Chapter 3), there exists a sequence of Lipschitz continuous functions λ_n^g such that for every T:

$$\sup_{t \in [0,T]} \|S_n(\lambda_n^g(t))g - S(g)(t)\| \to 0 \tag{3.50}$$

$$\lim_{n \to \infty} \sup_{s > t \geq 0} \left| \ln\left(\frac{\lambda_n^g(s) - \lambda_n^g(t)}{s - t}\right) \right| = 0. \tag{3.51}$$

There exists an N such that $\lambda_n^g(t) < 2t$, $\forall t \in \mathbb{R}_+$, $\forall n \geq N$. For $n \geq N$ and $t \in \mathbb{R}_+$ we have:

$$\|S(g)(t)\| \leq \|S(g)(t) - S_n(\lambda_n(t))g\| + \|S_n(\lambda_n(t))g\| \tag{3.52}$$

$$\leq \|S(g)(t) - S_n(\lambda_n(t))g\| + M_{2t}\|g\|. \tag{3.53}$$

Taking the limit as $n \to \infty$ we get $\|S(g)(t)\| \leq M_{2t}\|g\|$, which completes the proof that $S \in \mathbb{D}$ on Ω^*. By Proposition 3.2.3, $\{S_n\}$ converges to S in $(\mathbb{D}, d)$ on Ω^*. By Proposition 3.2.5 we immediately get the first upper bound for $\|S(t)\|_{op}$ on Ω^*.

Now that we know that $S \in \mathbb{D}$ on Ω^*, we note that S is indeed a $\mathbb{D}-$valued random variable by Proposition 3.2.6, because $S(f_n)$ are $D(\mathbb{R}_+, Y)$ valued ran-dom variables as almost sure limits of $D(\mathbb{R}_+, Y)$ valued random variables.

The fact that S doesn't depend on the choice of the countable dense set $\mathbf{f}$ is a direct consequence of Proposition 3.2.3. Indeed, by the latter proposition and if $\mathbf{g}$ is another dense countable subset of Y, we get on Ω^* that $\{S_n\}$ converges to S as $n \to \infty$ in $(\mathbb{D}, d_{\mathbf{g}})$, where $d_{\mathbf{g}}$ is the metric of (3.5) associated with the family $\mathbf{g}$.

Finally, let the probability one set:

$$\Omega_1^* := \bigcap_{j \in \mathbb{N}} \{S f_j \in C(\mathbb{R}_+, Y)\} \cap \Omega^*. \tag{3.54}$$

By Proposition 3.2.5 we immediately get $S \in \mathbb{C}$ on Ω_1^* as well as the second upper bound for $\|S(t)\|_{op}$ on Ω_1^*.

3.4 Weak Convergence of $\mathbb{D}$−Valued Random Variables

The other main result of this book is the weak convergence equivalent of Theorem 3.3.1. Assume that we proved that $S_n \Rightarrow S$ in $(\mathbb{D}, d)$ (where $\Rightarrow$ denotes weak convergence). We cannot hope to get estimations of $||S(t)||_{op}$ similar to those of Theorem 3.3.1, simply because S and S_n might be defined on different probability spaces and, therefore, it would make no sense to get a pathwise comparison of $||S(t)||_{op}$ and $||S_n(t)||_{op}$. If we still want some comparison results, the best we can do is to assume $||S_n(t)||_{op}$ is almost surely bounded by some deterministic function and derive a corresponding upper bound for $||S(t)||_{op}$.

Theorem 3.4.1. *Let $\{S_n\}$ be a sequence of $\mathbb{D}$−valued random variables defined on $(\Omega, \mathcal{F}, \mathbb{P})$, and define for each $T \in \mathbb{R}_+$:*

$$M_T^n(\omega) := \sup_{t \in [0,T]} ||S_n(\omega)(t)||_{op}. \tag{3.55}$$

If $\{S_n\mathbf{f}\}$ converges weakly in $D(\mathbb{R}_+, Y)^\infty$ as $n \to \infty$, and the sequence of real-valued random variables $\{M_T^n(\omega)\}_{n \in \mathbb{N}}$ is tight for each $T \in \mathbb{Q}_+$, then $\{S_n\}$ converges weakly to some $\mathbb{D}$−valued random variable S as $n \to \infty$ in $(\mathbb{D}, d)$. In this case, S doesn't depend on the sequence $\mathbf{f}$ in the sense that for every countable dense subset $\mathbf{g}$ of Y, $\{S_n\}$ are $(\mathbb{D}, d_\mathbf{g})$−valued random variables by Proposition 3.2.6 and $\{S_n\}$ converges weakly to S as $n \to \infty$ in $(\mathbb{D}, d_\mathbf{g})$, where $d_\mathbf{g}$ is the metric of (3.5) associated to the family $\mathbf{g}$. Furthermore, assume there exists a deterministic $m_n(t)$ such that for each n:

$$\mathbb{P}\left[||S_n(t)||_{op} \leq m_n(t), \forall t \in \mathbb{R}_+\right] = 1. \tag{3.56}$$

Then we have almost surely:

$$||S(t)||_{op} \leq \liminf_{n \to \infty} \sup_{u \in [t-\epsilon t, t+\epsilon t]} m_n(u), \quad \forall t \in \mathbb{R}_+, \forall \epsilon \in (0, 1]. \tag{3.57}$$

In addition, if $\mathbb{P}\left[S_n f_j \in C(\mathbb{R}_+, Y)\right] = 1$ for every j, n then we have almost surely:

$$S \in \mathbb{C} \text{ and } ||S(t)||_{op} \leq \liminf_{n \to \infty} m_n(t), \quad \forall t \in \mathbb{R}_+. \tag{3.58}$$

Proof. By assumption, the sequence $\{S_n\mathbf{f}, \{M_T^n\}_{T \in \mathbb{Q}}\}$ is tight in $D(\mathbb{R}_+, Y)^\infty \times \mathbb{R}^\infty$. Take a weakly converging subsequence:

$$\{S_{k_n}\mathbf{f}, \{M_T^{k_n}\}_{T \in \mathbb{Q}}\} \Rightarrow \{S(\mathbf{f}), \{M_T\}_{T \in \mathbb{Q}}\}. \tag{3.59}$$

By the Skorokhod representation theorem, we can consider this convergence to be almost sure, i.e. there exists a probability space $(\Omega', \mathcal{F}', \mathbb{P}')$ and random *variables with the same distributions as the previous ones (denoted by the subscript '), such that:*

$$\{S_{k_n}\mathbf{f}, \{M_T^{k_n}\}_{T\in\mathbb{Q}}\} \stackrel{d}{=} \{S_n'(\mathbf{f}), \{M_{n,T}'\}_{T\in\mathbb{Q}}\} \text{ for each } n,$$
$$S_n'(\mathbf{f}) \stackrel{a.s.}{\to} S'(\mathbf{f}), \tag{3.60}$$
$$\{M_{n,T}'\}_{T\in\mathbb{Q}} \stackrel{a.s.}{\to} \{\bar{M}_T'\}_{T\in\mathbb{Q}}.$$

We will prove that S_n' satisfies the conditions of Theorem 3.3.1, and apply the latter to conclude. For this, we need to prove that S_n' is a $\mathbb{D}$–valued random variable, and that it is uniformly bounded in the operator norm on compact intervals almost surely.

Since for each $T \in \mathbb{Q}_+$, $M_{n,T}' \stackrel{a.s.}{\to} \bar{M}_T'$, the random variable $M_T' := \sup_n M_{n,T}'$ is a.s. finite for each $T \in \mathbb{Q}_+$. Without loss of generality restrict Ω' to the set of probability one $\cap_{T\in\mathbb{Q}_+}\{M_T' < \infty\}$.

Because $\mathbf{f}$ is closed under finite rational combinations and using the equality in distribution:

$$\{S_{k_n}\mathbf{f}, \{M_T^{k_n}\}_{T\in\mathbb{Q}}\} \stackrel{d}{=} \{S_n'(\mathbf{f}), \{M_{n,T}'\}_{T\in\mathbb{Q}}\} \text{ for each } n, \tag{3.61}$$

we have for each $\lambda \in \mathbb{Q}, T \in \mathbb{Q}_+, i, j, n \in \mathbb{N}$:

$$\mathbb{P}'\left[d(S_n'(f_j), S_n'(f_i)) \le M_{n,T}'||f_i - f_j|| + e^{-T}\right] = 1,$$
$$\mathbb{P}'\left[S_n'(f_j) + \lambda S_n'(f_i) - S_n'(f_j + \lambda f_i) = 0\right] = 1, \tag{3.62}$$
$$\mathbb{P}'\left[\sup_{t\in[0,T]} ||S_n'(f_j)(t)|| \le M_{n,T}'||f_j||\right] = 1.$$

Let Ω_0' the corresponding probability one subset of Ω' (taking intersection over all n, i, j, T, λ). Therefore on Ω_0' we have in fact for each $T \in \mathbb{Q}_+$:

$$\mathbb{P}'\left[d(S_n'(f_j), S_n'(f_i)) \le M_T'||f_i - f_j|| + e^{-T}\right] = 1,$$
$$\mathbb{P}'\left[\sup_{t\in[0,T]} ||S_n'(f_j)(t)|| \le M_T'||f_j||\right] = 1. \tag{3.63}$$

Let $g \in Y$. There exists a sequence $\mathbf{g} \subseteq \mathbf{f}$ such that $g_j \to g$ as $j \to \infty$. On Ω_0' we have for every i, j, n and $T \in \mathbb{Q}_+$:

$$d(S_n'(g_i), S_n'(g_j)) \le M_T'||g_i - g_j|| + e^{-T}. \tag{3.64}$$

This shows that the sequence $S_n'(\mathbf{g})$ is Cauchy in $D(\mathbb{R}_+, Y)$ (on Ω_0') and therefore converges to some $S_n'(g)$ in $D(\mathbb{R}_+, Y)$. To show that S_n' is linear on Ω_0', take $g, h \in Y$ and $\lambda \in \mathbb{R}$. There exists $\mathbf{g}, \mathbf{h} \subseteq \mathbf{f}$ and $\lambda \subseteq \mathbb{Q}$ such that these sequences tend respectively to g, h, λ. The only possible limit for $S_n'(g_j) + \lambda_j S_n'(h_j)$ as $j \to \infty$ is $S_n'(g) + \lambda S_n'(h)$. By definition of Ω_0' we have $S_n'(g_j) + \lambda_j S_n'(h_j) = S_n'(g_j + \lambda_j h_j)$. And we have shown that

$S'_n(g_j + \lambda_j h_j) \to S'_n(g + \lambda h)$, *since* $\mathbf{f}$ *is closed under finite rational combinations. Therefore* $S'_n(g + \lambda h) = S'_n(g) + \lambda S'_n(h)$.

To finish the proof that S'_n *is a* $\mathbb{D}$−*valued random variable, it remains to show the boundedness of* S'_n *on* Ω'_0. *By [3] (Proposition 5.3, Chapter 3), there exists a sequence of Lipschitz continuous functions* λ_j *such that for every* T:

$$\sup_{t \in [0,T]} ||S'_n(g_j)(\lambda_j(t)) - S'_n(g)(t)|| \to 0 \ (as \ j \to \infty) \tag{3.65}$$

$$\lim_{j \to \infty} \sup_{s > t \geq 0} \left| \ln\left(\frac{\lambda_j(s) - \lambda_j(t)}{s - t} \right) \right| = 0. \tag{3.66}$$

There exists a N *such that* $\lambda_j(t) < 2t$, $\forall t \in \mathbb{R}_+$, $\forall j \geq N$. *For* $j \geq N$ *and* $t \in \mathbb{R}_+$ *we have:*

$$||S'_n(g)(t)|| \leq ||S'_n(g)(t) - S'_n(g_j)(\lambda_j(t))|| + ||S'_n(g_j)(\lambda_j(t))||. \tag{3.67}$$

Denote $\lceil t \rceil$ *the smallest integer greater than t. We have (on* Ω'_0, *by definition of it) for* $j \geq N$:

$$\sup_{u \in [0, \lceil 2t \rceil]} ||S'_n(g_j)(u)|| \leq M'_{\lceil 2t \rceil} ||g_j|| \tag{3.68}$$

$$\Rightarrow ||S'_n(g_j)(\lambda_j(t))|| \leq M'_{\lceil 2t \rceil} ||g_j||. \tag{3.69}$$

And therefore:

$$||S'_n(g)(t)|| \leq ||S'_n(g)(t) - S'_n(g_j)(\lambda_j(t))|| + M'_{\lceil 2t \rceil} ||g_j||. \tag{3.70}$$

Taking the limit as $j \to \infty$ *we get* $||S'_n(g)(t)|| \leq M'_{\lceil 2t \rceil} ||g||$, *which completes the proof that* $S'_n \in \mathbb{D}$ *on* Ω'_0 *for each n, and that we have for each* T:

$$\sup_{\substack{t \in [0,T] \\ n \in \mathbb{N}}} ||S'_n(t)||_{op} \leq M'_{\lceil 2T \rceil} < \infty, \quad on \ \Omega'_0. \tag{3.71}$$

We can now use Theorem 3.3.1 to get that S'_n *converges almost surely (in our case, on* Ω'_0) *to a* $\mathbb{D}$−*valued random variable* S' *in* $(\mathbb{D}, d)$, *which doesn't depend on the choice of the countable dense family* $\mathbf{f}$. *To see that we have* $S_n \Rightarrow S'$ *in* $(\mathbb{D}, d)$, *fix* $S^* \in \mathbb{D}$ *and* $\delta > 0$ *and observe that by the Portmanteau theorem, using the fact that* $S_{k_n} \mathbf{f} \overset{d}{=} S'_n(\mathbf{f})$ *in* $D(\mathbb{R}_+, Y)^\infty$ *(and therefore in* $D(\mathbb{R}_+, Y^\infty)$ *by Proposition 3.2.6) together with (3.5):*

$$\limsup_{n \to \infty} \mathbb{P}\left[d(S^*, S_n) \leq \delta \right] = \limsup_{n \to \infty} \mathbb{P}\left[d(S^*\mathbf{f}, S_n\mathbf{f}) \leq \delta \right]$$

$$= \limsup_{n \to \infty} \mathbb{P}\left[d(S^*\mathbf{f}, S_{k_n}\mathbf{f}) \leq \delta \right] = \limsup_{n \to \infty} \mathbb{P}'\left[d(S^*\mathbf{f}, S'_n(\mathbf{f})) \leq \delta \right] \tag{3.72}$$

$$\leq \mathbb{P}'\left[d(S^*\mathbf{f}, S'\mathbf{f}) \leq \delta \right] = \mathbb{P}'\left[d(S^*, S') \leq \delta \right].$$

The second equality in the previous formula comes from the unicity of the limit in distribution of $S_n\mathbf{f}$ in $D(\mathbb{R}_+, Y^\infty)$, by assumption. Now, assume that there exists a deterministic $m_n(t)$ such that

$$\mathbb{P}\left[||S_n(t)||_{op} \leq m_n(t), \forall t \in \mathbb{R}_+\right] = 1. \tag{3.73}$$

Denote the set:

$$B_{n,j} := \{x \in D(\mathbb{R}_+, Y) : ||x(t)|| \leq m_n(t)||f_j||, \forall t \in \mathbb{R}_+\}. \tag{3.74}$$

By equality in distribution, we have for every j, n:

$$\mathbb{P}'\left[S_n'(f_j) \in B_{n,j}\right] = 1, \tag{3.75}$$

We can modify Ω_0' in (3.62) to incorporate the latter probability one sets (taking the intersection over j, n). Now, let $\epsilon \in (0, 1]$. In (3.67), we can find a N_ϵ such that $-\frac{\epsilon}{3}t < \lambda_j(t) - t < \frac{\epsilon}{3}t$, $\forall t \in \mathbb{R}_+$, $\forall j \geq N_\epsilon$ and, therefore, we get instead of (3.70) $\forall t \in \mathbb{R}_+$, $\forall j \geq N_\epsilon$:

$$||S_n'(g)(t)|| \leq ||S_n'(g)(t) - S_n'(g_j)(\lambda_j(t))|| + m_n(\lambda_j(t))||g_j|| \tag{3.76}$$

$$\leq ||S_n'(g)(t) - S_n'(g_j)(\lambda_j(t))|| + \sup_{u \in [t-\frac{\epsilon}{3}t, t+\frac{\epsilon}{3}t]} m_n(u)||g_j||. \tag{3.77}$$

Taking the limit as $j \to \infty$ we get:

$$||S_n'(g)(t)|| \leq \sup_{u \in [t-\frac{\epsilon}{3}t, t+\frac{\epsilon}{3}t]} m_n(u)||g||. \tag{3.78}$$

And therefore $||S_n'(t)||_{op} \leq \sup_{u \in [t-\frac{\epsilon}{3}t, t+\frac{\epsilon}{3}t]} m_n(u)$. Now, we can apply Proposition 3.2.5 and get:

$$||S'(t)||_{op} \leq \liminf_{n \to \infty} \sup_{u \in [t-\frac{\epsilon}{3}t, t+\frac{\epsilon}{3}t]} ||S_n'(u)||_{op}, \quad \forall t \in \mathbb{R}_+, \forall \epsilon \in (0,1] \tag{3.79}$$

$$\leq \liminf_{n \to \infty} \sup_{u \in [t-\frac{\epsilon}{3}t, t+\frac{\epsilon}{3}t]} \sup_{v \in [u-\frac{\epsilon}{3}u, u+\frac{\epsilon}{3}u]} m_n(v) \tag{3.80}$$

$$= \liminf_{n \to \infty} \sup_{v \in [t-2\frac{\epsilon}{3}t-\frac{\epsilon^2}{9}t, t+2\frac{\epsilon}{3}t+\frac{\epsilon^2}{9}t]} m_n(v) \tag{3.81}$$

$$\leq \liminf_{n \to \infty} \sup_{u \in [t-\epsilon t, t+\epsilon t]} m_n(u). \tag{3.82}$$

Now assume that $\mathbb{P}\left[S_n f_j \in C(\mathbb{R}_+, Y)\right] = 1$ for every j, n. This implies $\mathbb{P}'\left[S_n'(f_j) \in C(\mathbb{R}_+, Y)\right] = 1$ for every j, n. Incorporate the latter probability one sets (taking intersection over j, n) into Ω_0'. Because $S_n'(g)$ was defined to be the limit of $S_n'(g_j)$ in $D(\mathbb{R}_+, Y)$, we get $S_n'(g) \in C(\mathbb{R}_+, Y)$ for every $g \in Y$, and, therefore, (3.78) becomes:

$$||S_n'(g)(t)|| \leq m_n(t)||g||, \quad \forall t \in \mathbb{R}_+. \tag{3.83}$$

$$\Rightarrow ||S_n'(t)||_{op} \leq m_n(t), \quad \forall t \in \mathbb{R}_+. \tag{3.84}$$

The application of Proposition 3.2.5 yields (on Ω_0'):

$$||S'(t)||_{op} \leq \liminf_{n \to \infty} ||S_n'(t)||_{op} \leq \liminf_{n \to \infty} m_n(t), \quad \forall t \in \mathbb{R}_+. \tag{3.85}$$

Remark 3.4.2. *In the last proof (Theorem 3.4.1), for the last estimation (3.58), we need the condition $S_n f_j \in C(\mathbb{R}_+, Y)$ a.e. for every j, n, and not only the weaker $S f_j \in C(\mathbb{R}_+, Y)$ a.e. for every j as in Theorem 3.3.1. This is because in the proof of Theorem 3.4.1, we need to construct the $\mathbb{D}-valued$ random variable S'_n, whereas we already have S_n in Theorem 3.3.1. So if we only assume $S f_j \in C(\mathbb{R}_+, Y)$ a.e. for every j, we will only get almost surely:*

$$||S'_n(t)||_{op} \leq \sup_{u \in [t-\epsilon t, t+\epsilon t]} m_n(u), \quad \forall t \in \mathbb{R}_+, \tag{3.86}$$

as in the non-continuous case, and we will not be able to get an estimation of type (3.58).

Bibliography

[1] Billingsley, P. *Convergence of Probability Measures*, John Wiley & Sons, Inc., 1999.

[2] Ethier, S., Kurtz, T. *Markov Processes: Characterization and Convergence*, John Wiley, 1986.

[3] Gulisashvili, A., van Casteren, J. *Non Autonomous Kato Classes and Feynman-Kac Propagators*, World Scientific Publishing Co. Pte. Ltd, 1986.

[4] Ledoux, M., Talagrand, M. *Probability in Banach Spaces: Isoperimetry and Processes*, Springer-Verlag, 1991

[5] Skorokhod, A. V. *Random Linear Operators*, Reidel Publishing Company, Dordrecht, 1984.

[6] Thang, D.H. *Random operators in Banach spaces*, Probab. Math. Statist. 8 (1987), 155-157.

[7] Thang, D.H. *The adjoint and the composition of random operators on a Hilbert space.*, Stochastic and Stochastic Reports 54 (1995), 53-73.

[8] Thang, D.H. *On the convergence of random mappings*, Vietnam Journal of Mathematics, no. 28, p. 71-80, 2000.

[9] Thang, D.H. and Thinh, N. *Random bounded operators and their extension*, Kyushu J. Math. 58 (2004), 257-276.

[10] Thang, D.H. *Transforming random operators into random bounded operators*, Random Operators / Stochastic Eqs. 16 (2008), 293-302.

[11] N. Vadori. *Semi-Markov Driven Models: Limit Theorems and Financial Applications*. PhD Thesis, University of Calgary, Calgary, AB, Canada, 2015.

[12] Vadori, N. and Swishchuk, A. Convergence of random bounded linear operators in the Skorokhod space. *Random Operators and Stochastic Equations*, 2019, 27 (3), 1-13 (https://doi.org/10.1515/rose-2019-2011).

[13] Watkins, J. *A CLT in random evolution.* Ann. Prob. 12. 2, 480-513, 1984

[14] Watkins, J. *A stochastic integral representation for random evolution.* Ann. Prob. 13. 2, 531-557, 1985

[15] Watkins, J. *Limit theorems for stationary random evolutions.* Stoch. Pr. Appl. 19, 189-224, 1985

Part II

Homogeneous and Inhomogeneous Random Evolutions

4

Homogeneous Random Evolutions (HREs) and their Applications

Chapter 4 is devoted to the definitions and properties of *homogeneous random evolutions* (HREs), limit theorems for REs, such as LLN and CLT, and also presents many examples of HREs. This Chapter 4 is a bridge between *homogeneous* and *inhomogeneous* REs (described in the following Chapter 5), and is to show the main differences and some similarities between these two random dynamics.

4.1 Random Evolutions

4.1.1 Definition and Classification of Random Evolutions

Let $(\Omega, \mathcal{F}, \mathcal{F}_t, \mathcal{P})$ be a probability space, $t \in R_+ := [0, +\infty]$, let (X, Ξ) be a measurable phase space, and let $(B, \mathcal{B}, \| \cdot \|)$ be a separable Banach space.

Let us consider a Markov renewal process $(x_n, \theta_n; n \geq 0)$, $x_n \in X$, $\theta_n \in R_+$, $n \geq 0$, with stochastic kernel

$$Q(x, A, t) := P(x, A)G_x(t),$$
$$P(x, A) := \mathcal{P}(x_{n+1} \in A/x_n = x), \tag{4.1}$$
$$G_x(t) := \mathcal{P}(\theta_{n+1}/x_n = x),$$

$x \in X, a \in \Xi, t \in R_+$. Process $x_t := x_{\nu(t)}$ is called a semi-Markov process, where $\nu(t) := \max\{n : \tau_n \leq t\}$, $\tau_n := \sum_{k=0}^{n} \theta_k$, $x_n = x_{\tau_n}$, $\mathcal{P}\{\nu(t) < +\infty, \forall t \in R_+\} = 1$. We note, that if $G_x(t) = 1 - e^{-\lambda(x)t}$, where $\lambda(x)$ is a measurable and bounded function on X, then x_t is called a jump Markov process.

Let $\{\Gamma(x); x \in X\}$ be a family of operators on the dense subspace $B_0 \in B$, which is common domain for $\Gamma(x)$, independent of x, noncommuting and unbounded in general, such that map $\Gamma(x)f : X \to B$ is strongly $\Xi/\mathcal{B}$-measurable for all $f \in B$, $\forall t \in R_+$; also, let $\{\mathcal{D}(x, y); x, y \in X\}$ be a family of bounded linear operators on B, such that map $\mathcal{D}(x, y)f : X \times X \to B$ is $\Xi \times \Xi/\mathcal{B}$-measurable, $\forall f \in B$.

Random Evolution (RE) is defined by the solution of stochastic operator integral equation in separable Banach space B:

$$V(t)f = f + \int_0^t V(s)\Gamma(x_s)f\,ds + \sum_{k=1}^{\nu(t)} V(\tau_k-)[\mathcal{D}(x_{k-1}, x_k) - I]f, \qquad (4.2)$$

where I is an identity operator on B, $\tau_k- := \tau_k - 0, f \in B$.

If x_t in (4.1) is a Markov or semi-Markov process, then RE in (4.2) is called a **Markov or semi-Markov RE**, respectively.

If $\mathcal{D}(x,y) \equiv I$, $\forall x,y \in X$, then $V(t)$ in (4.2) is called **a continuous RE**.

If $\Gamma(x) \equiv 0$, $\forall x \in X$, is a zero operator on B, then $V(t)$ in (4.2) is called a **jump RE**.

RE $V_n := V(\tau_n)$ is called a **discrete RE** .

Operators $\Gamma(x)$, $x \in X$, describe a continuous component $V^c(t)$ of RE $V(t)$ in (4.2), and operators $\mathcal{D}(x,y)$ describe a jump component $V^d(t)$ of RE $V^d(t)$ in (4.2).

In such a way, RE is described by two objects: 1) operator dynamical system $V(t)$; 2) random process x_t.

We note, that it turned out to be [5, 6]

$$V(t) = \Gamma_{x_t}(t - \tau_{\nu(t)}) \prod_{k=1}^{\nu(t)} \mathcal{D}(x_{k-1}, x_k)\Gamma_{x_{k-1}}(\theta_k), \qquad (4.3)$$

where $\Gamma_x(t)$ are the semigroups of operators (see Section 1.3) of t generated by the operators $\Gamma(x)$, $\forall x \in X$. We also note, that RE in (4.2) is usually called a **discontinuous RE**. Under the above introduced conditions the solution $V(t)$ of the equation (4.2) is unique and can be represented by product (4.3), that can be proved by constructive method [5].

Remark. *From the definition of random evolutions it follows that they are other examples of MOFs, as they satisfy all the conditions for MOFs.*

4.1.2   Some Examples of RE

Connection of RE with applied problems is explained by the generality of definition (4.2) of RE. It includes any homogeneous linear evolutionary system. If, for example,

$$\Gamma(x) := v(x)\frac{d}{dz}, \mathcal{D}(x,y) \equiv I, B = C^1(R),$$

then the equation (4.2) is a transport equation, which describes a motion of particle with random velocity $v(x_t)$. In such a way, various interpretations of operators $\Gamma(x)$ and $\mathcal{D}(x,y)$ give us many realizations of RE.

Example 1. Impulse traffic process. Let $B = C(R)$ and operators $\Gamma(x)$ and $\mathcal{D}(x, y)$ are defined by the following way:

$$\Gamma(x)f(z) := v(z, x)\frac{d}{dz}f(z), \mathcal{D}(x, y)f(z) := f(z + a(x, y)), \qquad (4.4)$$

where functions $v(z, x)$ and $a(x, y)$ are continuous and bounded on $R \times X$ and $X \times X$ respectively, $\forall z \in R$, $\forall x, y \in X$, $f(z) \in C^1(R) := B_0$. Then the equation (4.2) takes the form:

$$f(z_t) = f(z) + \int_0^t v(z_s, x_s)\frac{d}{dz}f(z_s)ds + \sum_{k=1}^{\nu(t)}[f(z_{\tau_k-} + a(x_{k-1}, x_k)) - f(z_{\tau_k-})],$$

$$(4.5)$$

and RE $V(t)$ is defined by the relation:

$$V(t)f(z) = f(z_t),$$

$z_0 = z$. Equation (4.5) is a functional one for **impulse traffic process** z_t, which satisfies the equation:

$$z_t = z + \int_0^t v(z_s, x_s)ds + \sum_{k=1}^{\nu(t)} a(x_{k-1}, x_k). \qquad (4.6)$$

We note that impulse traffic process z_t in (4.6) is a realization of discontinuous RE.

Example 2. Summation on a Markov chain. Let us put $v(z, x) \equiv 0$, $\forall z \in R$, $\forall x \in X$, in (4.6). Then the process

$$z_t = z + \sum_{k=1}^{\nu(t)} a(x_{k-1}, x_k) \qquad (4.7)$$

is a summation on a Markov chain $(x_n; n \geq 0)$, and it is a realization of a jump RE. Let $z_n := z_{\tau_n}$ in (4.7). Then discrete process

$$z_n = z + \sum_{k=1}^{n} a(x_{k-1}, x_k)$$

is a realization of a discrete RE.

Example 3. Diffusion process in random media. Let $B = C(R)$, $B_0 = C^2(R)$, $P_x(t, z, A)$ be a Markov continuous distribution function, which respects to the diffusion process $\xi(t)$, that is the solution of the stochastic differential equation in R with semi-Markov switchings:

$$d\xi(t) = \mu(\xi(t), x_t)dt + \sigma(\xi(t), x_t)dw_t, \xi(0) = z, \qquad (4.8)$$

where x_t is a semi-Markov process independent on a standard Wiener process w_t, coefficients $\mu(z,x)$ and $\sigma(z,x)$ are bounded and continuous functions on $R \times X$. Let us define the following contraction semigroups of operators on B:

$$\Gamma_x(t)f(z) := \int_R P_x(t,z,dy)f(y), f(y) \in B, x \in X. \tag{4.9}$$

Their infinitesimal operators $\Gamma(x)$ have the following kind:

$$\Gamma(x)f(z) = \mu(z,x)\frac{d}{dz}f(z) + 2^{-1}\sigma^2(z,x)\frac{d^2}{dz^2}f(z), f(z) \in B_0.$$

The process $\xi(t)$ is continuous one, that is why the operators $\mathcal{D}(x,y) \equiv I$, $\forall x,y \in X$, are identity operators. Then the equation (4.2) takes the form:

$$f(\xi(t)) = f(z) + \int_0^t [\mu(\xi(s),x_s)\frac{d}{dz} + 2^{-1}\sigma^2(\xi(s),x_s)\frac{d^2}{dz^2}]f(\xi(s))ds, \tag{4.10}$$

and RE $V(t)$ is defined by the relation

$$V(t)f(z) = E[f(\xi(t))/x_s; 0 \le s \le t; \xi(0) = z].$$

Equation (4.10) is a functional one for diffusion process $\xi(t)$ in (4.8) semi-Markov random media x_t. We note that diffusion process $\xi(t)$ in (4.8) is a realization of continuous RE.

Example 4. The Geometric Markov Renewal Process (GMRP) [12]. Let $(x_n, \theta_n)_{n \in \mathbf{Z}^+}$ be a Markov renewal process on the phase space $X \times \mathbf{R}^+$ with the semi-Markov kernel $Q(x,A,t)$ and $x(t) := x_{\nu(t)}$ be a semi-Markov process. Let $\rho(x)$ be a bounded continuous function on X such that $\rho(x) > -1$. We define a stochastic functional S_t with Markov renewal process $(x_n; \theta_n)_{n \in Z_+}$ as follows:

$$S_t := S_0 \prod_{k=0}^{\nu(t)} (1 + \rho(x_k)), \tag{4.11}$$

where $S_0 > 0$ is the initial value of S_t. We call the process $(S_t)_{t \in \mathbf{R}_+}$ in (4.11) a geometric Markov renewal process (GMRP). This process $(S_t)_{t \in \mathbf{R}_+}$ we call such by analogy with the geometric compound Poisson process

$$S_t = S_0 \prod_{k=1}^{N(t)} (1 + Y_k), \tag{4.12}$$

where $S_0 > 0$, $N(t)$ is a standard Poisson process, $(Y_k)_{k \in Z_+}$ are iid random variable, which is a trading model in many financial applications as a pure jump model (see [12] and Chapter 6).

Let $B : C_0(\mathbf{R}_+)$ be a space of continuous functions on $\mathbf{R}_+$, vanishing at infinity, and let us define a family of bounded contracting operators $D(x)$ on $C_0(\mathbf{R}_+)$:

$$D(x)f(s) := f(s(1 + \rho(x)), \ x \in X, s \in \mathbf{R}_+. \tag{4.13}$$

With these contraction operators $D(x)$ we define the following jump semi-Markov random evolution (JSMRE) $V(t)$ of geometric Markov renewal process in (4.11)

$$V(t) = \prod_{k=0}^{\nu(t)} D(x_k) := D(x_{\nu(t)}) \circ D(x_{\nu(t)-1}) \circ \ldots \circ D(x_1) \circ D(x_0). \tag{4.14}$$

Using (4.13) we obtain from (4.14)

$$V(t)f(s) = \prod_{k=0}^{\nu(t)} D(x_k)f(s) = f(s \prod_{k=0}^{\nu(t)} (1 + \rho(x_k)) = f(S_t), \tag{4.15}$$

where S_t is defined in (4.11) and $S_0 = s$.

4.1.3   Martingale Characterization of Random Evolutions

The main approaches to the study of RE are martingale methods (see Chapter 2). The main idea is that process

$$M_n := V_n - I - \sum_{k=0}^{n-1} E[V_{k+1} - V_k/\mathcal{F}_k], V_0 = I, \tag{4.16}$$

is an $\mathcal{F}_n$-martingale in B, where

$$\mathcal{F}_n := \sigma x_k, \tau_k; 0 \le k \le n, V_n := V(\tau_n),$$

E is an expectation by probability $\mathcal{P}$. Representation of the martingale M_n (see(4.4)) in the form of martingale-difference

$$M_n = \sum_{k=0}^{n-1} [V_{k+1} - E(V_{k+1}/\mathcal{F}_k)] \tag{4.17}$$

gives us the possibility to calculate the weak quadratic variation:

$$< l(M_n f) > := \sum_{k=0}^{n-1} E[l^2((V_{k+1} - V_k)f)/\mathcal{F}_k], \tag{4.18}$$

where $l \in B^*$, and B^* is a dual space to B, dividing points of B. The martingale method of obtaining of the limit theorems for the sequence of RE is founded on the solution of the following problems: 1) weak compactness of the family of measures generated by the sequences of RE; 2) any limiting point of this family of measures is the solution of martingale problem; 3) the solution of martingale problem is unique. The conditions 1)-2) quarantee the existence of weakly converging subsequence, and condition 3) gives the uniqueness of the weak limit. It follows from 1)-3) that the consequence of RE converges weak to the unique solution of martingale problem. The weak convergence of RE in series scheme we obtain from the criterion of weak compactness of the processes with values in separable Banach space [5]. The limit RE we obtain from the solution of some martingale problem in form of some integral operator equations in Banach space B. We also use the representation

$$V_{k+1} - V_k = [\Gamma_{x_k}(\theta_{k+1})\mathcal{D}(x_k, x_{k+1}) - I]V_k, V_k := V(\tau_k), \qquad (4.19)$$

and the following expression for semigroups of operators $\Gamma_x(t)$ [5]:

$$\Gamma_x(t)f = f + \sum_{k=1}^{n-1} \frac{t^k}{k!}\Gamma^k(x)f + (n-1)^{-1}\int_0^t (t-s)^n\Gamma_x(s)\Gamma^n(x)fds, \forall x \in X,$$

$$(4.20)$$

$\forall f \in \cap_{x \in X} Dom(\Gamma^n(x))$. Taking into account (4.4)-(4.8) we obtain the limit theorems for RE. In the previous subsection we considered the evolution equation assosiated with random evolutions by using the jump structure of the semi-Markov process or jump Markov process. In order to deal with more general driving processes and to consider other applications, it is useful to re-formulate the treatment of random evolution in terms of a **martingale problem.** It has been shown by Stroock and Varadhan that the entire theory of multidimentional diffusion processes (and many other continuous-parameter Markov processes) can be so formulated. Suppose that we have an evolution equation of the form:

$$\frac{df}{dt} = Gf. \qquad (4.21)$$

The **martingale problem** is to find a Markov process $x(t), t \geq 0$, and RE $V(t)$ so that for all smooth functions

$$V(t)f(x(t)) - \int_0^t V(s)Gf(x(s))ds \quad is \quad a \quad martingale. \qquad (4.22)$$

It is immediate that this gives the required solution. Indeed, the operator

$$f \to T(t)f := E_x[V(t)f(x(t))]$$

defines a semigroup of operators on the Banach space B, whose infinitesimal generator can be computed by taking the expectation:

$$E_x[V(t)f(x(t))] - f(x) = E_x[\int_0^t V(s)Gf(x(s))ds],$$

and

$$\lim_{t \to 0} t^{-1}[E_x[V(t)f(x(t))] - f(x)] = \lim_{t \to 0} t^{-1}E_x[\int_0^t V(s)Gf(x(s))ds] = Gf(x).$$

Remark. *In case $V(t) \equiv I$-identity operator, the above reduces to the usual martingale problem for Markov process [3].*

Remark. *In case $B = R$ the problem reduces to the determination of a real-valued multiplicative functional, which is related to a Feynman-Kac type formula. In the case of the one-dimensional Wiener process, a wide class of multiplicative functionals is provided by*

$$V(t) = \exp \int_0^t a(x(s))ds + \int_0^t b(x(s))dw(s),$$

where $w(t)$ is a standard Wiener process.

Let us illustrate the martingale problem for discontinuous RE over a jump Markov process, diffusion process, etc.

Martingale problem for discontinuous RE over a jump Markov process. Let $x(t), t \geq 0$, be a conservative regular jump Markov process on a measurable state space (X, Ξ) with rate function $\lambda(x) > 0$ and a family of probability measures $P(x, dy)$. Let also $V(t)$ be a discontinuous RE in (4.2). For any Borel function f we have the sum:

$$f(x(t)) = f(x(0)) + \sum_{0 \leq s \leq t} [f(x(s+0)) - f(x(s-0))]. \qquad (4.23)$$

From this we see that the product $V(t)f(x(t))$ satisfies the differential equation:

$$\frac{dV(t)f(x(t))}{dt} = V(t)\Gamma(x(t))f(x(t)), if \tau_k < t < \tau_{k+1},$$

and the jump across $t = \tau_k$ is evaluated as

$$V(t)f(x(t))|_{\tau_k-}^{\tau_k+} = V(\tau_k-)\mathcal{D}(x(\tau_k-), x(\tau_k+))f(x(\tau_k+0)) - f(x(\tau_k - 0))$$

leading to the equation:

$$\begin{aligned} V(t)f(x(t)) &= f(x) + \int_0^t V(s)\Gamma(x(s))f(x(s))ds \\ &+ \sum_{0 \leq \tau_k \leq t} V(\tau_k-)[\mathcal{D}(x(\tau_k-), x(\tau_k+))f(x(\tau_k+)) \qquad (4.24) \\ &- f(x(\tau_k-))], x(0) = x, \tau_k\pm := \tau_k \pm 0. \end{aligned}$$

To put this in the appropriate form of the martingale problem, we use the following identity from the theory of Markov processes:for any positive Borel-measurable function $\phi(.,.)$:

$$E_x[\sum_{0 \leq \tau_k \leq t} \phi(x(\tau_k-), x(\tau_k+))] = E_x[\int_0^t \lambda(x(s)) \int_X \phi(x(s), y)P(x(s), dy)ds].$$

$$(4.25)$$

We note, that the difference

$$\sum_{0 \leq \tau_k \leq t} \phi(x(\tau_k-), x(\tau_k+)) - \int_0^t \lambda(x(s))(P\phi)(x(s))ds$$

is a martingale, where P is an operator generated by $P(x, A)$, $x \in X$, $A \in \Xi$. Applying this to the above computations we see that

$$V(t)f(x(t)) = f(x) + \int_0^t V(s)Gf(x(s))ds + Z(t), \qquad (4.26)$$

where $Z(t), t \geq 0$, is a martingale and

$$Gf(x) = \Gamma(x)f + \lambda(x) \int_X [\mathcal{D}(x, y)f(y) - f(x)]P(x, dy).$$

Martingale problem for discontinuous RE over semi-Markov process. It is known, that process $(x(t), \gamma(t))$ (with $\gamma(t) := t - \tau_{\nu(t)}$ and $x(t)$ as semi-Markov process) is a Markov process in $X \times R_+$ with infinitesimal operator

$$\hat{Q} := \frac{d}{dt} + \frac{g_x(t)}{\bar{G}_x(t)}[P - I],$$

where $g_x(t) := dG_x(t)/dt$, $\bar{G}_x(t) := 1 - G_x(t)$, P is an operator generated by $P(x, A)$, $x \in X$, $A \in \Xi$, $P(x, A)$ and $G_x(t)$ are defined in (4.1). We note, that in Markov case, $G_x(t) = 1 - \exp -\lambda(x)t$, $g_x(t) = \lambda(x)\exp -\lambda(x)t$, $\hat{G}_x(t) = \exp -\lambda(x)t$, and $g_x(t)/\hat{G}_x(t) = \lambda(x)$, $\forall x \in X$. Hence, $\hat{Q} = \lambda(x)[P - I]$ is an infinitesimal operator of a jump Markov process $x(t)$ in X. Using the reasonings (4.23)-(4.26) of the previous example for Markov process $y(t) := (x(t), \gamma(t))$ in $X \times R_+$ we obtain that the solution of the martingale problem is operator

$$Gf(x, t) = \frac{d}{dt}f(x, t) + \Gamma(x)f(x, t) + \frac{g_x(t)}{\hat{G}_x(t)} \int_X [\mathcal{D}(x, y)f(y, t) - f(x, 0)]P(x, dy),$$

and the process $y(t)$.

Martingale problem for RE over Wiener process.

Let $w(t), t \geq 0$, be the Wiener process in R^d and consider the linear stochastic equation:

$$V(t) = I + \int_0^t V(s)\Gamma_0(w(s))ds + \sum_{j=1}^d \int_0^t V(s)\Gamma_j(w(s))dw_j(s),$$

where the final term is a stochastic integral of the Ito variety and $\Gamma_0, ..., \Gamma_d$ are bounded operators on a Banach space B. If f is any C^2 function Ito's formula gives

$$f(w(t)) = f(w(0)) + 2^{-1} \int_0^t \Delta f(w(s))ds + \sum_{j=1}^d \int_0^t \frac{\partial f}{\partial w_j}(w(s))dw_j(s).$$

Using the stochastic product rule

$$d(Mf) = Mdf + (dM)f + (dM)df \qquad (4.27)$$

and re-arranging terms, we have:

$$V(t)f(w(t)) = f(w(0)) + \int_0^t V(s)(2^{-1}\Delta f + \sum_{j=1}^d \Gamma_j \frac{\partial f}{\partial w_j} + \Gamma_0 f)(w(s))ds + Z(t),$$

where $Z(t) := \sum_{j=1}^d \int_0^t V(s)(\frac{\partial f}{\partial w_j}(w(s)) + \Gamma_j(w(s))f(w(s)))dw_j(s)$, which is a martingale. Therefore we have obtained the solution of the martingale problem, with the infinitesimal generator

$$Gf = 2^{-1}\Delta f(w) + \sum_{j=1}^d \Gamma_j(w)\frac{\partial f}{\partial w_j}(w) + \Gamma_0(w)f(w).$$

This corresponds to the stochastic solution of the parabolic system

$$\frac{\partial u}{\partial t} = Gu.$$

Martingale problem for RE over diffusion process. Let $\xi(t), t \geq 0$, be the diffusion process in R:

$$d\xi(t) = a(\xi(t))dt + \sigma(\xi(t))dw(t)$$

and consider the linear stochastic equation:

$$V(t) = I + \int_0^t V(s)\Gamma_0(\xi(s))ds + \int_0^t V(s)\Gamma_1(\xi(s))d\xi(s),$$

with the bounded operators Γ_0 and Γ_1 on B. If f is any C^2 function Ito's formula gives:

$$
\begin{aligned}
f(\xi(t)) &= f(\xi(0)) + \int_0^t [a(\xi(s))\frac{df(\xi(s))}{d\xi} + 2^{-1}\sigma^2(\xi(s))\frac{d^2 f(\xi(s))}{d\xi^2}]ds \\
&+ \int_0^t \frac{\partial f(\xi(s))}{\partial \xi}\sigma(\xi(s))dw(s).
\end{aligned}
$$

Using the stochastic product rule (4.27) we have:

$$V(t)f(\xi(t)) = f(\xi(0)) + \int_0^t V(s)(a\frac{df}{d\xi} + 2^{-1}\sigma^2\frac{d^2 f}{d\xi^2} + \Gamma_1\frac{df}{d\xi} + \Gamma_0 f)(\xi(s))ds + Z(t),$$

where

$$Z(t) := \int_0^t V(s)(\sigma\frac{df}{d\xi} + \Gamma_1 f)(\xi(s))dw(s),$$

which is a martingale. Therefore, we have obtained the solution of the martingale problem with the operator

$$Gf = a\frac{df}{d\xi} + 2^{-1}\sigma^2\frac{d^2 f}{d\xi^2} + \Gamma_1\frac{df}{d\xi} + \Gamma_0 f.$$

Other solutions of martingale problems for RE we will obtain in the limit theorems for RE.

4.1.4   Analogue of Dynkin's Formula for RE (see Chapter 2)

Let $x(t), t \geq 0$, be a strongly measurable strong Markov process, let $V(t)$ be a multiplicative operator functionals (MOF) of $x(t)$ [7, 10], let A be the infinitesimal operator of semigroup

$$(T(t)f)(x) := E_x[V(t)f(x(t))], \qquad (4.28)$$

and let τ be a stopping time for $x(t)$. It is known [10], that if $Ah = g$ and $E_x\tau < +\infty$, then

$$E_x[V(\tau)h(x(\tau)) - h(x) = E_x \int_0^\tau V(t)Ah(x(t))dt. \qquad (4.29)$$

Formula (4.28) is an analogue of Dynkin's formula for MOF [10]. In fact, if w set $V(t) \equiv I$-identity operator, then from (4.29) we obtain:

$$E_x[h(x(\tau))] - h(x) = E_x \int_0^\tau Qh(x(t))dt, \qquad (4.30)$$

where Q is an infinitesimal operator of $x(t)$ (see (4.28)). Formula (4.30) is the well-known Dynkin's formula. Let $x(t), t \geq 0$, be a continuous Markov process on (X, Ξ) and $V(t)$ be a continuous RE:

$$dV(t)/dt = V(t)\Gamma(x(t)), V(0) = I. \qquad (4.31)$$

We note, that the function $u(t, x) := E_x[V(t)f(x(t))]$ satisfies the following equation [10]:

$$du(t, x)/dt = Qu(t, x) + \Gamma(x)u(t, x), u(0, x) = f(x), \qquad (4.32)$$

where Q is an infinitesimal operator of $x(t)$. From (4.29) and (4.32) we obtain the **analogue of Dynkin's formula for continuous Markov RE** $V(t)$ in (4.31):

$$E_x[V(\tau)h(x(\tau))] - h(x) = E_x \int_0^\tau V(t)[Q + \Gamma(x(t))]h(x(t))dt. \qquad (4.33)$$

Let $x(t), t \geq 0$, be a jump Markov process with infinitesimal operator Q and $V(t)$ be a discontinuous Markov RE in (4.2). In this case the function $u(t, x) := E_x[V(t)f(x(t))]$ satisfies the equation [10]:

$$du(t, x)/dt = Qu(t, x) + \Gamma(x)u(t, x) + \lambda(x)$$
$$\int_X P(x, dy)[\mathcal{D}(x, y) - I]u(t, y), u(0, x) = f(x). \qquad (4.34)$$

From (4.29) and (4.34) we obtain the **analogue of Dynkin's formula for discontinuous Markov RE** in (4.2):

$$E_x[V(\tau)f(x(\tau))] - f(x) = E_x \int_0^\tau V(s)[Q + \Gamma(x(t)) + \lambda(x)$$
$$\int_X P(x(t), dy)(\mathcal{D}(x(t), y) - I)]f(x(t))dt. \qquad (4.35)$$

Let finally $x(t), t \geq 0$, be a semi-Markov process, and $V(t)$ be a semi-Markov random evolution in (4.2). Let us define the process

$$\gamma(t) := t - \tau_{\nu(t)}). \tag{4.36}$$

Then the process

$$y(t) := (x(t), \gamma(t)) \tag{4.37}$$

is a Markov process in $X \times R_+$ with infinitesimal operator [6]

$$\hat{Q} := \frac{d}{dt} + \frac{g_x(t)}{\bar{G}_x(t)}[P - I], \tag{4.38}$$

where $g_x(t) := dG_x(t)/dt$, $\bar{G}_x(t) := 1 - G_x(t)$, P is an operator generated by the kernel $P(x, A)$. Hence, the process $(V(t)f; x(t); \gamma(t); t \geq 0) \equiv (V(t)f; y(t); t \geq 0)$ in $B \times X \times R_+$ is a Markov process with infinitesimal operator

$$L(x) := \hat{Q} + \Gamma(x) + \frac{g_x(t)}{\bar{G}_x(t)} \int_X P(x, dy)[\mathcal{D}(x, y) - I], \tag{4.39}$$

where $\hat{Q}$ is defined in (4.38).

Let $f(x, t)$ be a function on $X \times R_+$ bounded by x and differentiable by t, and let τ be a stopping time for $y(t) = (x(t), \gamma(t))$. Then for semi-Markov RE $V(t)$ in (4.2) we have from (4.29), (4.36)-(4.39) the following **analogue of Dynkin's formula:**

$$E_y[V(\tau)f(y(\tau))] - f(y) = E_y \int_0^\tau V(s)[\hat{Q} + \Gamma(x(t)) + \frac{g_x(t)}{\bar{G}_x(t)}$$
$$\int_X P(x(t), dy)[\mathcal{D}(x(t), y) - I]f(y(t))dt, \tag{4.40}$$

where $y := y(0) = (x, 0), f(y) = f(x, 0)$.

4.1.5 Boundary Value Problems for RE (see Chapter 2)

Let $x(t), t \geq 0$, be a continuous Markov process in semicompact state space (X, Ξ). Let $V(t)$ be a continuous Markov RE in (4.31), and let G be an open set satisfying the following conditions:

$$\forall x \in G, \exists U : E_x \tau_U < +\infty, U \in \Xi,$$
$$\tau_U := \inf_t t : x(t) \notin U, P_x \tau_G = +\infty = 0, \forall x \in X. \tag{4.41}$$

If $f(x)$ is a bounded measurable function on ∂G (boundary of G) and function

$$b(x) := E_x[V(\tau_G)f(x(\tau_G))] \tag{4.42}$$

is continuous on X, then function $b(x)$ is the solution of the equation [10]:

$$Qb(x) + \Gamma(x)b(x) = 0, \forall x \in G, \tag{4.43}$$

where Q is an infinitesimal operator of $x(t)$. If function

$$H(x) := E_x[\int_0^{\tau_G} V(t)g(x(t))dt] \tag{4.44}$$

is continuous and bounded, then this function satisfies the following equation [10]:

$$QH(x) + \Gamma(x)H(x) = -g(x), \forall x \in X. \tag{4.45}$$

It follows from (4.41)-(4.44) that the boundary value problem

$$QH(x) + \Gamma(x)H(x) = -g(x), H(x)|_{\partial G} = f(x) \tag{4.46}$$

has the following solution:

$$H(x) = E_x \int_0^{\tau_G} [V(s)g(x(s))ds] + E_x[V(\tau_G f(x(\tau_G))]. \tag{4.47}$$

Let $x(t), t \geq 0$, be a jump Markov process in (X, Ξ), let $V(t)$ be a discontinuous Markov RE in (4.2), and let conditions (4.41) be satisfied. It follows from (4.44)-(4.47), that the boundary value problem

$$QH(x) + \Gamma(x)H(x) + \int_X P(x, dy)[\mathcal{D}(x, y) - I]H(y) = -g(x), H(x)|_{\partial G} = f(x)$$

has the following solution:

$$H(x) = E_x \int_0^{\tau_G} V(s)g(x(s))ds + E_x[V(\tau_G)f(x(\tau_G))].$$

4.2 Limit Theorems for Random Evolutions

The main approach to the investigation of SMRE in the limit theorems is a martingale method.

The martingale method of obtaining the limit theorems (averaging and diffusion approximation) for the sequence of SMRE is bounded on the solution of the following problems:

1) weak compactness of the family of measures generated by the sequence of SMRE;

2) any limiting point of this family of measures is the solution of martingale problem;

3) the solution of the martingale problem is unique.

The conditions 1) - 2) guarantee the existence of weakly converging subsequence, and condition 3) gives the uniqueness of a weak limit.

From 1) - 3) it follows that the consequence of SMRE converges weakly to the unique solution of martingale problem.

4.2.1 Weak Convergence of Random Evolutions (see Chapter 2 and 3)

A **weak convergence of SMRE** in a series scheme we obtain from the criterion of weak compactness of the process with values in separable Banach spaces [5]. The **limit SMRE** we obtain from the solution of some martingale problem in kind of some integral operator equations in Banach space B.

The main idea is that process

$$M_n := V_n - I - \sum_{k=0}^{n-1} E\left[V_{k+1} - V_k/\mathcal{F}_k\right], \quad V_0 = I, \tag{4.48}$$

is an $\mathcal{F}_n$–martingale in B, where

$$\mathcal{F}_n := \sigma\{x_k, \tau_k; 0 \le k \le n\}, \quad V_n := V(\tau_n),$$

E is an expectation of probability $\mathcal{P}$ on a probability space $(\Omega, \mathcal{F}, \mathcal{P})$.

Representation of the martingale M_n in the form of **martingale - differences**

$$M_n = \sum_{k=0}^{n-1} [V_{k+1} - E(V_{k+1}/\mathcal{F}_k)] \tag{4.49}$$

gives us the possibility to calculate the **weak quadratic variation**:

$$< l(M_n f) >:= \sum_{k=0}^{n-1} E\left[l^2((V_{k+1} - V_k)f)/\mathcal{F}_k\right], \tag{4.50}$$

where $l \in B^*$, and B^* is a dual space to B, dividing points of B.

From (4.19) it follows that

$$V_{k+1} - V_k = [\Gamma_{x_k}(\theta_{k+1})D(x_k, x_{k+1}) - I] \cdot V_k. \tag{4.51}$$

We note that the following expression for a semigroup of operators $\Gamma_x(t)$ is fulfilled:

$$\Gamma_x(t)f = I + \sum_{k=1}^{n-1} \frac{t^k}{k!}\Gamma_{(x)}^k f + \frac{1}{(n-1)!}\int_0^t (t-s)^n \Gamma_x(s)\Gamma_{(x)}^n f ds,$$

$$\forall x \in X, \quad \forall f \in \bigcap_x Dom(\Gamma^n(x)). \tag{4.52}$$

Taking into account (4.48)–(4.52) we obtain the mentioned above results.

Everywhere we suppose that the following conditions will be satisfied:

A) there exists Hilbert spaces H and H^* such that compactly imbedded in Banach spaces B and B^* respectively, $H \subset B$, $H^* \subset B^*$, where B^* is a dual space to B, that divides points of B;

B) operators $\Gamma(x)$ and $(\Gamma(x))^*$ are dissipative on any Hilbert space H and H^* respectively;

C) operators $D(x,y)$ and $D^*(x,y)$ are contractive on any Hilbert space H and H^* respectively;

D) $(x_n; \ n \geq 0)$ is a uniformly ergodic Markov chain with stationary distribution $\rho(A), \quad A \in \mathcal{X}$;

E) $m_i(x) := \int_0^\infty t^i G_x(dt)$ are uniformly integrable, $\forall i = 1, 2, 3$, where

$$G_x(t) := \mathcal{P}\{\omega : \theta_{n+1} \leq t/x_n = x\}; \tag{4.53}$$

F)

$$\int_X \rho(dx)\|\Gamma(x)f\|^k < +\infty; \ \int_X \rho(dx)\|PD_j(x,\cdot)f\|^k < +\infty;$$

$$\int_X \rho(dx)\|\Gamma(x)f\|^{k-1} \cdot \|PD_j(x,\cdot)f\|^{k-1} < +\infty; \quad \forall k = 1,2,3,4, f \in B, \tag{4.54}$$

where P is an operator generated by the transition probabilities $P(x, A)$ of Markov chain $(x_n; \ n \geq 0)$:

$$P(x, A) := \mathcal{P}\{\omega : x_{n+1} \in A/x_n = x\}, \tag{4.55}$$

and $\{D_j(x, y); \quad x, y \in X, \quad j = 1, 2\}$ is a family of some closed operators.

If $B := C_0(R)$, then $H := W^{l,2}(R)$ is a Sobolev space [8], and $W^{l,2}(R) \subset C_0(R)$ and this imbedding is compact. For the spaces $B := L_2(R)$ and $H := W^{l,2}(R)$ it is the same.

It follows from the conditions A) - B) that operators $\Gamma(x)$ and $(\Gamma(x))^*$ generate a strongly continuous contractive semigroup of operators $\Gamma_x(t)$ and $\Gamma_x^*(t), \quad \forall x \in X$, in H and H^* respectively. From the conditions A–C it follows that SMRE $V(t)$ in (1) is a contractive operator in $H, \quad \forall t \in R_+$, and $\|V(t)f\|_H$ is a semimartingale $\forall f \in H$. In such a way, the conditions A) - C) supply the following result:

SMRE $V(t)f$ is a tight process in B, namely, $\forall \Delta > 0$ there exists a compact set K_Δ:

$$\mathcal{P}\{V(t)f \in K_\Delta; 0 \leq t \leq T\} \geq 1 - \Delta. \tag{4.56}$$

This result follows from Kolmogorov - Doob inequality [4] for semimartingale $\|V(t)f\|_H$ [5].

Condition (4.56) is the main step in the proving of limit theorems and rates of convergence for the sequence of SMRE in a series scheme.

4.2.2 Averaging of Random Evolutions

Let's consider a SMRE in series scheme:

$$V_\varepsilon(t) = f + \int_0^t \Gamma(x(s/\varepsilon))V_\varepsilon(s)f ds + \sum_{k=1}^{\nu(t/\varepsilon)} [D^\varepsilon(x_{k-1}, x_k) - I] V_\varepsilon(\varepsilon\tau_k-)f, \tag{4.57}$$

where

$$D^\varepsilon(x,y) = I + \varepsilon D_1(x,y) + 0(\varepsilon), \qquad (4.58)$$

$\{D_1(x,y); x,y \in X\}$ is a family of closed linear operators, $\|0(\varepsilon)f\|/\varepsilon \to 0$ $\varepsilon \to 0$, ε is a small parameter,

$$f \in B_0 := \bigcap_{x,y \in X} Dom(\Gamma^2(x)) \cap Dom(D_1^2(x,y)). \qquad (4.59)$$

Another form for $V_\varepsilon(t)$ in (4.57) is:

$$V_\varepsilon(t) = \Gamma_{x(t/\varepsilon)}(t - \varepsilon\tau_{\nu(t/\varepsilon)}) \prod_{k=1}^{\nu(t/\varepsilon)} D^\varepsilon(x_{k-1}, x_k)\Gamma_{k-1}(\varepsilon\theta_k). \qquad (4.60)$$

Under conditions A) - C) the sequence of SMRE $V_\varepsilon(t)f$ is tight (see (4.56)).
$\rho - a.s.$

Under conditions D), E), $i = 2, F$), $k = 2, j = 1$, the sequence of SMRE $V_\varepsilon(t)f$ is weakly compact $\rho - a.s.$ in $D_B[0, +\infty)$ with limit points in $C_B[0, +\infty)$, $f \in B_0$.

Let's consider the following process in $D_B[0, +\infty)$:

$$M_{\nu(t/\varepsilon)}^\varepsilon f^\varepsilon := V_{\nu(t/\varepsilon)}^\varepsilon f^\varepsilon - f^\varepsilon - \sum_{k=0}^{\nu(t/\varepsilon)-1} E_\rho[V_{k+1}^\varepsilon f_{k+1}^\varepsilon - V_k^\varepsilon f_k^\varepsilon / \mathcal{F}_k], \qquad (4.61)$$

where $V_n^\varepsilon := V_\varepsilon(\varepsilon\tau_n)$ (see (4.19)),

$$f^\varepsilon := f + \varepsilon f_1(x(t/\varepsilon)),$$

$$f_k^\varepsilon := f^\varepsilon(x_k),$$

function $f_1(x)$ is defined from the equation

$$(P - I)f_1(x) = \left[(\hat\Gamma + \hat D) - (m(x)\Gamma(x) + PD_1(x,\cdot))\right] f,$$

$$\hat\Gamma := \int_x \rho(dx)m(x)\Gamma(x), \quad \hat D := \int_x \rho(dx)PD_1(x,\cdot),$$

$$m(x) := m_1(x) \qquad (4.62)$$

(see E), $f \in B_0$.

The process $M_{\nu(t/\varepsilon)}^\varepsilon f^\varepsilon$ is an $\mathcal{F}_t^\varepsilon$-martingale with respect to the σ-algebra $\mathcal{F}_t^\varepsilon := \sigma\{x(s/\varepsilon); 0 \le s \le t\}$.

The martingale $M_{\nu(t/\varepsilon)}^\varepsilon f^\varepsilon$ in (4.61) has the asymptotic representation:

$$M_{\nu(t/\varepsilon)}^\varepsilon f^\varepsilon = V_{\nu(t/\varepsilon)}^\varepsilon f - f - \varepsilon \sum_{k=0}^{\nu(t/\varepsilon)} (\hat\Gamma + \hat D)V_k^\varepsilon f + 0_f(\varepsilon), \qquad (4.63)$$

where $\hat{\Gamma}, \hat{D}, f, f^{\varepsilon}$ are defined in (4.61)–(4.62) and

$$\|0_f(\varepsilon)\|/\varepsilon \to const \quad as \varepsilon \to 0, \quad \forall f \in B_0.$$

We've used (4.19), (4.20) *as* $n = 2$, and representation (4.51) and (4.61) in (4.63).

The families $l(M^{\varepsilon}_{\nu(t/\varepsilon)} f^{\varepsilon})$ and

$$l\left(\sum_{k=0}^{\nu(t/\varepsilon)} E_{\rho}[(V^{\varepsilon}_{k+1} f^{\varepsilon}_{k+1} - V^{\varepsilon}_k f^{\varepsilon}_k)/\mathcal{F}_k]\right)$$

are weakly compact for all $l \in B_0^*$ is a some dense subset from B^*. Let $V_0(t)$ be a limit process for $V_{\varepsilon}(t) as \quad \varepsilon \to 0$.

Since (see (4.60))

$$[V_{\varepsilon}(t) - V^{\varepsilon}_{\nu(t/\varepsilon)}] = [\Gamma_{x(t/\varepsilon)}(t - \varepsilon\tau_{\nu(t/\varepsilon)}) - I] \cdot V^{\varepsilon}_{\nu(t/\varepsilon)} \tag{4.64}$$

and the righthand side in (4.64) tends to zero *as* $\varepsilon \to 0$, then it's clearly that the limits for $V_{\varepsilon}(t)$ and $V^{\varepsilon}_{\nu(t/\varepsilon)}$ are the same, namely, $V_0(t) \quad \rho - a.s.$

The sum $\varepsilon \cdot \sum_{k=0}^{\nu(t/\varepsilon)}(\hat{\Gamma}+\hat{D})V^{\varepsilon}_k f$ converges strongly as $\quad \varepsilon \to 0$ to the integral

$$m^{-1} \cdot \int_0^t (\hat{\Gamma} + \hat{D})V_0(s)f ds.$$

The quadratic variation of the martingale $l(M^{\varepsilon}_{\nu(t/\varepsilon)} f^{\varepsilon})$ tends to zero, and, hence,

$$M^{\varepsilon}_{\nu(t/\varepsilon)} f^{\varepsilon} \to 0 \quad as \quad \varepsilon \to 0, \quad \forall f \in B_0, \quad \forall e \in B_0^*.$$

Passing to the limit in (4.63) *as* $\varepsilon \to 0$ and taking into account all previous reasonings we obtain that the limit process $V_0(t)$ satisfies the equation:

$$0 = V_0(t)f - f - m^{-1} \int_0^t (\hat{\Gamma} + \hat{D})V_0(s)f ds, \tag{4.65}$$

where

$$m := \int_X \rho(dx)m(x), \quad f \in B_0, \quad t \in [0, T].$$

4.2.3 Diffusion Approximation of Random Evolutions

Let us consider SMRE $V_{\varepsilon}(t/\varepsilon)$, where $V_{\varepsilon}(t)$ is defined in (4.57) or (4.60), with the operators

$$D^{\varepsilon}(x, y) := I + \varepsilon D_1(x, y) + \varepsilon^2 D_2(x, y) + 0(\varepsilon^2), \tag{4.66}$$

$\{D_i(x,y); \quad x,y \in X, \quad i = 1,2\}$ are closed linear operators and
$\|0(\varepsilon^2)f\|/\varepsilon^2 \to 0, \varepsilon \to 0$

$$\forall f \in B_0 := \bigcap_{x,y \in X} Dom(\Gamma^4(x)) \bigcap Dom(D_2(x,y)),$$

$$Dom(D_2(x,y)) \subseteq Dom(D_1(x,y)); \quad D_1(x,y) \subseteq Dom(D_1(x,y)),$$
$$\forall x,y \in X, \quad \Gamma^i(x) \subset Dom(D_2(x,y)), \quad i = \overline{1,3}. \quad (4.67)$$

In such a way

$$V_\varepsilon(t/\varepsilon) = \Gamma_{x(t/\varepsilon^2)}(t/\varepsilon - \varepsilon\tau_{\nu(t/\varepsilon^2)}) \prod_{k=1}^{\nu(t/\varepsilon^2)} D^\varepsilon(x_{k-1}, x_k)\Gamma_{x_{k-1}}(\varepsilon, \theta_k), \quad (4.68)$$

where $D^\varepsilon(x,y)$ are defined in (4.66).

Under conditions A) - C) the sequence of SMRE $V_\varepsilon(t/\varepsilon)f$ is tight (see (4.56)) $\rho - a.s.$

Under conditions D), E), $i = 3, F), k = 4$, the sequence of SMRE $V_\varepsilon(t/\varepsilon)f$ is weakly compact $\rho - a.s.$ in $D_B[0, +\infty)$ with limit points in $C_B[o, +\infty)$, $f \in B_0$.

Let us the **balance condition** be satisfied:

$$\int_X \rho(dx)[m(x)\Gamma(x) + PD_1(x,\cdot)]f = 0, \quad \forall f \in B_0. \quad (4.69)$$

Let us consider the following process in $D_B[0, +\infty)$:

$$M^\varepsilon_{\nu(t/\varepsilon^2)}f^\varepsilon := V^\varepsilon_{\nu(t/\varepsilon^2)}f^\varepsilon - f^\varepsilon - \sum_{k=0}^{\nu(t/\varepsilon^2)-1} E_\rho[V^\varepsilon_{k+1}f^\varepsilon_{k+1} - V^\varepsilon_k f^\varepsilon_k/\mathcal{F}_k], \quad (4.70)$$

where $f^\varepsilon := f + \varepsilon f_1(x(t/\varepsilon^2)) + \varepsilon^2 f_2(x(t/\varepsilon^2))$, and functions f_1 and f_2 are defined from the following equations:

$$\begin{aligned} (P-I)f_1(x) &= -[m(x)\Gamma(x) + PD_1(x,\cdot)]f, \\ (P-I)f_2(x) &= [\hat{L} - L(x)]f, \\ \hat{L} : &= \int_X \rho(dx)L(x), \end{aligned} \quad (4.71)$$

$$\begin{aligned} L(x) := & (m(x)\Gamma(x) + PD_1(x,\cdot))(R_0 - I)(m(x)\Gamma(x) + PD_1(x,\cdot)) + \\ & +m_2(x)\Gamma^2(x)/2 + m(x)PD_1(x,\cdot)\Gamma(x) + PD_2(x,\cdot), \end{aligned}$$

R_0 is a potential operator of $(x_n; \quad n \geq 0)$.

The balance condition (4.69) and condition $\prod(\hat{L} - L(x)) = 0$ give the solvability of the equations in (4.71).

The process $M^\varepsilon_{\nu(t/\varepsilon^2)}f^\varepsilon$ is an $\mathcal{F}^\varepsilon_t$–martingale with respect to the σ–algebra $\mathcal{F}^\varepsilon_t := \sigma\{x(s/\varepsilon^2); 0 \le s \le t\}$.

This martingale has the asymptotic representation:

$$M^\varepsilon_{\nu(t/\varepsilon^2)}f^\varepsilon = V^\varepsilon_{\nu(t/\varepsilon^2)}f - f - \varepsilon^2 \sum_{k=0}^{\nu(t/\varepsilon^2)-1} \hat{L}V^\varepsilon_k f - 0_f(\varepsilon t), \qquad (4.72)$$

where $\hat{L}$ is defined in (4.72) and

$$\|0_f(\varepsilon)\|/\varepsilon \to const \quad \varepsilon \to 0, \quad \forall f \in B_0.$$

We have used (4.19), (4.20) *as* $n = 3$, and representation (4.70) and (4.71) in (4.72).

The families $l(M^\varepsilon_{\nu(t/\varepsilon^2)}f^\varepsilon)$ and $l(\sum_{k=0}^{\nu(t/\varepsilon^2)} E_\rho[(V^\varepsilon_{k+1}f^\varepsilon_{k+1} - V^\varepsilon_k f^\varepsilon_k)/\mathcal{F}_k])$ are weakly compact for all $l \in B_0^*$, $f \in B_0$.

Set $V^0(t)$ for the limit process for $V_\varepsilon(t/\varepsilon)as$ $\varepsilon \to 0$.

From (4.60) we obtain that the limits for $V_\varepsilon(t/\varepsilon)$ and $V^\varepsilon_{\nu(t/\varepsilon^2)}$ are the some, namely, $V^0(t)$.

The sum $\varepsilon^2 \sum_{k=0}^{\nu(t/\varepsilon^2)} \hat{L}V^\varepsilon_k f$ converges strongly *as* $\varepsilon \to 0$ to the integral $m^{-1} \int_0^t \hat{L}V^0(s)fds$.

Set $M^0(t)f$ be a limit martingale for $M^\varepsilon_{\nu(t/\varepsilon^2)}f^\varepsilon$ *as* $\varepsilon \to 0$.

Then, from (4.71)–(4.72) and previous reasonings we have *as* $\varepsilon \to 0$:

$$M^0(t)f = V^0(t)f - f - m^{-1} \cdot \int_0^t \hat{L}V^0(s)fds. \qquad (4.73)$$

The quadratic variation of the martingale $M^0(t)f$ has the form:

$$< l(M^0(t)f) >= \int_0^t \int_X l^2(\sigma(x)\Gamma(x)V^0(s)f)\sigma(dx)ds, \qquad (4.74)$$

where

$$\sigma^2(x) := [m_2(x) - m^2(x)]/m.$$

The solution of the martingale problem for $M^0(t)$ (namely, to find the representation of $M^0(t)$ with quadratic variation (4.73)) is expressed by the integral over Wiener orthogonal martingale measure $W(dx, ds)$ with quadratic variation $\rho(dx) \cdot ds$:

$$M^0(t)f = \int_0^t \int_x \sigma(x)\Gamma(x)V^0(s)fW(dx, ds). \qquad (4.75)$$

In such a way, the limit process $V^0(t)$ satisfies the following equation (see (4.73) and (4.74)):

$$V^0(t)f = f + m^{-1} \cdot \int_0^t \hat{L} \cdot V^0(s)fds + \int_0^t \int_X \sigma(x)\Gamma(x)V^0(s)fW(dx, ds).$$

$$(4.76)$$

If the operator $\hat{L}$ generates the semigroup $U(t)$ then the process $V^0(t)f$ in (4.76) satisfied equation:

$$V^0(t)f = U(t)f + \int_0^t \int_x \sigma(x)U(t-s)\Gamma(x)V^0(s)fW(dx, ds). \qquad (4.77)$$

The **uniqueness** of the limit evolution $V_0(t)f$ in **averaging** scheme follows from the equation (4.77) and the fact that if the operator $\hat{\Gamma} + \hat{D}$ (see (4.62)) generates a semigroup, then $V_0(t)f = \exp\{(\hat{\Gamma}+\hat{D})\cdot t\}f$ and this representation is unique.

The **uniqueness** of the limit evolution $V^0(t)f$ in **diffusion approximation** scheme follows from the uniqueness of the solution of the martingale problem for $V^0(t)f$ (see (4.73)–(4.74)) [9]. The latter is proved by **dual SMRE** in a series scheme by the construction of the limit equation in diffusion approximation and by using a dual identify [5].

4.2.4   Averaging of Random Evolutions in Reducible Phase Space. Merged Random Evolutions

Suppose that the following conditions hold true:

a) **decomposition** of phase space X (**reducible** phase space):

$$X = \bigcup_{u \in U} X_u, \quad X_u \bigcap X_{u'} = \emptyset, \quad u \neq u' : \qquad (4.78)$$

where $(U, \mathcal{U})$ is a some measurable phase space (**merged** phase space);

b) Markov renewal process $(x_n^\varepsilon, \theta_n; n \geq 0)$ on $(X, \mathcal{X})$ has the **semi-Markov kernel**:

$$Q_\varepsilon(x, A, t) := P_\varepsilon(x, A)G_x(t), \qquad (4.79)$$

where $P_\varepsilon(x, A) = P(x, A) - \varepsilon^l P_1(x, A)$, $x \in X$, $A \in \mathcal{X}$, $= 1, 2$; $P(x, A)$ are the transition probabilities of the **supporting nonperturbed** Markov chain $(x_n; n \geq 0)$;

c) the stochastic kernel $P(x, A)$ is adapted to the decomposition (38) in the following form:

$$P(x, X_u) = \begin{cases} 1, x \in X_u \\ 0, x \overline{\in} X_u, \end{cases} \quad u \in U;$$

d) the Markov chain $(x_n; n \geq 0)$ is uniformly ergodic with stationary distributions $\rho_u(B)$:

$$\rho_u(B) = \int_{X_u} P(x, B)\rho_u(dx), \quad \forall u \in U, \quad \forall B \in \mathcal{X}. \qquad (4.80)$$

e) there is a family $\{\rho_u^\varepsilon(A); u \in U, A \in \chi, \varepsilon > 0\}$ of stationary distributions of perturbed Markov chain $(x_n^\varepsilon; n \geq 0)$;

f)

$$b(u) := \int_{X_u} \rho_u(dx) P_1(x, X_u) > 0, \quad \forall u \in U,$$

$$b(u, \Delta) := - \int_{X_u} \rho_u(dx) P_1(x, X_\Delta) > 0, \quad \forall u \bar{\in} \Delta, \quad \Delta \in U; \qquad (4.81)$$

g) the operators $\Gamma(u) := \int_{X_u} \rho_u(dx) m(x) \Gamma(x)$ and

$$\hat{D}(u) := \int_{X_u} \rho_u(dx) \int_{X_u} P(x, dy) D_1(x, y) \qquad (4.82)$$

are closed $\forall u \in U$ with common domain B_0, and operators $\hat{\Gamma}(u) + \hat{D}(u)$ generate the semigroup of operators $\forall u \in U$.

Decomposition (4.78) in a) defines the **merging** function

$$u(x) = u \quad \forall x \in Xu, \quad u \in U. \qquad (4.83)$$

We note that σ–algebras $\mathcal{X}$ and $\mathcal{U}$ are coordinated such that

$$X_\Delta = \bigcup_{u \in \Delta} Xu, \quad \forall u \in U, \quad \Delta \in \mathcal{U}. \qquad (4.84)$$

We set $\prod_u f(u) := \int_{X_u} \rho_u(dx) f(x)$ and $x^\varepsilon(t) := x^\varepsilon_{\nu(t/\varepsilon^2)}$.

SMRE in reducible phase space X is defined by the solution of the equation:

$$
\begin{aligned}
V_\varepsilon(t) &= I + \int_0^t \Gamma(x^\varepsilon(s/\varepsilon)) V_\varepsilon(s) ds \\
&+ \sum_{k=0}^{\nu(t/\varepsilon)} [D^\varepsilon(x^\varepsilon_{k-1}, x^\varepsilon_k) - I] V_\varepsilon(\varepsilon \tau_k^-), \qquad (4.85)
\end{aligned}
$$

where $D^\varepsilon(x, y)$ are defined in (4.58).

Let's consider the martingale

$$
\begin{aligned}
M^\varepsilon_{\nu(t/\varepsilon)} f^\varepsilon(x^\varepsilon(t/\varepsilon)) &:= V^\varepsilon_{\nu(t/\varepsilon)} f^\varepsilon(x^\varepsilon(t/\varepsilon)) - f^\varepsilon(x) \\
&- \sum_{k=0}^{\nu(t/\varepsilon)-1} E_{\rho_u^\varepsilon}[V^\varepsilon_{k+1} f^\varepsilon_{k+1} - V^\varepsilon_k f^\varepsilon_k / \mathcal{F}^\varepsilon_k], \quad (4.86)
\end{aligned}
$$

where

$$\mathcal{F}^\varepsilon_n := \sigma\{x^\varepsilon_k, \theta_k; 0 \le k \le n\},$$

$$f^\varepsilon(x) := \hat{f}(u(x)) + \varepsilon f^1(x), \quad \hat{f}(u) := \int_{X_u} \rho_u(dx) f(x), \qquad (4.87)$$

$$(P - I)f_1(x) = [-(m(x)\Gamma(x) + PD_1(x, \cdot)) + \hat{\Gamma}(u)$$
$$+ \hat{D}(u) + (\Pi_u - I)P_1]\hat{f}(u), \tag{4.88}$$

$$f_k^\varepsilon := f^\varepsilon(x_k^\varepsilon), \quad V_n^\varepsilon := V_\varepsilon(\varepsilon\tau_n),$$

and $V_\varepsilon(t)$ is defined in (4.85), P_1 is an operator generated by $P_1(x, A)$ (see (4.79)).

The follows representation is true [5]:

$$\Pi_u^\varepsilon = \Pi_u - \varepsilon^r \Pi_u P_1 R_0 + \varepsilon^{2r} \Pi_u^\varepsilon (P_1 R_0)^2, \quad r = 1, 2, \tag{4.89}$$

where $\prod_u^\varepsilon, \prod_u, P_1$ are the operators generated by $\rho_u^\varepsilon, \quad \rho_u$ and $P_1(x, A)$ respectively, $x \in X, \quad A \in \mathcal{X}, \quad u \in U$.

It follows from (4.89) that for any continuous and bounded function $f(x)$

$$E_{\rho_u^\varepsilon} f(x) \to \varepsilon \to 0 E_{\rho_u} f(x), \quad \forall u \in U,$$

and the all calculations in Section 4 we use in this section replacing E_{ρ_u} by $E_{\rho_u^\varepsilon}$ that reduce to the calculations by $E_{\rho_u} \quad as \varepsilon \to 0$.

Under conditions $A) - C)$ the sequence of SMRE $V_\varepsilon(t)f$ in (4.85), $f \in B_0$, is tight $\rho_u - a.s., \quad \forall u \in U$.

Under conditions $D), E), i = 2, F), k = 2, j = 1$, the sequence of SMRE $V_\varepsilon(t)f$ is weakly compact $\rho_u - a.s., \quad \forall u \in U$, in $D_B[0, +\infty)$ with limit points in $C_B[0, +\infty)$.

We note that $u(x^\varepsilon(t/\varepsilon)) \to \hat{x}(t) \quad as \varepsilon \to 0$, where $\hat{x}(t)$ is a **merged** jump Markov process in $(U, \mathcal{U})$ with **infinitesimal operator** $\Lambda(\hat{P} - I)$,

$$\Lambda\hat{f}(u) := [b(u)/m(u)]\hat{f}(u),$$
$$\hat{P}\hat{f}(u) := \int_U [b(u, du')/b(u)]\hat{f}(u),$$
$$m(u) := \int_{X_u} \rho_u(dx)m(x), \tag{4.90}$$

$b(u)$ and $b(u, \Delta)$ are defined in (4.81). We also note that

$$\Pi_u P_1 = \Lambda(\hat{P} - I), \tag{4.91}$$

where $\prod_u$ is defined in (4.89), P_1–in (4.89), Λ and $\hat{P}$–in (4.90).

Using (4.19), (4.20) $as \quad n = 2$, and (4.87)–(4.88), (4.89) $as \quad r = 1$, (4.91), we obtain the following representation:

$$M_{\nu(t/\varepsilon)}^\varepsilon f^\varepsilon(x^\varepsilon(t/\varepsilon)) = V_{\nu(t/\varepsilon)}^\varepsilon \hat{f}(u(x^\varepsilon(t/\varepsilon))) - \hat{f}(u(x)) -$$

$$\varepsilon \sum_{k=0}^{\nu(t/\varepsilon)} [m(u)\hat{\Gamma}(u) + m(u)\hat{D}(u) + m(u)\Lambda(\hat{P} - I)]V_k^\varepsilon \hat{f}(u(x_k^\varepsilon)) + 0_f(\varepsilon), \tag{4.92}$$

where $\|0_f(\varepsilon)\|/\varepsilon \to const \quad \varepsilon \to 0, \quad \forall f \in B_0$. Since the third term in (4.92) tends to the integral

$$\int_0^t [\Lambda(\hat{P} - I) + \hat{\Gamma}(\hat{x}(s)) + \hat{D}(\hat{x}(s))] \times \hat{V}_0(s)\hat{f}(\hat{x}(s))ds$$

and the quadratic variation of the martingale $l(M_{\nu(t/\varepsilon)}^\varepsilon f^\varepsilon(x^\varepsilon(t/\varepsilon)))$ tends to zero $as\varepsilon \to 0($ and, hence, $M_{\nu(t/\varepsilon)}^\varepsilon f^\varepsilon(x^\varepsilon(t/\varepsilon)) \to 0, \varepsilon \to 0)$,　$\forall l \in B_0^*$, then we obtain from (4.92) that the limit evolution $\hat{V}_0(t)$ satisfies equation:

$$\hat{V}_0(t)\hat{f}(\hat{x}(t)) = \hat{f}(u) + \int_0^t [\Lambda(\hat{P} - I) + \hat{\Gamma}(\hat{x}(s)) + \hat{D}(\hat{x}(s))]\hat{V}_0(s)\hat{f}(\hat{x}(s))ds.$$

$$(4.93)$$

RE　$\hat{V}_0(t)$ is called a **merged** RE in averaging scheme.

4.2.5　Diffusion Approximation of Random Evolutions in Reducible Phase Space

Let us consider SMRE $V_\varepsilon(t/\varepsilon)$ with expansion (4.66), where $V_\varepsilon(t)$ is defined in (4.85), and conditions $A) - F)(as$　$i = 3, k = 4, j = 1, 2)$ and conditions $a) - f)(e = 2)$ be satisfied.

Let us consider the balance condition

$$\int_{X_u} \rho_u(dx)[m(x)\Gamma(x) + PD_1(x, \cdot)]f = 0, \quad \forall u \in U,$$

$$(4.94)$$

be also satisfied and operator

$$L(u) := \int_{X_u} \rho_u(dx)L(x)/m(u),$$

$$(4.95)$$

generates the semigroup of operators, where $L(x)$ is defined in (4.71) and $m(u)$ in (4.90).

Let us also consider the martingale

$$M_{\nu(t/\varepsilon^2)}^\varepsilon f^\varepsilon(x^\varepsilon(t/\varepsilon^2)) = V_{\nu(t/\varepsilon^2)}^\varepsilon f^\varepsilon(x^\varepsilon(t/\varepsilon^2)) - f^\varepsilon(x)$$

$$- \sum_{k=0}^{\nu(t/\varepsilon^2)} E\rho_u^\varepsilon[V_{k+1}^\varepsilon f_{k+1}^\varepsilon - V_k^\varepsilon f_k^\varepsilon/\mathcal{F}_k^\varepsilon], \quad (4.96)$$

where

$$f^\varepsilon(x) := \hat{f}(u(x)) + \varepsilon f^1(x) + \varepsilon^2 f^2(x),$$
$$(P - I)f^1(x) = [m(x)\Gamma(x) + PD_1(x, \cdot)]\hat{f}(u),$$
$$(P - I)f^2(x) = [m(u)L(u) - L(x) + (\Pi_u - I)P_1]\hat{f}(u), \quad (4.97)$$

where $L(u)$ is defined in (4.95).

From the balance condition (4.94) and from the condition

$$\Pi_u[L(u) - L(x) + (\Pi_u - I)P_1] = 0$$

it follows that functions $f^i(x), i = 1, 2$, are defined unique.

Set $\hat{V}^0(t)$ for the limit of $V_\varepsilon(t/\varepsilon)$ *as* $\varepsilon \to 0$. From (4.64) we obtain that the limit for $V_\varepsilon(t/\varepsilon)$ and $V^\varepsilon_{\nu(t/\varepsilon^2)}$ are the same, namely, $\hat{V}^0(t)$.

Weak compactness of $V_\varepsilon(t/\varepsilon)$ is analogical to the one in Section 2.3 with the use of (4.79) *as* $l = 2$ and (4.88) *as* $r = 2$. That is why all calculations in Section 5 we use in this section replacing E_{ρ_u} by $E_{\rho^\varepsilon_u}$ that reduce to the rates by E_{ρ_u} *as* $\varepsilon \to 0$.

Using (4.19), (4.20) *as* $n = 3$, and representations (4.66) and (4.96)–(4.97) we have the following representation for $M^\varepsilon f^\varepsilon$:

$$M^\varepsilon_{\nu(t/\varepsilon^2)} f^\varepsilon = V^\varepsilon_{\nu(t/\varepsilon^2)} \hat{f}(u(x^\varepsilon(t/\varepsilon^2))) - \hat{f}(u)(x) - \varepsilon^2$$

$$\sum_{k=0}^{\nu(t/\varepsilon^2)} [m(u)L(u(x^\varepsilon_k) + \Pi_u P_1]V^\varepsilon_k \hat{f}(u(x^\varepsilon_k)) + 0_f(\varepsilon), \qquad (4.98)$$

where $L(u)$ is defined in (4.95), $\|0_f(\varepsilon)\|/\varepsilon \to const \;\; \varepsilon \to 0$. The sum in (4.98) converges strongly *as* $\varepsilon \to 0$ to the integral

$$\int_0^t [\Lambda(\hat{P} - I) + L(\hat{x}(s))]\hat{V}^0(s)\hat{f}(\hat{x}(s))ds, \qquad (4.99)$$

because of the relation (4.90), where $\hat{x}(t)$ is a jump Markov process in (U, U) with infinitesimal operator $\Lambda(\hat{P} - I)$, $\hat{x}(0) = u \in U$.

Let $\hat{M}^0(t)f$ be a limit martingale for

$$M^\varepsilon_{\nu(t/\varepsilon^2)} f^\varepsilon(x^\varepsilon(t/\varepsilon^2)) \quad as \quad \varepsilon \to 0.$$

In such a way from (4.93)–(4.98) we have the equation *as* $\varepsilon \to 0$:

$$\hat{M}^0(t)\hat{f}(\hat{x}(t)) = \hat{V}^0(t)\hat{f}(\hat{x}(t)) - \hat{f}(u)$$

$$- \int_0^t [\Lambda(\hat{P} - I) + L(\hat{x}(s))]\hat{V}^0(s)\hat{f}(\hat{x}(s))ds. \quad (4.100)$$

The quadratic variation of the martingale $\hat{M}^0(t)$ has the form:

$$< l(\hat{M}^0(t)\hat{f}(u)) >= \int_0^t \int_{X_u} l^2(\sigma(x, u)\Gamma(x)\hat{V}^0(s)\hat{f}(u))\rho_u(dx)ds, \quad (4.101)$$

where

$$\sigma^2(x, u) := [m_2(x) - m^2(x)]/m(u).$$

The solution of martingale problem for $\hat{M}^0(t)$ is expressed by integral:

$$\hat{M}^0(t)\hat{f}(\hat{x}(t)) = \int_0^t \hat{W}(ds, \hat{x}(s))\hat{V}^0(s)\hat{f}(\hat{x}(s)), \qquad (4.102)$$

where

$$\hat{W}(t, u)f := \int_{X_u} W_{\rho_u}(t, dx)\sigma(x, u)\Gamma(x)f.$$

Finally, from (4.99)–(4.101) it follows that the limit process $\hat{V}^0(t)$ satisfies the following equation:

$$\hat{V}^0(t)\hat{f}(\hat{x}(t)) = \hat{f}(u) + \int_0^t [\Lambda(\hat{P} - I) + L(\hat{x}(s))]\hat{V}^0(s)\hat{f}(\hat{x}(s))ds$$
$$+ \int_0^t \hat{W}(ds, \hat{x}(s))\hat{V}^0(s)\hat{f}(\hat{x}(s)). \qquad (4.103)$$

RE $\hat{V}^0(t)$ in (4.103) is called a **merged** *RE* in a diffusion approximation scheme. If the operator $\hat{U}^0(t)$ be a solution of Cauchy problem:

$$\begin{cases} d\hat{U}^0(t)dt = \hat{U}^0(t)L(\hat{x}(t)) \\ \hat{U}^0(0) = I, \end{cases}$$

then the operator process $\hat{V}^0\hat{f}(\hat{x}(t))$ satisfies equation:

$$\hat{V}^0(t)\hat{f}(\hat{x}(t)) = \hat{U}^0(t)\hat{f}(u) + \int_0^t \hat{U}^0(t - s)\Lambda(\hat{P} - I)\hat{V}^0(s)\hat{f}(\hat{x}(s))ds$$
$$+ \int_0^t \hat{U}^0(t - s)\hat{W}(ds, \hat{x}(s))\hat{V}^0(s)\hat{f}(\hat{x}(s)). \qquad (4.104)$$

The uniqueness of the limit *RE* $\hat{V}^0(t)$ is established by dual SMRE.

4.2.6 Normal Deviations of Random Evolutions

The averaged evolution obtained in averaging and merging schemes can be considered as the first approximation to the initial evolution. The diffusion approximation of the SMRE determines the second approximation to the initial evolution, since the first approximation under balance condition - the averaged evolution - appears to be trivial.

Here we consider the **double approximation** to the SMRE - the averaged and the diffusion approximation - provided that the balance condition failed. We introduce the **deviation process** as the normalized difference between the initial and averaged evolutions. In the limit we obtain the **normal deviations** of the initial SMRE from the averaged one.

Let us consider the SMRE $V_\varepsilon(t)$ in (4.57) and the averaged evolution $V_0(t)$ in (4.65). Let's also consider the deviation of the initial evolution $V_\varepsilon(t)f$ from the averaged one $V_0(t)f$:

$$W_\varepsilon(t)f := \varepsilon^{-1/2} \cdot [V_\varepsilon(t) - V_0(t)]f, \quad \forall f \in B_0. \tag{4.105}$$

Taking into account the equations (4.57) and (4.105) we obtain the relation for $W_\varepsilon(t)$:

$$
\begin{aligned}
W_\varepsilon(t)f &= \varepsilon^{-1/2} \int_0^t (\Gamma(x(s/\varepsilon)) - \hat{\Gamma}) V_\varepsilon(s) f \, ds \\
&+ \int_0^t \hat{\Gamma} W_\varepsilon(s) f \, ds + \\
&\varepsilon^{-1/2} [V_\varepsilon^d(t) - \int_0^t \hat{D} \cdot V_0(s) ds] f, \quad \forall f \in B_0, \tag{4.106}
\end{aligned}
$$

where

$$V_\varepsilon^d(t)f := \sum_{k=1}^{\nu(t/\varepsilon)} [D^\varepsilon(x_{k-1}, x_k) - I] V_\varepsilon(\varepsilon \tau_k^-) f,$$

and $\hat{\Gamma}, \hat{D}$ are defined in (4.62).

If the process $W_\varepsilon(t)f$ has the weak limit $W_0(t)f$ *as* $\varepsilon \to 0$ then we obtain:

$$\int_0^t \hat{\Gamma} W_\varepsilon(s) f \, ds \to \int_0^t \hat{\Gamma} W_0(s) f \, ds, \varepsilon \to 0. \tag{4.107}$$

Since the operator $\Gamma(x) - \hat{\Gamma}$ satisfies to the balance condition

$$(\Pi(\Gamma(x) - \hat{\Gamma})f = 0),$$

then the diffusion approximation of the first term in the righthand side of (4.106) gives:

$$\varepsilon^{-1/2} \int_0^t e((\Gamma(x(s/\varepsilon)) - \hat{\Gamma})f) ds \to l(\sigma_1 f) w(t), \varepsilon \to 0 \tag{4.108}$$

where

$$
\begin{aligned}
l^2(\sigma_1 f) &= \int_X \rho(dx)[m(x)l((\Gamma(x) - \hat{\Gamma})f)(R_0 - I)m(x)l((\Gamma(x) - \hat{\Gamma})f) \\
&+ 2^{-1} \cdot m_2(x)l^2((\Gamma(x) - \hat{\Gamma})f)]/m,
\end{aligned}
$$

$\forall l \in B_0, \quad w(t)$ is a standard Wiener process.

Since $\prod(PD_1(x,\cdot) - \hat{D})f = 0$, then the diffusion approximation of the third term in the righthand side of (65) gives the following limit:

$$\varepsilon^{-1/2} \cdot l(V_\varepsilon^d(t)f - \int_0^t \hat{D}V_0(s)fds) \to l(\sigma_2 f) \cdot w(t), \varepsilon \to 0, \qquad (4.109)$$

where

$$l^2(\sigma_2 f) := \int_X \rho(dx)l((PD_1(x,\cdot) - \hat{D})f)(R_0 - I) \cdot l((PD_1(x,\cdot) - \hat{D})f).$$

The passage to the limit *as* $\varepsilon \to 0$ in the representation (4.106) by encountering (4.107)–(4.109) arrives at the equation for $W_0(t)f$:

$$W_0(t)f = \int_0^t \hat{\Gamma}W_0(s)fds + \sigma fw(t), \qquad (4.110)$$

where the variance operator σ is determined from the relation:

$$l^2(\sigma f) := l^2(\sigma_1 f) + l^2(\sigma_2 f), \quad \forall l \in B_0, \quad \forall l \in B_0^*, \qquad (4.111)$$

where operators σ_1 and σ_2 are defined in (4.108) and (4.109) respectively.

Double approximation of the SMRE has the form:

$$V_\varepsilon(t)f \approx V_0(t)f + \sqrt{\varepsilon}W_0(t)f$$

for small ε, which perfectly fits the standart form of the *CLT* with non - zero limiting mean value.

4.2.7 Rates of Convergence in the Limit Theorems for RE

The rates of convergence in the averaging and diffusion approximation scheme for the sequence of SMRE are considered in this section.

Averaging Scheme. The problem is to estimate the value

$$\|E_\rho[V_\varepsilon(t)f^\varepsilon(x(t/\varepsilon)) - V_0(t)f]\|, \quad \forall f \in B_0, \qquad (4.112)$$

where $V_0(t), V_\varepsilon(t), f^\varepsilon, f$ and B_0 are defined in (4.65), (4.57), (4.61), (4.59), respectively.

We use the following representation

$$\|E_\rho[V_\varepsilon(t)f^\varepsilon(x(t/\varepsilon)) \quad - \quad V_0(t)f]\| \le \|E_\rho[V_\varepsilon(t)f - V_\varepsilon(\tau_{\nu(t/\varepsilon)})f]\| + $$
$$+\|E_\rho[V_\varepsilon(\tau_{\nu(t/\varepsilon)})f \quad - \quad V_0(t)f]\| + \varepsilon\|E_\rho V_\varepsilon(t)f_1(x(t/\varepsilon))\| \qquad (4.113)$$

that follows from (4.107) and (4.65), (4.61), (4.59).

For the first term in the righthand side of (4.113) we obtain (see (4.64) and (4.52) *as* $n = 2$):

$$\|E_\rho[V_\varepsilon(t)f - V_\varepsilon(\tau_{\nu(t/\varepsilon)}))f]\| \le \varepsilon \cdot C_1(T, f), \quad \forall t \in [0, T], \qquad (4.114)$$

where

$$C_1(T, f) := \int_X \rho(dx)[C_0(T, x, f) + C_0^2(T, x, f)],$$

$$C_0(T, x, f) := T \cdot m_2(x) \|\Gamma(x)f\|/2m, \quad \forall f \in B_0.$$

For the second term in the righthand side of (4.113) we have from (4.63) and (4.112) (since $E_\rho M_{\nu(t/\varepsilon)}^\varepsilon f^\varepsilon(x(t/\varepsilon)) = 0$):

$$\|E_\rho[V_\varepsilon(\tau_{\nu(t/\varepsilon)})f - V_0(t)f]\| \leq \varepsilon \|E_\rho[V_{\nu(t/\varepsilon)}^\varepsilon - I]f_1(x(t/\varepsilon))\|$$

$$+ \varepsilon \|E_\rho[\sum_{k=0}^{\nu(t/\varepsilon)-1} (\hat{\Gamma} + \hat{D})V_k^\varepsilon f - \varepsilon^{-1}m^{-1}\int_0^1 (\hat{\Gamma} + \hat{D})V_0(s)f ds]\|$$

$$+ \varepsilon \cdot C_2(T, f), \tag{4.115}$$

where constant $C_2(T, f)$ is expressed by algebraic sum of $\int_X m_i(x)\|\Gamma^i(x)f\|\rho(dx)$ and

$$\int_X m_i(x)\|PD_1(x, \cdot) \cdot \Gamma^i(x)f\|\rho(dx), \quad i = 1, 2, \quad f \in B_0,$$

and $\|R_0\|$, R_0 is a potential of Markov chain $(x_n; n \geq 0)$.

For the third term in righthand side of (4.113) we obtain:

$$E_\rho\|f_1(x)\| \leq 2C_3(f), \tag{4.116}$$

where

$$C_3(f) := \|R_0\| \cdot \int_X p(dx)[m(x)\|\Gamma(x)f\| + \|PD_1(x, \cdot)f\|].$$

Finally, from (4.113)–(4.116) we obtain the estimate of the value in (4.112), namely, **rate of convergence in averaging scheme for SMRE**:

$$\|E_\rho[V_\varepsilon(t)f^\varepsilon(x(t/\varepsilon)) - V_0(t)f]\| \leq \varepsilon \cdot C(T, f), \tag{4.117}$$

where constant $C(T, f)$ is expressed by $C_i(T, f)$, $i = \overline{1, 3}$.

Diffusion Approximation. The problem is to estimate the value:

$$\|E_\rho[V_\varepsilon(t/\varepsilon)f^\varepsilon(x(t/\varepsilon^2)) - V^0(t)f]\|, \quad \forall f \in B_0, \tag{4.118}$$

where $V_\varepsilon(t/\varepsilon)$, f^ε, $V^0(t)$, f, B_0 are defined in (4.68),(4.70),(4.76),(4.67) respectively.

Here, we use the following representation:

$$\|E_\rho[V_\varepsilon(t/\varepsilon)f^\varepsilon(x(t/\varepsilon^2)) - V^0(t)f]\| \leq \|E_\rho[V_\varepsilon(t/\varepsilon)f - V_\varepsilon(\tau_{\nu(t/\varepsilon^2)})f]\|$$

$$+ \|E_\rho[V_\varepsilon(\tau_{\nu(t/\varepsilon^2)})f - V^0(t)f]\| + \varepsilon\|E_\rho[V_\varepsilon(t/\varepsilon)f_1(x(t/\varepsilon^2))]\|$$

$$+ \varepsilon^2\|E_\rho[V_\varepsilon(t/\varepsilon)f_2(x(t/\varepsilon^2))]\|, \tag{4.119}$$

that follows from (4.117) and (4.70), (4.64), respectively.

First of all we have for the fourth term in the righthand side of (4.118):

$$\varepsilon^2 \|E_\rho[V_\varepsilon(t/\varepsilon)f_2(x(t/\varepsilon^2))]\| \leq \varepsilon^2 \cdot 2\|R_0\| \cdot \int_X \rho(dx)\|L(x)f\| := \varepsilon^2 d_1(f),$$
(4.120)

where $L(x)$ is defined in (4.71).

For the third term in the righthand side of (4.118) we obtain:

$$\varepsilon \|E_\rho[V_\varepsilon(t/\varepsilon)f_1(x(t/\varepsilon^2))]\| \leq \varepsilon \cdot d_2(f),$$
(4.121)

where

$$d_2(f) := 2\|R_0\| \cdot \int_X \rho(dx)[m(x)\|\Gamma(x)f\| + \|PD_1(x,\cdot)f\|], \quad f \in B_0.$$

For the first term in the righthand side of (4.118) we have from (4.117):

$$\|E_\rho[V_\varepsilon(t/\varepsilon)f - V_\varepsilon(\tau_{\nu(t/\varepsilon^2)})f]\| \leq \varepsilon \cdot C_1(T,f),$$
(4.122)

where $C_1(T,f)$ is defined in (4.117).

For the second term in the righthand side of (4.123) we use the asymptotic representation (4.72) for the martingale $M^\varepsilon_{\nu(t/\varepsilon^2)}f^\varepsilon$ and the conditions

$$E_\rho M^\varepsilon f^\varepsilon = 0, \quad E_\rho M^0(t)f = 0, \quad \forall f \in B_0$$
(4.123)

$$\|E_\rho[V_\varepsilon(\tau_{\nu(t/\varepsilon^2)})f - V^0(t)f]\| \leq \varepsilon\|E_\rho[V^\varepsilon(\tau_{\nu(t/\varepsilon^2)})f_1 - f_1(x)]\| +$$

$$+\varepsilon^2\|E_\rho[V_\varepsilon(\tau_{\nu(t/\varepsilon^2)})f_2 - f_2(x)]\| + \varepsilon^2\|E_\rho[\sum_{k=0}^{\nu(t/\varepsilon^2)-1} \hat{L}V^\varepsilon_k f$$

$$-\varepsilon^{-2}m^{-1}\int_0^t \hat{L}V^0(s)fds]\| + \varepsilon \cdot d_3(f), \quad (4.124)$$

where constant $d_3(f)$ is expressed by algebraic sum of

$$\int_X m_i(x)\|\Gamma^j(x)PD_e(x,\cdot)f\|\rho(dx), \quad i = \overline{1,3}, \quad j = \overline{0,3}, \quad e = \overline{1,2}.$$

We note that

$$\|E_\rho[\sum_{k=0}^{\nu(t/\varepsilon^2)-1} \hat{L}V^\varepsilon_k f - \varepsilon^{-2}m^{-1}\int_0^t \hat{L}V^0(s)fds]\| \leq d_4(T,f).$$
(4.125)

Finally, from (4.119)–(4.125) we obtain the estimate of the value in (4.118), namely, **rate of convergence in diffusion approximation scheme for SMRE**:

$$\|E_\rho[V_\varepsilon(t/\varepsilon)f^\varepsilon(x(t/\varepsilon^2)) - V^0(t)f]\| \leq \varepsilon \cdot d(T,f),$$
(4.126)

where constant $d(T,f)$ is expressed by $d_i, i = \overline{1,4}$, and $C_1(T,f)$, $f \in B_0$.

Bibliography

[1] Arnold, L. *Random Dynamical Systems.* Springer-Verlag, 1998.

[2] Bhattacharya, R. and Majumdar, M. *Random Dynamical Systems: Theory and Applications.* Cambridge University Press, 2007.

[3] Dynkin, E. B. *Markov Processes.* Springer-Verlag, 1991.

[4] Jacod, J. and Shiryaev, A. N. *Limit Theorems for Stochastic Processes.* Springer-Verlag, 2010.

[5] Korolyuk, V. S. and Swishchuk A. V. *Evolution of Systems in Random Media.* Chapman & Hall CRC, 1995.

[6] Korolyuk, V. S. and Swishchuk A. V. *Semi-Markov Random Evolutions.* Kluwer AP, 1995.

[7] Pinsky, M. *Lectures on Random Evolutions.* World Scientific Publishers, 1991.

[8] Sobolev, S. L. *Some Applications of Functional Analysis in Mathematical Physics.* American Mathematical Society, 1991.

[9] Strook, D. and Varadhan, S.R.S. *Multidimensional Diffusion Processes.* Springer-Verlag, 1979.

[10] Swishchuk, A. V. *Random Evolutions and Their Applications.* Kluwer AP, 1997.

[11] Swishchuk, A. V. *Random Evolutions and Their Applications. New Trends.* Kluwer AP, 2000.

[12] Swishchuk, A. and Islam, S. The geometric Markov renewal processes with applications to finance. *Stochastic Analysis and Applications.* v. 29, N. 4, 684-705, 2010.

5

Inhomogeneous Random Evolutions (IHREs)

This chapter is devoted to the *inhomogeneous (semi-Markov) random evolutions* (IHREs). We first introduce *propagators* and then give definitions and properties of IHRE. *Weak law of large numbers* (WLLN) and *central limit theorem* (CLT) are proved for IHREs, and they represent the main results of Chapter 5 (see [26], [27]). We apply the results from Chapter 3 here. introduce the concept of *regular propagators* in Section 1, which we will characterize as unique solutions to well-posed Cauchy problems, and this will be of crucial importance for both our main WLLN and CLT results, in order to get the *unicity of the limiting process*. In Section 2 we introduce inhomogeneous random evolutions and present some of their properties. In Sections 3 and 4 we prove respectively a WLLN and a CLT, which are the main results of the chapter (Theorems 5.3.19 and 5.4.7). In particular, for the CLT, we obtain a precise (and new) characterization of the limiting process using weak Banach-valued stochastic integrals and so-called *orthogonal martingale measures*.

Let us explain here the main ideas of the main results of the book, Theorems 4.19 (LLN) and 5.7 (FCLT), using one of the applications, namely, a regime-switching inhomogeneous Lévy-based stock price model. We will consider a regime-switching inhomogeneous Lévy-based stock price model, very similar in the spirit to the recent article [4]. In short, an inhomogeneous Lévy process differs from a classical Lévy process in the sense that it has time-dependent (and absolutely continuous) characteristics. We let $\{L^x\}_{x \in \mathbb{J}}$ a collection of such $\mathbb{R}^d$–valued inhomogeneous Lévy processes with characteristics $(b_t^x, c_t^x, \nu_t^x)_{x \in \mathbb{J}}$, and we define:

$$\Gamma_x(s,t)f(z) := \mathbb{E}[f(L_t^x - L_s^x + z)], \quad z \in \mathbb{R}^d, \, x \in \mathbb{J}, \tag{5.1}$$

$$D^\epsilon(x,y)f(z) := f(z + \epsilon\alpha(x,y)), \quad z \in \mathbb{R}^d, \, x, y \in \mathbb{J}, \tag{5.2}$$

for some bounded function α. We will give in Section 6 a financial interpretation of this function α, as well as reasons why we consider a regime-switching model. In this setting, f represents a contingent claim on a (d–dimensional) risky asset S having regime-switching inhomogeneous Lévy dynamics driven by the processes $\{L^x\}_{x \in \mathbb{J}}$: on each random time interval $[T_k^{\epsilon,s}(s), T_{k+1}^{\epsilon,s}(s))$, the risky asset is driven by the process $L^{x_k(s)}$. Indeed, we have the following representation, for $\omega' \in \Omega$ (to make clear that the expectation below is taken w.r.t. ω embedded in the process L and not ω'):

$$V_\epsilon(s,t)(\omega')f(z) =$$

$$\mathbb{E}\left[f\left(z + \sum_{k=1}^{N_s\left(t^{\frac{1}{\epsilon},s}\right)(\omega')+1} \Delta L_k + \sum_{k=1}^{N_s\left(t^{\frac{1}{\epsilon},s}\right)(\omega')} \epsilon\alpha(x_{k-1}(s)(\omega'), x_k(s)(\omega'))\right)\right],$$

$$(5.3)$$

where we have denoted for clarity:

$$\Delta L_k = \Delta L_k(\epsilon, \omega') := L_{T_k^{\epsilon,s}(s)(\omega')\wedge t}^{x_{k-1}(s)(\omega')} - L_{T_{k-1}^{\epsilon,s}(s)(\omega')}^{x_{k-1}(s)(\omega')}. \qquad (5.4)$$

The random evolution $V_\epsilon(s,t)f$ represents in this case the present value of the contingent claim f of maturity t on the risky asset S, *conditionally on the regime switching process* $(x_n, T_n)_{n\geq 0}$: indeed, remember that $V_\epsilon(s,t)f$ is random, and that its randomness (only) comes from the Markov renewal process. Our main results Theorems 5.3.19 and 5.4.7 allow to approximate the impact of the regime-switching on the present value $V_\epsilon(s,t)f$ of the contingent claim. Indeed, we get the following normal approximation, for small ϵ:

$$V_\epsilon(s,t)f \approx \underbrace{\widehat{\Gamma}(s,t)f}_{\substack{1^{st}\text{ order}\\ \text{regime-switching approx.}}} + \underbrace{\sqrt{\epsilon}I_\sigma(s,t)f}_{\substack{\text{noise due to}\\ \text{regime-switching}}} \qquad (5.5)$$

The above approximation allows quantifying the risk inherent in regime-switchings occurring at a high frequency governed by ϵ. The parameter ϵ reflects the frequency of the regime-switchings and can therefore be calibrated to market data by the risk manager. For market practitioners, because of the computational cost, it is often convenient to have asymptotic formulas that allow them to approximate the present value of a given derivative, and by extent the value of their whole portfolio. In addition, the asymptotic normal form of the regime-switching cost allows the risk manager to derive approximate confidence intervals for his portfolio, as well as other quantities of interest such as reserve policies linked to a given model.

5.1 Propagators (Inhomogeneous Semigroup of Operators)

This section aims at presenting some results on propagators, which will be used in what follows. Most of them (as well as the corresponding proofs) are similar to what can be found in [16] Chapter 5, [7] Chapter 2 or [6], but to the best of our knowledge, they don't appear in the literature in the form presented below. In particular, the main result of this section is Theorem 5.1.10 which characterizes so-called *regular* propagators as unique solutions to well-posed Cauchy problems.

Let $(Y, ||\cdot||)$ be a real separable Banach space. Let Y^* the dual space of Y. $(Y_1, ||\cdot||_{Y_1})$ is assumed to be a real separable Banach space which is continuously embedded in Y (this idea was used in [16], Chapter 5), i.e. $Y_1 \subseteq Y$ and $\exists c_1 \in \mathbb{R}^+ \colon ||f|| \leq c_1 ||f||_{Y_1} \ \forall f \in Y_1$. Unless mentioned otherwise, limits are taken in the $Y-$norm, normed vector spaces are equipped with the norm topology and subspaces of normed vector spaces are equipped with the subspace topology. Limits in the Y_1 norm will be denoted $Y_1-\lim$, for example. In the following, J will refer either to $\mathbb{R}^+$ or to $[0, T_\infty]$ for some $T_\infty > 0$ and $\Delta_J := \{(s,t) \in J^2 : s \leq t\}$. Let also, for $s \in J \colon J(s) := \{t \in J : s \leq t\}$ and $\Delta_J(s) := \{(r,t) \in J^2 : s \leq r \leq t\}$. We start by a few introductory definitions:

Definition 5.1.1. *A function* $\Gamma : \Delta_J \to \mathcal{B}(Y)$ *is called a Y-(backward) propagator if:*

i) $\forall t \in J \colon \Gamma(t,t) = I$

ii) $\forall (s,r), (r,t) \in \Delta_J \colon \Gamma(s,r)\Gamma(r,t) = \Gamma(s,t)$

If in addition, $\forall (s,t) \in \Delta_J \colon \Gamma(s,t) = \Gamma(0, t - s)$, Γ *is called a* $Y-$*semigroup.*

Note that we focus our attention on backward propagators as many applications only fit the backward case, as it will be shown later on. Forward propagators differ from backward propagators in the way that they satisfy $\Gamma(t,r)\Gamma(r,s) = \Gamma(t,s)$ $(s \leq r \leq t)$. We now introduce the generator of the propagator:

Definition 5.1.2. *For* $t \in int(J)$ *define:*

$$\mathcal{D}(A_\Gamma(t)) := \left\{ f \in Y : \lim_{\substack{h\downarrow 0 \\ t+h\in J}} \frac{(\Gamma(t, t+h) - I)f}{h} = \lim_{\substack{h\downarrow 0 \\ t-h\in J}} \frac{(\Gamma(t-h, t) - I)f}{h} \in Y \right\}$$
(5.6)

and for $f \in \mathcal{D}(A_\Gamma(t))$:

$$A_\Gamma(t)f := \lim_{\substack{h\downarrow 0 \\ t+h\in J}} \frac{(\Gamma(t, t+h) - I)f}{h} = \lim_{\substack{h\downarrow 0 \\ t-h\in J}} \frac{(\Gamma(t-h, t) - I)f}{h}$$
(5.7)

Define similarly for $t = 0$:

$$\mathcal{D}(A_\Gamma(0)) := \left\{ f \in Y : \lim_{\substack{h\downarrow 0 \\ h\in J}} \frac{(\Gamma(0, h) - I)f}{h} \in Y \right\}$$
(5.8)

and for $f \in \mathcal{D}(A_\Gamma(0)) \colon A_\Gamma(0)f := \lim_{\substack{h\downarrow 0 \\ h\in J}} \frac{(\Gamma(0, h) - I)f}{h}$,
(5.9)

and define $A_\Gamma(T_\infty)$ *similarly to* $A_\Gamma(0)$. *Let* $\mathcal{D}(A_\Gamma) := \bigcap_{t\in J} \mathcal{D}(A_\Gamma(t))$. *Then* $A_\Gamma : J \to L(\mathcal{D}(A_\Gamma), Y)$ *is called the infinitesimal generator of the Y-propagator* Γ.

In the following definitions, which deal with continuity and boundedness of propagators, $(E_1, || \cdot ||_{E_1})$ and $(E_2, || \cdot ||_{E_2})$ represent Banach spaces such that $E_2 \subseteq E_1$ (possibly $E_1 = E_2$).

Definition 5.1.3. *A E_1-propagator Γ is $\mathcal{B}(E_2, E_1)$-bounded if* $\sup_{(s,t) \in \Delta_J} ||\Gamma(s,t)||_{\mathcal{B}(E_2,E_1)} < \infty$. *It is a $\mathcal{B}(E_2, E_1)$-contraction if* $\sup_{(s,t) \in \Delta_J} ||\Gamma(s,t)||_{\mathcal{B}(E_2,E_1)} \leq 1$. *It is $\mathcal{B}(E_2, E_1)$-locally bounded if* $\sup_{(s,t) \in K} ||\Gamma(s,t)||_{\mathcal{B}(E_2,E_1)} < \infty$ *for every compact $K \subseteq \Delta_J$.*

Definition 5.1.4. *Let $F \subseteq E_2$. A E_1-propagator Γ is $(F, || \cdot ||_{E_2})$-strongly continuous if $\forall (s,t) \in \Delta_J$, $\forall f \in F$:*

$$\Gamma(s,t)F \subseteq E_2 \quad and \quad \lim_{\substack{(h_1,h_2) \to (0,0) \\ (s+h_1,t+h_2) \in \Delta_J}} ||\Gamma(s+h_1,t+h_2)f - \Gamma(s,t)f||_{E_2} = 0.$$

$$(5.10)$$

When $E_1 = E_2 = Y$, we will simply write that it is F-strongly continuous.

We use the terminologies t-continuity and s-continuity for the continuity of the partial applications. By [7], strong joint continuity is equivalent to strong separate continuity together with local boundedness of the propagator.

Definition 5.1.5. *Let $F \subseteq E_2$. The generator A_Γ or the E_1-propagator Γ is $(F, || \cdot ||_{E_2})$-strongly continuous if $\forall t \in J$, $\forall f \in F$:*

$$A_\Gamma(t)F \subseteq E_2 \quad and \quad \lim_{\substack{h \to 0 \\ t+h \in J}} ||A_\Gamma(t+h)f - A_\Gamma(t)f||_{E_2} = 0. \quad (5.11)$$

When $E_1 = E_2 = Y$, we will simply write that it is F-strongly continuous.

The following results give conditions under which the propagator is differentiable in s and t.

Theorem 5.1.6. *Let Γ be a Y-propagator. Assume that $\forall (s,t) \in \Delta_J$, $\Gamma(s,t)Y_1 \subseteq \mathcal{D}(A_\Gamma)$. Then:*

$$\frac{\partial^-}{\partial s}\Gamma(s,t)f = -A_\Gamma(s)\Gamma(s,t)f, \quad \forall (s,t) \in \Delta_J, \forall f \in Y_1. \quad (5.12)$$

If in addition Γ is $(Y_1, || \cdot ||_{Y_1})$-strongly s-continuous, Y_1-strongly t-continuous, then:

$$\frac{\partial}{\partial s}\Gamma(s,t)f = -A_\Gamma(s)\Gamma(s,t)f \quad \forall (s,t) \in \Delta_J, \forall f \in Y_1. \quad (5.13)$$

Proof of Theorem 5.1.6. Let $(s,t) \in \Delta_J$, $f \in Y_1$.

$$\frac{\partial^-}{\partial s}\Gamma(s,t)f = \lim_{\substack{h \downarrow 0 \\ (s-h,t) \in \Delta_J}} \frac{\Gamma(s,t)f - \Gamma(s-h,t)f}{h} \quad (5.14)$$

$$= - \lim_{\substack{h \downarrow 0 \\ (s-h,t) \in \Delta_J}} \frac{\Gamma(s-h,s) - I}{h}\Gamma(s,t)f = -A_\Gamma(s)\Gamma(s,t)f \quad (5.15)$$

since $\Gamma(s,t)f \in \mathcal{D}(A_\Gamma)$. We have for $s < t$:

$$\frac{\partial^+}{\partial s}\Gamma(s,t)f = \lim_{\substack{h\downarrow 0 \\ (s+h,t)\in\Delta_J}} \frac{\Gamma(s+h,t)f - \Gamma(s,t)f}{h} \tag{5.16}$$

$$= -\lim_{\substack{h\downarrow 0 \\ (s+h,t)\in\Delta_J}} \frac{\Gamma(s,s+h) - I}{h}\Gamma(s+h,t)f. \tag{5.17}$$

Let $h \in (0, t-s]$:

$$\left\|\frac{(\Gamma(s,s+h) - I)}{h}\Gamma(s+h,t)f - A_\Gamma(s)\Gamma(s,t)f\right\|$$

$$\leq \left\|\frac{(\Gamma(s,s+h) - I)}{h}\Gamma(s,t)f - A_\Gamma(s)\Gamma(s,t)f\right\| \tag{5.18}$$

$$+ \left\|\frac{(\Gamma(s,s+h) - I)}{h}\right\|_{\mathcal{B}(Y_1,Y)} \|\Gamma(s+h,t)f - \Gamma(s,t)f\|_{Y_1},$$

the last inequality holding because $\forall(s,t) \in \Delta_J$: $\Gamma(s,t)Y_1 \subseteq Y_1$. We are going to apply the uniform boundedness principle to show that $\sup_{h\in(0,t-s]} \left\|\frac{(\Gamma(s,s+h)-I)}{h}\right\|_{\mathcal{B}(Y_1,Y)} < \infty$. Y_1 is Banach. We have to show that $\forall g \in Y_1$: $\sup_{h\in(0,t-s]} \left\|\frac{(\Gamma(s,s+h)-I)}{h}g\right\| < \infty$. Let $g \in Y_1$. We have $\left\|\frac{(\Gamma(s,s+h)-I)}{h}g\right\| \overset{h\downarrow 0}{\to} \|A_\Gamma(s)g\|$ since $Y_1 \subseteq \mathcal{D}(A_\Gamma)$. $\exists\delta(g) \in (0, t-s) : h \in (0,\delta) \Rightarrow \left\|\frac{(\Gamma(s,s+h)-I)}{h}g\right\| < 1 + \|A_\Gamma(s)g\|$. Then, by Y_1-strong t−continuity of Γ, $h \to \left\|\frac{(\Gamma(s,s+h)-I)}{h}g\right\| \in C([\delta, t-s], \mathbb{R})$. Let $M := \max_{h\in[\delta,t-s]} \left\|\frac{(\Gamma(s,s+h)-I)}{h}g\right\|$. Then we get $\left\|\frac{(\Gamma(s,s+h)-I)}{h}g\right\| \leq \max(M, 1 + \|A_\Gamma(s)g\|) \; \forall h \in (0, t-s]$ $\Rightarrow \sup_{h\in(0,t-s]} \left\|\frac{(\Gamma(s,s+h)-I)}{h}g\right\| < \infty$.

Further, by $(Y_1, \|\cdot\|_{Y_1})$−strong s−continuity of Γ, $\|\Gamma(s+h,t)f - \Gamma(s,t)f\|_{Y_1} \overset{h\downarrow 0}{\to} 0$. Finally, since $\Gamma(s,t)f \in \mathcal{D}(A_\Gamma)$, $\left\|\frac{(\Gamma(s,s+h)-I)}{h}\Gamma(s,t)f - A_\Gamma(s)\Gamma(s,t)f\right\| \overset{h\downarrow 0}{\to} 0$.

Therefore we get $\frac{\partial^+}{\partial s}\Gamma(s,t)f = -A_\Gamma(s)\Gamma(s,t)f$ for $s < t$, which shows that $\frac{\partial}{\partial s}\Gamma(s,t)f = -A_\Gamma(s)\Gamma(s,t)f$ for $(s,t) \in \Delta_J$.

Theorem 5.1.7. *Let Γ be a Y-propagator. Assume that $Y_1 \subseteq \mathcal{D}(A_\Gamma)$. Then we have:*

$$\frac{\partial^+}{\partial t}\Gamma(s,t)f = \Gamma(s,t)A_\Gamma(t)f \qquad \forall(s,t) \in \Delta_J, \forall f \in Y_1. \tag{5.19}$$

If in addition Γ *is* Y-*strongly t-continuous, then we have:*

$$\frac{\partial}{\partial t}\Gamma(s,t)f = \Gamma(s,t)A_\Gamma(t)f \qquad \forall(s,t)\in\Delta_J, \forall f\in Y_1. \tag{5.20}$$

Proof of Theorem 5.1.7. Let $(s,t)\in\Delta_J$, $f\in Y_1$. We have:

$$\frac{\partial^+}{\partial t}\Gamma(s,t)f = \lim_{\substack{h\downarrow 0 \\ (s,t+h)\in\Delta_J}} \frac{\Gamma(s,t+h)f - \Gamma(s,t)f}{h} \tag{5.21}$$

$$= \lim_{\substack{h\downarrow 0 \\ (s,t+h)\in\Delta_J}} \Gamma(s,t)\frac{(\Gamma(t,t+h)-I)f}{h}. \tag{5.22}$$

And for $h\in J: t+h\in J$:

$$\left\|\Gamma(s,t)\frac{(\Gamma(t,t+h)-I)f}{h} - \Gamma(s,t)A_\Gamma(t)f\right\|$$
$$\leq \|\Gamma(s,t)\|_{\mathcal{B}(Y)}\left\|\frac{(\Gamma(t,t+h)-I)f}{h} - A_\Gamma(t)f\right\| \overset{h\downarrow 0}{\to} 0. \tag{5.23}$$

since $f\in\mathcal{D}(A_\Gamma)$. Therefore $\frac{\partial^+}{\partial t}\Gamma(s,t)f = \Gamma(s,t)A_\Gamma(t)f$. Now if $s<t$:

$$\frac{\partial^-}{\partial t}\Gamma(s,t)f = \lim_{\substack{h\downarrow 0 \\ (s,t-h)\in\Delta_J}} \frac{\Gamma(s,t)f - \Gamma(s,t-h)f}{h} \tag{5.24}$$

$$= \lim_{\substack{h\downarrow 0 \\ (s,t-h)\in\Delta_J}} \Gamma(s,t-h)\frac{(\Gamma(t-h,t)-I)f}{h}. \tag{5.25}$$

For $h\in(0,t-s]$:

$$\left\|\Gamma(s,t-h)\frac{(\Gamma(t-h,t)-I)f}{h} - \Gamma(s,t)A_\Gamma(t)f\right\|$$
$$\leq \|\Gamma(s,t-h)\|_{\mathcal{B}(Y)}\left\|\frac{(\Gamma(t-h,t)-I)f}{h} - A_\Gamma(t)f\right\| \tag{5.26}$$
$$+ \|(\Gamma(s,t-h)-\Gamma(s,t))A_\Gamma(t)f\|.$$

Since $f\in\mathcal{D}(A_\Gamma)$, $\left\|\frac{(\Gamma(t-h,t)-I)f}{h} - A_\Gamma(t)f\right\| \overset{h\downarrow 0}{\to} 0$. By Y-strong t-continuity of Γ: $\|(\Gamma(s,t-h)-\Gamma(s,t))A_\Gamma(t)f\| \overset{h\downarrow 0}{\to} 0$. By the principle of uniform boundedness together with the Y-strong t-continuity of Γ, we have $\sup_{h\in(0,t-s]}\|\Gamma(s,t-h)\|_{\mathcal{B}(Y)} \leq \sup_{h\in[0,t-s]}\|\Gamma(s,t-h)\|_{\mathcal{B}(Y)} < \infty$. Therefore we get $\frac{\partial^-}{\partial t}\Gamma(s,t)f = \Gamma(s,t)A_\Gamma(t)f$ for $s<t$, which shows $\frac{\partial}{\partial t}\Gamma(s,t)f = \Gamma(s,t)A_\Gamma(t)f$ for $(s,t)\in\Delta_J$.

In general, we will want to use the *evolution equation*: $\Gamma(s,t)f = f + \int_s^t\Gamma(s,u)A_\Gamma(u)fdu$, and, therefore, we will need that $u\to\Gamma(s,u)A_\Gamma(u)f$ is in $L^1_Y([s,t])$. The following result gives sufficient conditions for which it is the case.

Theorem 5.1.8. *Assume that Theorem 5.1.7 holds true, that $\forall t \in J$, $A_\Gamma(t) \in \mathcal{B}(Y_1, Y)$ and $\forall (s,t) \in \Delta_J$, $u \to \|A_\Gamma(u)\|_{\mathcal{B}(Y_1, Y)} \in L^1_{\mathbb{R}}([s,t])$. Then $\forall f \in Y_1$, $(s,t) \in \Delta_J$:*

$$\Gamma(s,t)f = f + \int_s^t \Gamma(s,u)A_\Gamma(u)f\,du. \tag{5.27}$$

Proof of Theorem 5.1.8. Let $f \in Y_1$, $(s,t) \in \Delta_J$. First $u \to \Gamma(s,u)A_\Gamma(u)f \in B_Y([s,t])$ as the derivative of $u \to \Gamma(s,u)f$. By the principle of uniform boundedness together with the Y−strong t−continuity of Γ, we have $M := \sup_{u \in [s,t]} \|\Gamma(s,u)\|_{\mathcal{B}(Y)} < \infty$. We then observe that for $u \in [s,t]$:

$$\|\Gamma(s,u)A_\Gamma(u)f\| \leq M\|A_\Gamma(u)f\| \leq M\|A_\Gamma(u)\|_{\mathcal{B}(Y_1, Y)}\|f\|_{Y_1}. \tag{5.28}$$

The following definition introduces the concept of *regular* propagator, which in short means that it is differentiable with respect to both variables, and that its derivatives are integrable.

Definition 5.1.9. *A Y-propagator Γ is said to be regular if it satisfies Theorems 5.1.6, 5.1.7 and $\forall (s,t) \in \Delta_J$, $\forall f \in Y_1$, $u \to \|A_\Gamma(u)\Gamma(u,t)f\|$ and $u \to \|\Gamma(s,u)A_\Gamma(u)f\|$ are in $L^1_{\mathbb{R}}([s,t])$.*

Now we are ready to characterize a *regular* propagator as the unique solution of a well-posed Cauchy problem, which will be needed in the sequel. Note that the proof of the theorem below requires that Γ satisfies both Theorems 5.1.6, 5.1.7 (hence our above definition of regular propagators).

Theorem 5.1.10. *Let A_Γ the generator of a a regular Y-propagator Γ and $s \in J$, $G_s \in \mathcal{B}(Y)$. A solution operator $G : J(s) \to \mathcal{B}(Y)$ to the Cauchy problem:*

$$\begin{cases} \frac{d}{dt}G(t)f = G(t)A_\Gamma(t)f & \forall t \in J(s), f \in Y_1 \\ G(s) = G_s \end{cases} \tag{5.29}$$

is said to be regular if it is Y−strongly continuous. If G is such a regular solution, then we have $G(t)f = G_s\Gamma(s,t)f$, $\forall t \in J(s)$, $\forall f \in Y_1$.

Proof of Theorem 5.1.10. Let $(s,u),(u,t) \in \Delta_J$, $f \in Y_1$. Consider the function $\phi : u \to G(u)\Gamma(u,t)f$. We are going to show that $\phi'(u) = 0$ $\forall u \in [s,t]$ and, therefore, that $\phi(s) = \phi(t)$. We have for $u < t$:

$$\frac{d^+\phi}{du}(u) = \lim_{\substack{h\downarrow 0 \\ h \in (0, t-u]}} \frac{1}{h}[G(u+h)\Gamma(u+h,t)f - G(u)\Gamma(u,t)f]. \tag{5.30}$$

Let $h \in (0, t - u]$. We have:

$$\left\| \frac{1}{h}[G(u+h)\Gamma(u+h,t)f - G(u)\Gamma(u,t)f] \right\| \leq$$

$$\underbrace{\left\| \frac{1}{h}G(u+h)\Gamma(u,t)f - \frac{1}{h}G(u)\Gamma(u,t)f - G(u)A_\Gamma(u)\Gamma(u,t)f \right\|}_{(1)}$$

$$+ \|G(u+h)\|_{\mathcal{B}(Y)} \underbrace{\left\| \frac{1}{h}\Gamma(u+h,t)f - \frac{1}{h}\Gamma(u,t)f + A_\Gamma(u)\Gamma(u,t)f \right\|}_{(2)}$$

$$+ \underbrace{\|G(u+h)A_\Gamma(u)\Gamma(u,t)f - G(u)A_\Gamma(u)\Gamma(u,t)f\|}_{(3)} \tag{5.31}$$

And we have:

- $(1) \to 0$ as G satisfies the initial value problem and $\Gamma(u,t)Y_1 \subseteq Y_1$.

- $(2) \to 0$ as $\frac{\partial}{\partial u}\Gamma(u,t)f = -A_\Gamma(u)\Gamma(u,t)f$.

- $(3) \to 0$ by Y-strong continuity of G.

Further, by the principle of uniform boundedness together with the Y-strong continuity of G, we have $\sup_{h \in (0, t-u]} \|G(u+h)\|_{\mathcal{B}(Y)} \leq \sup_{h \in [0, t-u]} \|G(u+h)\|_{\mathcal{B}(Y)} < \infty$. We therefore get $\frac{d^+\phi}{du}(u) = 0$. Now for $u > s$:

$$\frac{d^-\phi}{du}(u) = \lim_{\substack{h \downarrow 0 \\ h \in (0, u-s]}} \frac{1}{h}[G(u)\Gamma(u,t)f - G(u-h)\Gamma(u-h,t)f] \tag{5.32}$$

Let $h \in (0, u - s]$:

$$\left\| \frac{1}{h}[G(u)\Gamma(u,t)f - G(u-h)\Gamma(u-h,t)f] \right\| \leq$$

$$\underbrace{\left\| \frac{1}{h}G(u)\Gamma(u,t)f - \frac{1}{h}G(u-h)\Gamma(u,t)f - G(u)A_\Gamma(u)\Gamma(u,t)f \right\|}_{(4)}$$

$$+ \|G(u-h)\|_{\mathcal{B}(Y)} \underbrace{\left\| -\frac{1}{h}\Gamma(u-h,u)\Gamma(u,t)f + \frac{1}{h}\Gamma(u,t)f + A_\Gamma(u)\Gamma(u,t)f \right\|}_{(5)}$$

$$+ \underbrace{\|G(u)A_\Gamma(u)\Gamma(u,t)f - G(u-h)A_\Gamma(u)\Gamma(u,t)f\|}_{(6)}$$

$$\tag{5.33}$$

By the principle of uniform boundedness together with the $Y-$strong t-continuity of G, we have $\sup_{h \in (0,u-s]} ||G(u-h)||_{\mathcal{B}(Y)} \leq \sup_{h \in [0,u-s]} ||G(u-h)||_{\mathcal{B}(Y)} < \infty$. And:

- $(4) \to 0$ as G satisfies the initial value problem and $\Gamma(u,t)Y_1 \subseteq Y_1$.

- $(5) \to 0$ as $\Gamma(u,t)Y_1 \subseteq Y_1$.

- $(6) \to 0$ by Y-strong continuity of G.

We therefore get $\frac{d^-\phi}{du}(u) = 0$.

The following corollary expresses the fact that equality of generators implies equality of propagators.

Corollary 5.1.11. *Assume that Γ_1 and Γ_2 are regular Y-propagators and that $\forall f \in Y_1$, $\forall t \in J$, $A_{\Gamma_1}(t)f = A_{\Gamma_2}(t)f$. Then $\forall f \in Y_1$, $\forall (s,t) \in \Delta_J$: $\Gamma_1(s,t)f = \Gamma_2(s,t)f$. In particular, if Y_1 is dense in Y, then $\Gamma_1 = \Gamma_2$.*

We conclude this section with a second order Taylor formula for propagators. Let:

$$\mathcal{D}(A_\Gamma \in Y_1) := \{f \in \mathcal{D}(A_\Gamma) \cap Y_1 : A_\Gamma(t)f \in Y_1 \ \forall t \in J\}. \tag{5.34}$$

Theorem 5.1.12. *Let Γ be a regular Y-propagator, $(s,t) \in \Delta_J$. Assume that $\forall u \in J$, $A_\Gamma(u) \in \mathcal{B}(Y_1, Y)$ and $u \to ||A_\Gamma(u)||_{\mathcal{B}(Y_1,Y)} \in L^1_{\mathbb{R}}([s,t])$. Then we have for $f \in \mathcal{D}(A_\Gamma \in Y_1)$:*

$$\Gamma(s,t)f = f + \int_s^t A_\Gamma(u)f du + \int_s^t \int_s^u \Gamma(s,r)A_\Gamma(r)A_\Gamma(u)f dr du. \tag{5.35}$$

Proof of Theorem 5.1.12. Since Γ is regular and f, $A_\Gamma(u)f \in Y_1$ and $u \to ||A_\Gamma(u)||_{\mathcal{B}(Y_1,Y)}$ is integrable on $[s,t]$ we have by 5.1.8:

$$\Gamma(s,t)f = f + \int_s^t \Gamma(s,u)A_\Gamma(u)f du = f$$

$$+ \int_s^t \left[A_\Gamma(u)f + \int_s^u \Gamma(s,r)A_\Gamma(r)A_\Gamma(u)f dr \right] du \tag{5.36}$$

$$= f + \int_s^t A_\Gamma(u)f du + \int_s^t \int_s^u \Gamma(s,r)A_\Gamma(r)A_\Gamma(u)f dr du. \tag{5.37}$$

5.2 Inhomogeneous Random Evolutions (IHREs): Definitions and Properties

Let $(\Omega, \mathcal{F}, \mathbb{P})$ be a complete probability space, $\mathbb{J}$ a finite set and $(x_n, T_n)_{n \in \mathbb{N}}$ an inhomogeneous Markov renewal process on it, with associated inhomogeneous semi-Markov process $(x(t))_{t\in\mathbb{R}_+} := (x_{N_t})_{t\in\mathbb{R}_+}$ (as in [18]). In this section, we will use the same notations for the various kernels and cumulative distribution functions $(Q_s(i,j,t), F_s(i,t),$ etc.) as in [18] on inhomogeneous Markov renewal processes. Throughout the section we assume that the inhomogeneous Markov renewal process $(x_n, T_n)_{n\in\mathbb{N}}$ is regular (cf. definition in [18]), and that $Q_s(\cdot,\cdot,0) = 0$ for all $s \in \mathbb{R}_+$. Further assumptions on it will be made later on. We define the following random variables on Ω, for $s \leq t \in \mathbb{R}_+$:

- the number of jumps on $(s,t]$: $N_s(t) := N(t) - N(s)$,

- the jump times on (s,∞): $T_n(s) := T_{N(s)+n}$ for $n \in \mathbb{N}^*$, and $T_0(s) := s$,

- the states visited by the process on $[s,\infty)$: $x_n(s) := x(T_n(s))$, for $n \in \mathbb{N}$.

Consider a family of $Y-$propagators $(\Gamma_x)_{x\in\mathbb{J}}$, with respective generators $(A_x)_{x\in\mathbb{J}}$, satisfying:

$$\forall s \in J : (r,t,x,f) \to \Gamma_x(r \wedge t, r \vee t)f \tag{5.38}$$

$$\text{is } Bor(J(s)) \otimes Bor(J(s)) \otimes Bor(\mathbb{J}) \otimes Bor(Y) - Bor(Y) \text{ measurable}, \tag{5.39}$$

as well as a family $(D(x,y))_{(x,y)\in\mathbb{J}^2} \subseteq \mathcal{B}(Y)$ of $\mathcal{B}(Y)-$contractions, satisfying:

$$(x,y,f) \to D(x,y)f \text{ is } Bor(\mathbb{J}) \otimes Bor(\mathbb{J}) \otimes Bor(Y) - Bor(Y) \text{ measurable}. \tag{5.40}$$

We define the *inhomogeneous random evolution* the following way:

Definition 5.2.1. *The function $V : \Delta_J \times \Omega \to \mathcal{B}(Y)$ defined pathwise by:*

$$V(s,t)(\omega) = \left[\prod_{k=1}^{N_s(t)} \Gamma_{x_{k-1}(s)}\left(T_{k-1}(s), T_k(s)\right) D(x_{k-1}(s), x_k(s)) \right] \Gamma_{x(t)}$$
$$\left(T_{N_s(t)}(s), t\right) \tag{5.41}$$

is called a $(\Gamma, D, x)-$inhomogeneous Y-random evolution, or simply an inhomogeneous Y-random evolution. V is said to be continuous (resp. purely discontinuous) if $D(x,y) = I$ (resp. $\Gamma_x = I$), $\forall(x,y) \in \mathbb{J}^2$. V is said to be regular (resp. $\mathcal{B}(Y)-$contraction) if $(\Gamma_x)_{x\in\mathbb{J}}$ are regular (resp. $\mathcal{B}(Y)-$contraction).

Remark. *In the latter definition, we use as conventions that $\prod_{k=1}^{0} := I$ and $\prod_{k=1}^{n} A_k := A_1...A_{n-1}A_n$, that is, the product operator applies the product on the right. Further, if $N_s(t) > 0$, then $x_{N_s(t)}(s) = x(T_{N_s(t)}(s)) = x(T_{N(t)}) = x(t)$. If $N_s(t) = 0$, then $x(s) = x(t)$ and $x_{N_s(t)}(s) = x_0(s) = x(T_0(s)) = x(s) = x(t)$. Therefore in all cases $x_{N_s(t)}(s) = x(t)$. By Proposition 5.2.3 below, if $D = I$, we see that V is continuous and if $\Gamma_x = I$ (and therefore $A_x = 0$), we see that V has no continuous part, hence Definition 5.2.1.*

We have the following measurability result:

Proposition 5.2.2. *For* $s \in J$, $f \in Y$, *the stochastic process* $(V(s,t)(\omega)f)_{(\omega,t) \in \Omega \times J(s)}$ *is adapted to the (augmented) filtration:*

$$\mathcal{F}_t(s) := \sigma\left[x_{n \wedge N_s(t)}(s), T_{n \wedge N_s(t)}(s) : n \in \mathbb{N}\right] \vee \sigma(\mathbb{P} - null \ sets). \quad (5.42)$$

Proof of Proposition 5.2.2. Let $E \in Bor(Y)$, $(s,t) \in \Delta_J$, $f \in Y$. We have:

$$V(s,t)f^{-1}(E) = \bigcup_{n \in \mathbb{N}} \{V(s,t)f \in E\} \cap \{N_s(t) = n\}. \quad (5.43)$$

Denote the $\mathcal{F}_t(s) - Bor(\mathbb{R}^+)$ measurable (by construction) function $h_k := T_{(k+1) \wedge N_s(t)}(s) - T_{k \wedge N_s(t)}(s)$. Since $Q(\cdot,\cdot,0) = 0$, remark that $N_s(t)(\omega) = \sup_{m \in \mathbb{N}} \sum_{k=0}^{m} 1_{h_k^{-1}(\mathbb{R}^{+*})}(\omega)$ and is therefore $\mathcal{F}_t(s) - Bor(\mathbb{R}^+)$ measurable. Therefore $\{N_s(t) = n\} \in \mathcal{F}_t(s)$. Let:

$$\Omega_n := \{N_s(t) = n\} \qquad M := \{n \in \mathbb{N} : \Omega_n \neq \emptyset\}. \quad (5.44)$$

$M \neq \emptyset$ since $\Omega = \bigcup_{n \in \mathbb{N}} \Omega_n$, and for $n \in M$, let the sigma-algebra $\mathcal{F}_n := \mathcal{F}_t(s)|_{\Omega_n} := \{A \in \mathcal{F}_t(s) : A \subseteq \Omega_n\}$ ($\mathcal{F}_n$ is a sigma-algebra on Ω_n since $\Omega_n \neq \emptyset \in \mathcal{F}_t(s)$). Now consider the map $V_n(s,t)f : (\Omega_n, \mathcal{F}_n) \to (Y, Bor(Y))$:

$$V_n(s,t)f := \left[\prod_{k=1}^{n} \Gamma_{x_{k-1}(s)}(T_{k-1}(s), T_k(s)) D(x_{k-1}(s), x_k(s))\right] \Gamma_{x_n(s)}$$
$$(T_n(s), t) f. \quad (5.45)$$

We have:

$$V(s,t)f^{-1}(E) = \bigcup_{n \in \mathbb{N}} \{V_n(s,t)f \in E\} \cap \Omega_n = \bigcup_{n \in \mathbb{N}} \{\omega \in \Omega_n : V_n(s,t)f \in E\}$$
$$\quad (5.46)$$

$$= \bigcup_{n \in \mathbb{N}} V_n(s,t)f^{-1}(E). \quad (5.47)$$

Therefore it remains to show that $V_n(s,t)f^{-1}(E) \in \mathcal{F}_n$, since $\mathcal{F}_n \subseteq \mathcal{F}_t(s)$. First let $n > 0$. Notice that $V_n(s,t)f = \psi \circ \beta_n \circ \alpha_n ... \circ \beta_1 \circ \alpha_1 \circ \phi$, where:

$$\phi : \Omega_n \to J(s) \times \mathbb{J} \times \Omega_n \to Y \times \Omega_n \quad (5.48)$$
$$\omega \to (T_n(s)(\omega), x_n(s)(\omega), \omega) \to (\Gamma_{x_n(s)(\omega)}(T_n(s)(\omega), t)f, \omega). \quad (5.49)$$

The previous mapping holding since $T_k(s)(\omega) \in [s,t]$ $\forall \omega \in \Omega_n$, $k \in [[1,n]]$. ϕ is measurable iff each one of the coordinate mappings are. The canonical projections are trivially measurable. Let $A \in Bor(J(s))$, $B \in Bor(\mathbb{J})$. We have:

$$\{\omega \in \Omega_n : T_n(s) \in A\} = \Omega_n \cap T_{n \wedge N_s(t)}(s)^{-1}(A) \in \mathcal{F}_n \quad (5.50)$$
$$\{\omega \in \Omega_n : x_n(s) \in B\} = \Omega_n \cap x_{n \wedge N_s(t)}(s)^{-1}(B) \in \mathcal{F}_n. \quad (5.51)$$

Now, by measurability assumption, we have for $B \in Bor(Y)$:

$$\{(t_n, y_n) \in J(s) \times \mathbb{J} : \Gamma_{y_n}(t \wedge t_n, t \vee t_n)f \in B\} = C \in Bor(J(s)) \otimes Bor(\mathbb{J}) \tag{5.52}$$

Therefore $\{(t_n, y_n, \omega) \in J(s) \times \mathbb{J} \times \Omega_n : \Gamma_{y_n}(t \wedge t_n, t \vee t_n)f \in B\}$ (5.53)

$$= C \times \Omega_n \in Bor(J(s)) \otimes Bor(\mathbb{J}) \otimes \mathcal{F}_n. \tag{5.54}$$

Therefore ϕ is $\mathcal{F}_n - Bor(Y) \otimes \mathcal{F}_n$ measurable. Define for $i \in [1, n]$:

$$\alpha_i : Y \times \Omega_n \to \mathbb{J} \times \mathbb{J} \times Y \times \Omega_n \to Y \times \Omega_n \tag{5.55}$$

$$(g, \omega) \to (x_{n-i}(s)(\omega), x_{n-i+1}(s)(\omega), g, \omega)$$

$$\to (D(x_{n-i}(s)(\omega), x_{n-i+1}(s)(\omega))g, \omega). \tag{5.56}$$

Again, the canonical projections are trivially measurable. We have for $p \in [|0, n|]$:

$$\{\omega \in \Omega_n : x_p(s) \in B\} = \Omega_n \cap x_{p \wedge N_s(t)}(s)^{-1}(B) := C \in \mathcal{F}_n \tag{5.57}$$

Therefore $\{(g, \omega) \in Y \times \Omega_n : x_p(s) \in B\} = Y \times C \in Bor(Y) \otimes \mathcal{F}_n.$ (5.58)

Now, by measurability assumption, $\forall B \in Bor(Y)$, $\exists C \in Bor(\mathbb{J}) \otimes Bor(\mathbb{J}) \otimes Bor(Y)$:

$$\{(y_{n-i}, y_{n-i+1}, g, \omega) \in \mathbb{J} \times \mathbb{J} \times Y \times \Omega_n : D(y_{n-i}, y_{n-i+1})g \in B\} \tag{5.59}$$

$$= C \times \Omega_n \in Bor(\mathbb{J}) \otimes Bor(\mathbb{J}) \otimes Bor(Y) \otimes \mathcal{F}_n, \tag{5.60}$$

which proves the measurability of α_i. Then we define for $i \in [1, n]$:

$$\beta_i : Y \times \Omega_n \to J(s) \times J(s) \times \mathbb{J} \times Y \times \Omega_n \to Y \times \Omega_n \tag{5.61}$$

$$(g, \omega) \to (T_{n-i}(s)(\omega), T_{n-i+1}(s)(\omega), x_{n-i}(s)(\omega), g, \omega) \tag{5.62}$$

$$\to (\Gamma_{x_{n-i}(s)(\omega)}(T_{n-i}(s)(\omega), T_{n-i+1}(s)(\omega))g, \omega). \tag{5.63}$$

By measurability assumption, $\forall B \in Bor(Y)$, $\exists C \in Bor(J(s)) \otimes Bor(J(s)) \otimes Bor(\mathbb{J}) \otimes Bor(Y)$:

$$\{(t_{n-i}, t_{n-i+1}, y_{n-i}, g, \omega) \in J(s) \times J(s) \times \mathbb{J} \times Y \times \Omega_n :$$

$$\Gamma_{y_{n-i}}(t_{n-i} \wedge t_{n-i+1}, t_{n-i} \vee t_{n-i+1})g \in B\} \tag{5.64}$$

$$= C \times \Omega_n \in Bor(J(s)) \otimes Bor(J(s)) \otimes Bor(\mathbb{J}) \otimes Bor(Y) \otimes \mathcal{F}_n, \tag{5.65}$$

which proves the measurability of β_i. Finally, define the canonical projection:

$$\psi : Y \times \Omega_n \to Y \tag{5.66}$$

$$(g, \omega) \to g \tag{5.67}$$

which proves the measurability of $V_n(s, t)f$. For $n = 0$, we have $V_n(s, t)f = \Gamma_{x(s)}(s, t)f$ and the proof is similar.

The following result characterizes an inhomogeneous random evolution as a propagator, shows that it is right-continuous and that it satisfies an integral representation (which will be used extensively later on). It also clarifies why we used the terminology "continuous inhomogeneous $Y-$random evolution" when $D = I$.

Proposition 5.2.3. *Let V an inhomogeneous Y-random evolution and $(s,t) \in \Delta_J$, $\omega \in \Omega$. Then $V(\bullet, \bullet)(\omega)$ is a $Y-$propagator. If we assume that V is regular, then we have on Ω the following integral representation:*

$$V(s,t)f = f + \int_s^t V(s,u)A_{x(u)}(u)f\,du$$

$$+ \sum_{k=1}^{N_s(t)} V(s, T_k(s)^-)[D(x_{k-1}(s), x_k(s)) - I]f. \qquad (5.68)$$

Further $u \to V(s,u)(\omega)$ is $Y-$strongly RCLL on $J(s)$, i.e. $\forall f \in Y$, $u \to V(s,u)(\omega)f \in D(J(s), (Y, \|\cdot\|))$. More precisely, we have for $f \in Y$:

$$V(s, u^-)f = V(s,u)f \text{ if } u \notin \{T_n(s) : n \in \mathbb{N}\} \qquad (5.69)$$

$$V(s, T_{n+1}(s))f = V(s, T_{n+1}(s)^-)D(x_n(s), x_{n+1}(s))f \quad \forall n \in \mathbb{N}, \qquad (5.70)$$

where we denote $V(s, t^-)f := \lim_{u \uparrow t} V(s, u)f$.

Proof of Proposition 5.2.3. The fact that $V(s,t) \in \mathcal{B}(Y)$ is straightforward from the definition of V and using the fact that $(D(x,y))_{(x,y) \in \mathbb{J}^2}$ are $\mathcal{B}(Y)-$contractions. We can also obtain easily that V is a propagator by straightforward computations. We will now show that $u \to V(s,u)(\omega)$ is $Y-$strongly continuous on each $[T_n(s), T_{n+1}(s)) \cap J(s)$, $n \in \mathbb{N}$ and $Y-$strongly RCLL at each $T_{n+1}(s) \in J(s)$, $n \in \mathbb{N}$. Let $n \in \mathbb{N}$ such that $T_n(s) \in J(s)$. $\forall t \in [T_n(s), T_{n+1}(s)) \cap J(s)$, we have:

$$V(s,t) = \left[\prod_{k=1}^n \Gamma_{x_{k-1}(s)} (T_{k-1}(s), T_k(s)) \, D(x_{k-1}(s), x_k(s)) \right] \Gamma_{x_n(s)} (T_n(s), t).$$

$$(5.71)$$

Therefore by $Y-$strong $t-$continuity of Γ, we get that $u \to V(s,u)(\omega)$ is $Y-$strongly continuous on $[T_n(s), T_{n+1}(s)) \cap J(s)$. If $T_{n+1}(s) \in J(s)$, the fact that $V(s, \bullet)$ has a left limit at $T_{n+1}(s)$ also comes from the $Y-$strong $t-$continuity of Γ:

$$V(s, T_{n+1}(s)^-)f = \lim_{h \downarrow 0} G_n^s \Gamma_{x_n(s)}(T_n(s), T_{n+1}(s) - h)f$$

$$= G_n^s \Gamma_{x_n(s)}(T_n(s), T_{n+1}(s))f \qquad (5.72)$$

$$G_n^s = \prod_{k=1}^n \Gamma_{x_{k-1}(s)} (T_{k-1}(s), T_k(s)) \, D(x_{k-1}(s), x_k(s)). \qquad (5.73)$$

Therefore we get the relationship:

$$V(s, T_{n+1}(s))f = V(s, T_{n+1}(s)^-)D(x_n(s), x_{n+1}(s))f. \qquad (5.74)$$

To prove the integral representation, let $s \in J$, $\omega \in \Omega$, $f \in Y_1$. We are going to proceed by induction and show that $\forall n \in \mathbb{N}$, we have $\forall t \in [T_n(s), T_{n+1}(s)) \cap J(s)$:

$$\begin{aligned} V(s,t)f = f &+ \int_s^t V(s,u)A_{x(u)}(u)f\,du \\ &+ \sum_{k=1}^n V(s, T_k(s)^-)[D(x_{k-1}(s), x_k(s)) - I]f. \end{aligned} \qquad (5.75)$$

For $n = 0$, we have $\forall t \in [s, T_1(s)) \cap J(s)$: $V(s,t)f = \Gamma_{x(s)}(s,t)f$, and therefore $V(s,t)f = f + \int_s^t V(s,u)A_{x(u)}(u)f\,du$ by regularity of Γ. Now assume that the property is true for $n - 1$, namely: $\forall t \in [T_{n-1}(s), T_n(s)) \cap J(s)$, we have:

$$\begin{aligned} V(s,t)f = f &+ \int_s^t V(s,u)A_{x(u)}(u)f\,du \\ &+ \sum_{k=1}^{n-1} V(s, T_k(s)^-)[D(x_{k-1}(s), x_k(s)) - I]f. \end{aligned} \qquad (5.76)$$

Therefore it implies that (by continuity of the Bochner integral):

$$\begin{aligned} V(s, T_n(s)^-)f = f &+ \int_s^{T_n(s)} V(s,u)A_{x(u)}(u)f\,du \\ &+ \sum_{k=1}^{n-1} V(s, T_k(s)^-)[D(x_{k-1}(s), x_k(s)) - I]f. \end{aligned} \qquad (5.77)$$

Now, $\forall t \in [T_n(s), T_{n+1}(s)) \cap J(s)$ we have that:

$$V(s,t) = G_n^s \Gamma_{x_n(s)}(T_n(s), t) \qquad (5.78)$$

$$G_n^s := \prod_{k=1}^n \Gamma_{x_{k-1}(s)}(T_{k-1}(s), T_k(s)) D(x_{k-1}(s), x_k(s)), \qquad (5.79)$$

and therefore $\forall t \in [T_n(s), T_{n+1}(s)) \cap J(s)$, by Theorem 5.1.7 and regularity of Γ:

$$\begin{aligned} \frac{\partial}{\partial t} V(s,t)f = V(s,t)A_{x(t)}(t)f &\Rightarrow V(s,t)f = V(s, T_n(s))f \\ &+ \int_{T_n(s)}^t V(s,u)A_{x(u)}(u)f\,du. \end{aligned} \qquad (5.80)$$

Further, we already proved that $V(s, T_n(s))f = V(s, T_n(s)^-)D(x_{n-1}(s),$ $x_n(s))f$. Therefore combining these results we have:

$$V(s,t)f = V(s, T_n(s)^-)D(x_{n-1}(s), x_n(s))f + \int_{T_n(s)}^t V(s,u)A_{x(u)}(u)f du \qquad (5.81)$$

$$= V(s, T_n(s)^-)f + \int_{T_n(s)}^t V(s,u)A_{x(u)}(u)f du$$

$$+ V(s, T_n(s)^-)D(x_{n-1}(s), x_n(s))f - V(s, T_n(s)^-)f \qquad (5.82)$$

$$= f + \int_s^{T_n(s)} V(s,u)A_{x(u)}(u)f du + \sum_{k=1}^{n-1} V(s, T_k(s)^-)[D(x_{k-1}(s), x_k(s)) - I]f$$

$$\qquad (5.83)$$

$$+ \int_{T_n(s)}^t V(s,u)A_{x(u)}(u)f du + V(s, T_n(s)^-)D(x_{n-1}(s), x_n(s))f - V(s, T_n(s)^-)f$$

$$\qquad (5.84)$$

$$= f + \int_s^t V(s,u)A_{x(u)}(u)f du + \sum_{k=1}^n V(s, T_k(s)^-)[D(x_{k-1}(s), x_k(s)) - I]f. \qquad (5.85)$$

5.3 Weak Law of Large Numbers (WLLN)

In this section we introduce a rescaled random evolution V_ϵ, in which time is rescaled by a small parameter ϵ. The main result of this section is Theorem 5.3.19 in Subsection 5.3.5. In order to prove the weak convergence of V_ϵ to some regular propagator $\widehat{\Gamma}$, we prove in Subsection 5.3.3 that V_ϵ is relatively compact, which informally means that for any sequence $\epsilon_n \to 0$, there exists a subsequence $\{n_k\}$ along which $V_{\epsilon_{n_k}}$ converges weakly. To show the convergence of V_ϵ to $\widehat{\Gamma}$, we need to show that all limit points of the latter $V_{\epsilon_{n_k}}$ are equal to $\widehat{\Gamma}$. In order to prove relative compactness, we need among other things that V_ϵ satisfies the so-called compact containment criterion (abbreviated "CCC") - which in short requires that for every $f \in Y$, $V_\epsilon(s,t)f$ remains in a compact set of Y with an arbitrarily high probability as $\epsilon \to 0$. This compact containment criterion is the topic of Subsection 5.3.2. Subsection 5.3.1 introduces the rescaled random evolution V_ϵ as well as some regularity assumptions (condensed in assumption 5.3.1), which will be assumed to hold throughout the rest of the thesis. It also reminds the reader of some definitions and results on relative compactness in the Skorokhod space, which are mostly taken from the well-known book [3]. Finally, the main WLLN result Theorem 5.3.19 is proved using a martingale method similar in the spirit to what is done in [23] (Chapter 4, Section 4.2.1) for time-homogeneous random evolutions. This method is here adapted rigorously to the time-inhomogeneous setting: this is the topic of Subsection 5.3.4. The martingale representation presented in Lemma 5.3.17 of 5.3.4 will be used in Section 5.4 to prove a CLT for time-inhomogeneous random evolutions.

5.3.1　Preliminary Definitions and Assumptions

In this section we prove a weak law of large numbers for inhomogeneous random evolutions. We rescale both time and the jump operators D in a suitable way by a small parameter ϵ and study the limiting behavior of the rescaled random evolution. To this end, the same way we introduced inhomogeneous $Y-$random evolutions, we consider a family $(D^\epsilon(x,y))_{(x,y)\in\mathbb{J}^2,\epsilon\in(0,1]}$ of $\mathcal{B}(Y)-$contractions, satisfying $\forall\epsilon\in(0,1]$:

$$(x,y,f) \to D^\epsilon(x,y)f \text{ is } Bor(\mathbb{J}) \otimes Bor(\mathbb{J}) \otimes Bor(Y) - Bor(Y) \text{ measurable.}$$
(5.86)

and let $D^0(x,y) := I$. We define:

$$\mathcal{D}(D_1) := \bigcap_{\substack{\epsilon\in[0,1]\\(x,y)\in\mathbb{J}^2}} \left\{ f \in Y : \lim_{\substack{h\to 0\\\epsilon+h\in[0,1]}} \frac{D^{\epsilon+h}(x,y)f - D^\epsilon(x,y)f}{h} \in Y \right\}$$
(5.87)

and $\forall f \in \mathcal{D}(D_1)$:

$$D_1^\epsilon(x,y)f := \lim_{\substack{h\to 0\\\epsilon+h\in[0,1]}} \frac{D^{\epsilon+h}(x,y)f - D^\epsilon(x,y)f}{h}.$$
(5.88)

The latter operators correspond, in short, to the (first order) derivatives of the operators D^ϵ with respect to ϵ. We will need them in the following to be able to use the expansion $D^\epsilon \approx I + \epsilon D_1^\epsilon + \ldots$, which will prove useful when proving limit theorems for random evolutions. The same way, we introduce D_2^ϵ, corresponding to the second derivative. We also let:

$$\mathcal{D}(D_1^0 \in Y_1) := \left\{ f \in \mathcal{D}(D_1) \cap Y_1 : D_1^0(x,y)f \in Y_1 \ \forall(x,y) \in \mathbb{J}^2 \right\}.$$
(5.89)

For $x \in \mathbb{J}$, remembering the definition of $\mathcal{D}(A_x \in Y_1)$ in (5.34), we let:

$$\mathcal{D}(A_x') := \mathcal{D}(A_x \in Y_1)$$

$$\cap \left\{ f \in \mathcal{D}(A_x) \cap Y_1 : Y_1 - \lim_{\substack{h\to 0\\t+h\in J}} \frac{A_x(t+h)f - A_x(t)f}{h} \in Y_1 \ \forall t \in J \right\}$$
(5.90)

and for $t \in J$, $f \in \mathcal{D}(A_x')$:

$$A_x'(t)f := Y_1 - \lim_{\substack{h\to 0\\t+h\in J}} \frac{A_x(t+h)f - A_x(t)f}{h}.$$
(5.91)

Here $Y_1 - \lim$ simply indicates that the limit is taken in the Y_1 norm. We also introduce the space $\widehat{D}$ on which we will mostly be working:

$$\widehat{D} := \bigcap_{x \in \mathbb{J}} \mathcal{D}(A'_x) \cap \mathcal{D}(D_2) \cap \mathcal{D}(D_1^0 \in Y_1). \tag{5.92}$$

Throughout this section we make the following set of regularity assumptions, that we first state before commenting on them just afterwards. We recall that the various notions of continuity and regularity have been defined in Section 5.1.

Assumption 1. *Assumptions on the structure of spaces*:

1. The subset $\widehat{D}$ contains a countable family which is dense in both Y_1 and Y.

2. $Y_1 \subseteq \mathcal{D}(D_1)$.

Assumptions on the regularity of operators:

1. $(\Gamma_x)_{x \in \mathbb{J}}$ are regular Y−propagators.

2. A_x is Y_1−strongly continuous, $\forall x \in \mathbb{J}$.

Assumptions on the boundedness of operators:

1. $(\Gamma_x)_{x \in \mathbb{J}}$ are $\mathcal{B}(Y)$−exponentially bounded, i.e. $\exists \gamma \geq 0$ such that $\|\Gamma_x(s,t)\|_{\mathcal{B}(Y)} \leq e^{\gamma(t-s)}$, for all $x \in \mathbb{J}$, $(s,t) \in \Delta_J$.

2. $A_x(t) \in \mathcal{B}(Y_1, Y)$ and $\sup_{u \in [0,t]} \|A_x(u)\|_{\mathcal{B}(Y_1,Y)} < \infty$ $\forall t \in J$, $\forall x \in \mathbb{J}$.

3. $\sup_{\substack{t \in [0,T] \\ x \in \mathbb{J}}} \|A'_x(t)f\| < \infty$, $\sup_{\substack{t \in [0,T] \\ x \in \mathbb{J}}} \|A_x(t)f\|_{Y_1} < \infty$, $\forall f \in \bigcap_{x \in \mathbb{J}} \mathcal{D}(A'_x)$, for all $T \in J$.

4. $D_1^0(x,y) \in \mathcal{B}(Y_1, Y)$ $\forall x, y \in \mathbb{J}$.

5. $\sup_{\substack{\epsilon \in [0,1] \\ (x,y) \in \mathbb{J}^2}} \|D_1^\epsilon(x,y)f\| < \infty$, $\forall f \in \mathcal{D}(D_1)$.

6. $\sup_{\substack{\epsilon \in [0,1] \\ (x,y) \in \mathbb{J}^2}} \|D_2^\epsilon(x,y)f\| < \infty$, $\forall f \in \mathcal{D}(D_2)$.

Assumptions on the semi-Markov process:

1. (ergodicity) Assumptions from [18] hold true for the function $t \to t$, so that:

$$\lim_{t \to \infty} \frac{N(t)}{t} = \frac{1}{\Pi m} \text{ a.e.} \tag{5.93}$$

2. (uniform boundedness of sojourn increments) $\exists \bar{\tau} > 0$ such that:

$$\sup_{\substack{t \in \mathbb{R}_+ \\ i \in \mathbb{J}}} F_t(i, \bar{\tau}) = 1. \tag{5.94}$$

3. (regularity of the inhomogeneous Markov renewal process) The conditions for $F_t(i, \bar{\tau})$ are satisfied (see [18]), namely: there exists $\tau' > 0$ and $\beta' > 0$ such that:

$$\sup_{\substack{t \in \mathbb{R}_+ \\ i \in \mathbb{J}}} F_t(i, \tau') < 1 - \beta'. \tag{5.95}$$

Let's make a few comments on the previous assumptions. The assumptions regarding the regularity of operators mainly ensure that we will be able to use the results obtained on propagators in Section 5.1, for example Theorem 5.1.12. The (strong) continuity of A_x will also prove to be useful when working with convergence in the Skorokhod space. The assumptions on the boundedness of operators will be used to show that various quantities converge well. Finally, regarding the assumptions on the semi-Markov process, the a.e. convergence of $t^{-1}N(t)$ as $t \to \infty$ will be used very often. It is one of the fundamental requirement for the work below. The uniform boundedness of the sojourn increments is a mild assumption in practice. There might be a possibility to weaken it, but the proofs would become heavier, for example because the jumps of the martingales introduced below would not be uniformly bounded anymore.

Notation: in the following we let for $n \in \mathbb{N}$, $i \in \mathbb{J}$ and $t \in \mathbb{R}_+$ (their existence is guaranteed by assumption 5.3.1):

$$m_n(i, t) := \int_0^\infty s^n F_t(i, ds) \qquad m_n(i) := \int_0^\infty s^n F(i, ds). \tag{5.96}$$

We also let $\mathbf{J} := J$ if $J = \mathbb{R}_+$ and $\mathbf{J} := [0, T_\infty - \bar{\tau})$ if $J = [0, T_\infty]$. In the latter case it is assumed that $T_\infty > \bar{\tau}$. Similarly, we let for $s \in \mathbf{J}$: $\mathbf{J}(s) := \{t \in \mathbf{J} : s \le t\}$.

We now introduce the rescaled random evolution, with the notation $t^{\epsilon, s} := s + \epsilon(t - s)$:

Definition 5.3.1. Let V an inhomogeneous $Y-$random evolution. We define (pathwise on Ω) the rescaled inhomogeneous $Y-$random evolution V_ϵ for $\epsilon \in (0, 1]$, $(s, t) \in \Delta_J$ by:

$$V_\epsilon(s, t) := \left[\prod_{k=1}^{N_s\left(t^{\frac{1}{\epsilon}, s}\right)} \Gamma_{x_{k-1}(s)} \left(T_{k-1}^{\epsilon, s}(s), T_k^{\epsilon, s}(s)\right) D^\epsilon(x_{k-1}(s), x_k(s)) \right] \Gamma_{x\left(t^{\frac{1}{\epsilon}, s}\right)}$$

$$\times \left(T_{N_s\left(t^{\frac{1}{\epsilon}, s}\right)}^{\epsilon, s}(s), t \right). \tag{5.97}$$

Remark: we notice that V_ϵ is well-defined since on Ω:

$$T_{N_s\left(t^{\frac{1}{\epsilon}, s}\right)}^{\epsilon, s}(s) = s + \epsilon \left(T_{N_s\left(t^{\frac{1}{\epsilon}, s}\right)}(s) - s \right) \le s + \epsilon \left(t^{\frac{1}{\epsilon}, s} - s \right) = t, \tag{5.98}$$

and that it coincides with V for $\epsilon = 1$, i.e. $V_1(s, t) = V(s, t)$.

Our goal is to prove, as in [23], that for each f in some suitable subset of Y, $\{V_\epsilon(s, \bullet)f\}$ - seen as a family of elements of $D(\mathbf{J}(s), Y)$ - converges weakly to some continuous limiting process $V_0(s, \bullet)f$ to be determined. To this end, we will first prove that $\{V_\epsilon(s, \bullet)f\}$ is relatively compact with almost surely continuous weak limit points. This is equivalent to the notion of C−tightness in [10] (VI.3) because $\mathcal{P}(D(\mathbf{J}(s), Y))$ topologized with the Prohorov metric is a separable and complete metric space (Y being a separable Banach space), which implies that relative compactness and tightness are equivalent in $\mathcal{P}(D(\mathbf{J}(s), Y))$ (by Prohorov's theorem). Then we will identify the limiting operator-valued process V_0, using results of Chapter 4. We first need some elements that can be found in [23] (Section 1.4) and [3] (Sections 3.8 to 3.11):

Definition 5.3.2. *Let* $(\nu_n)_{n \in \mathbb{N}}$ *a sequence of probability measures on a metric space* (S, d). *We say that* ν_n *converges weakly to* ν, *and write* $\nu_n \Rightarrow \nu$ *iff* $\forall f \in C_b(S, \mathbb{R})$:

$$\lim_{n \to \infty} \int_S f d\nu_n = \int_S f d\nu. \tag{5.99}$$

Definition 5.3.3. *Let* $\{\nu_\epsilon\}$ *a family of probability measures on a metric space* (S, d). $\{\nu_\epsilon\}$ *is said to be relatively compact iff for any sequence* $(\nu_n)_{n \in \mathbb{N}} \subseteq \{\nu_\epsilon\}$, *there exists a weakly converging subsequence.*

Definition 5.3.4. *Let* $s \in \mathbf{J}$, $\{X_\epsilon\}$ *a family of stochastic processes with sample paths in* $D(\mathbf{J}(s), Y)$. *We say that* $\{X_\epsilon\}$ *is relatively compact iff* $\{\mathcal{L}(X_\epsilon)\}$ *is (in the metric space* $\mathcal{P}(D(\mathbf{J}(s), Y))$ *endowed with the Prohorov metric). We write that* $X_\epsilon \Rightarrow X$ *iff* $\mathcal{L}(X_\epsilon) \Rightarrow \mathcal{L}(X)$. *We say that* $\{X_\epsilon\}$ *is C-relatively compact iff it is relatively compact and if ever* $X_\epsilon \Rightarrow X$, *then* X *has a.e. continuous sample paths. If* $E_Y \subseteq Y$, *we say that* $\{V_\epsilon\}$ *is* E_Y−*relatively compact (resp.* E_Y−*C-relatively compact) iff* $\{V_\epsilon(s, \bullet)f\}$ *is* $\forall f \in E_Y$, $\forall s \in \mathbf{J}$.

Definition 5.3.5. *Let* $s \in \mathbf{J}$, $\{X_\epsilon\}$ *a family of stochastic processes with sample paths in* $D(\mathbf{J}(s), Y)$. *We say that* $\{X_\epsilon\}$ *satisfies the compact containment criterion (*$\{X_\epsilon\} \in \mathbf{CCC}$*) if* $\forall \Delta \in (0, 1]$, $\forall t \in \mathbf{J}(s) \cap \mathbb{Q}$, $\exists K \subseteq Y$ *compact set such that:*

$$\liminf_{\epsilon \to 0} \mathbb{P}[X_\epsilon(t) \in K] \geq 1 - \Delta. \tag{5.100}$$

We say that $\{V_\epsilon\}$ *satisfies the compact containment criterion in* $E_Y \subseteq Y$ *(*$\{V_\epsilon\} \in E_Y$-$\mathbf{CCC}$*), if* $\forall f \in E_Y$, $\forall s \in \mathbf{J}$, $\{V_\epsilon(s, \bullet)f\} \in CCC$.

Theorem 5.3.6. *Let* $s \in \mathbf{J}$, $\{X_\epsilon\}$ *a family of stochastic processes with sample paths in* $D(\mathbf{J}(s), Y)$. $\{X_\epsilon\}$ *is C-relatively compact iff it is relatively compact* and $j_s(X_\epsilon) \Rightarrow 0$, *where:*

$$j_s(X) := \int_{\mathbf{J}(s)} e^{-u}(j_s(X, u) \wedge 1)du \qquad (5.101)$$

$$j_s(X, u) := \sup_{t \in [s, u]} ||X(t) - X(t^-)||. \qquad (5.102)$$

Theorem 5.3.7. *Let $s \in \mathbf{J}$, $\{X_\epsilon\}$ a family of stochastic processes with sample paths in $D(J(s), Y)$. $\{X_\epsilon\}$ is relatively compact in $D(\mathbf{J}(s), Y)$ iff:*

  1. $\{X_\epsilon\} \in CCC$

  2. $\forall T \in \mathbf{J}(s)$, $\exists r > 0$ and a family $\{C_s(\epsilon, \eta) : (\epsilon, \eta) \in (0, 1] \times (0, 1)\}$ of nonnegative random variables such that $\forall(\epsilon, \eta) \in (0, 1] \times (0, \bar{\tau} \wedge 1)$, $\forall h \in [0, \eta]$, $\forall t \in [s, T]$:

$$\mathbb{E}\left[||X_\epsilon(t+h) - X_\epsilon(t)||^r | \mathcal{G}_t^{\epsilon,s}\right] \leq \mathbb{E}[C_s(\epsilon, \eta)|\mathcal{G}_t^{\epsilon,s}] \qquad (5.103)$$

$$\lim_{\eta \to 0} \limsup_{\epsilon \to 0} \mathbb{E}[C_s(\epsilon, \eta)] = 0, \qquad (5.104)$$

where $\mathcal{G}_t^{\epsilon,s} := \sigma[X_\epsilon(u) : u \in [s, t]]$.

If $\{X_\epsilon\}$ is relatively compact, then the stronger compact containment criterion holds: $\forall \Delta \in (0, 1]$, $\forall T \in \mathbf{J}(s)$, $\exists K \subseteq Y$ compact set such that:

$$\liminf_{\epsilon \to 0} \mathbb{P}[X_\epsilon(t) \in K \quad \forall t \in [s, T]] \geq 1 - \Delta. \qquad (5.105)$$

5.3.2　The Compact Containment Criterion (CCC)

We saw that in order to prove relative compactness, we need to prove that the compact containment criterion is satisfied. We give below some sufficient conditions for which it is the case, in particular for the space $C_0(\mathbb{R}^d)$ which is used in many applications. In [13] it is mentioned that there exists a compact embedding of a Hilbert space into $C_0(\mathbb{R}^d)$. Unfortunately, this is not true (to the best of our knowledge), and we show below in Proposition 5.3.9 how to overcome this problem. This latter proposition is applied in Section 6 to the time-inhomogeneous Lévy case, and the corresponding proof can easily be recycled for many other examples.

Proposition 5.3.8. *Assume that there exists a Banach space $(Z, || \cdot ||_Z)$ compactly embedded in Y, that $(\Gamma_x)_{x \in \mathbb{J}}$, are $\mathcal{B}(Z)-$exponentially bounded (uniformly in $\mathbb{J}$), and that $(D^\epsilon(x, y))_{\substack{\epsilon \in (0,1] \\ (x,y) \in \mathbb{J}^2}}$ are $\mathcal{B}(Z)-$contractions. Then $\{V_\epsilon\} \in Z\text{-}CCC$.*

Proof. *Let $f \in Z$, $(s, t) \in \Delta_\mathbb{J}$, and assume $||\Gamma_x(s, t)f||_Z \leq e^{r(t-s)}||f||_Z$ for some $r \geq 0$. Let $c := e^{r(t-s)}||f||_Z$ and $K := cl(Y) - S_c(Z)$, the $Y-$closure of the $Z-$closed ball of radius c. K is compact because of the compact embedding* *of Z into Y. Let $\epsilon \in (0,1]$. We have $\forall \omega \in \Omega$: $||V_\epsilon(s,t)(\omega)f||_Z \leq c$. Therefore $V_\epsilon(s,t)(\omega)f \in S_c(Z) \subseteq K$ and so $\mathbb{P}[V_\epsilon(s,t)f \in K] = \mathbb{P}(\Omega) = 1 \geq 1 - \Delta$.*

For example, we can consider the Rellich-Kondrachov compactness theorem: if $U \subseteq \mathbb{R}^d$ is an open, bounded Lipschitz domain, then the Sobolev space $W^{1,p}(U)$ is compactly embedded in $L^q(U)$, where $p \in [1,d)$ and $q \in [1, \frac{dp}{d-p})$.

For the space $C_0(\mathbb{R}^d)$, there is no well-known such compact embedding, therefore we have to proceed differently. The result below will be applied later on the time-inhomogeneous Lévy case (see Section 6), and the corresponding proof can easily be recycled for other examples.

Proposition 5.3.9. *Let $Y := C_0(\mathbb{R}^d)$, $E_Y \subseteq Y$. Assume that $\forall \Delta \in (0,1]$, $(s,t) \in \Delta_J$, $\epsilon \in (0,1]$, $f \in E_Y$, $\exists A_\epsilon \subseteq \Omega : \mathbb{P}(A_\epsilon) \geq 1 - \Delta$ and the family $\{V_\epsilon(s,t)(\omega)f : \epsilon \in (0,1], \omega \in A_\epsilon\}$ converge uniformly to 0 at infinity, is equicontinuous and uniformly bounded. Then $\{V_\epsilon\} \in E_Y\text{-CCC}$.*

Proof. *Let $f \in E_Y$, K the $Y-$closure of the set:*

$$K_1 := \{V_\epsilon(s,t)(\omega)f : \epsilon \in (0,1], \omega \in A_\epsilon\}. \tag{5.106}$$

K_1 is a family of elements of Y that are equicontinuous, uniformly bounded and that converge uniformly to 0 at infinity by assumption. Therefore it is well-known, using the Arzela-Ascoli theorem on the one-point compactification of $\mathbb{R}^d$, that K_1 is relatively compact in Y and therefore that K is compact in Y. And we have $\forall \epsilon \in (0,1]$:

$$\mathbb{P}[V_\epsilon(s,t)f \in K] \geq \mathbb{P}[\omega \in A_\epsilon : V_\epsilon(s,t)f \in K] = \mathbb{P}(A_\epsilon) \geq 1 - \Delta. \tag{5.107}$$

5.3.3 Relative Compactness of $\{V_\epsilon\}$

This section is devoted to proving that $\{V_\epsilon\}$ is relatively compact. In the following we assume that $\{V_\epsilon\}$ satisfies the compact containment criterion:

$$\{V_\epsilon\} \in Y_1 - \text{CCC}. \tag{5.108}$$

We first state an integral representation of $\{V_\epsilon\}$, proof of which is the same as the proof of Proposition 5.2.3.

Lemma 5.3.10. *Let assumption 5.3.1 hold true. Let $(s,t) \in \Delta_J$, $f \in Y_1$. Then V_ϵ satisfies on Ω:*

$$V_\epsilon(s,t)f = f + \int_s^t V_\epsilon(s,u) A_{x\left(u^{\frac{1}{\epsilon},s}\right)}(u)f\, du$$

$$+ \sum_{k=1}^{N_s\left(t^{\frac{1}{\epsilon},s}\right)} V_\epsilon(s, T_k^{\epsilon,s}(s)^-)[D^\epsilon(x_{k-1}(s), x_k(s)) - I]f. \tag{5.109}$$

We now prove that $\{V_\epsilon\}$ is relatively compact.

Lemma 5.3.11. *Let assumptions 5.3.1, 5.3.3 hold true. Then $\{V_\epsilon\}$ is Y_1-relatively compact.*

Proof. *We are going to use Theorem 5.3.7 to show this result. Using Lemma 5.3.10 we have for $h \in [0, \eta]$:*

$$\|V_\epsilon(s, t+h)f - V_\epsilon(s, t)f\| \tag{5.110}$$

$$\leq \left\| \int_t^{t+h} V_\epsilon(s, u) A_{x\left(\frac{1}{u\epsilon}, s\right)}(u) f du + \sum_{k=N_s\left(t^{\frac{1}{\epsilon}, s}\right)+1}^{N_s\left((t+h)^{\frac{1}{\epsilon}, s}\right)} V_\epsilon(s, T_k^{\epsilon, s}(s)^-)[D^\epsilon(x_{k-1}(s), x_k(s)) - I]f \right\| \tag{5.111}$$

$$\leq \eta M_1 + \epsilon e^{\gamma(T+1-s)} \sum_{k=N_s\left(t^{\frac{1}{\epsilon}, s}\right)+1}^{N_s\left((t+\eta)^{\frac{1}{\epsilon}, s}\right)} \frac{1}{\epsilon} \|D^\epsilon(x_{k-1}(s), x_k(s))f - f\| \tag{5.112}$$

$$\leq \eta M_1 + \epsilon M_2 \left[N\left((t+\eta)^{\frac{1}{\epsilon}, s}\right) - N\left(t^{\frac{1}{\epsilon}, s}\right)\right]. \tag{5.113}$$

where

$$M_1 := e^{\gamma(T+1-s)} \sup_{x \in \mathbb{J}, u \in [s, T+1 \wedge \bar{T}]} \|A_x(u)\|_{\mathcal{B}(Y_1, Y)} \|f\|_{Y_1}, \tag{5.114}$$

$$M_2 := e^{\gamma(T+1-s)} \sup_{\epsilon, x, y} \|D_1^\epsilon(x, y)f\|, \tag{5.115}$$

by assumption 5.3.1. Now, for $\epsilon \in (0, 1]$:

$$\epsilon \left[N\left((t+\eta)^{\frac{1}{\epsilon}, s}\right) - N\left(t^{\frac{1}{\epsilon}, s}\right)\right] \tag{5.116}$$

$$\leq \epsilon \sup_{t \in [s, s+\eta]} \left[N\left((t+\eta)^{\frac{1}{\epsilon}, s}\right) - N\left(t^{\frac{1}{\epsilon}, s}\right)\right] + \epsilon \sup_{t \in [s+\eta, T]} \left[N\left((t+\eta)^{\frac{1}{\epsilon}, s}\right) - N\left(t^{\frac{1}{\epsilon}, s}\right)\right] \tag{5.117}$$

$$\leq \epsilon N\left((s+2\eta)^{\frac{1}{\epsilon}, s}\right) + \epsilon \sup_{t \in [s+\eta, T]} \left[N\left((t+\eta)^{\frac{1}{\epsilon}, s}\right) - N\left(t^{\frac{1}{\epsilon}, s}\right)\right]. \tag{5.118}$$

Note that the supremums in the previous expression are a.e. finite as they are a.e. bounded by $N\left((T+1)^{\frac{1}{\epsilon}, s}\right)$. Now let:

$$C_s(\epsilon, \eta) := \eta M_1 + M_2 \epsilon N\left((s+2\eta)^{\frac{1}{\epsilon}, s}\right) + M_2 \epsilon \sup_{t \in [s+\eta, T]} \left[N\left((t+\eta)^{\frac{1}{\epsilon}, s}\right) - N\left(t^{\frac{1}{\epsilon}, s}\right)\right]. \tag{5.119}$$

We have to show that $\lim_{\eta \to 0} \lim_{\epsilon \to 0} \mathbb{E}[C_s(\epsilon, \eta)] = 0$. *We have:*

$$\lim_{\eta \to 0} \lim_{\epsilon \to 0} \eta M_1 + M_2 \epsilon \mathbb{E}\left[N\left((s + 2\eta)^{\frac{1}{\epsilon}, s}\right)\right] = \lim_{\eta \to 0} \eta M_1 + M_2 \frac{2\eta}{\Pi m} = 0. \quad (5.120)$$

Let $\{\epsilon_n\}$ any sequence converging to 0, and denote

$$Z_n := \epsilon_n \sup_{t \in [s+\eta, T]} \left[N\left((t + \eta)^{\frac{1}{\epsilon_n}, s}\right) - N\left(t^{\frac{1}{\epsilon_n}, s}\right)\right]. \quad (5.121)$$

We first want to show that $\{Z_n\}$ is uniformly integrable. By [3], it is sufficient to show that $\sup_n \mathbb{E}(Z_n^2) < \infty$. We have that $\mathbb{E}(Z_n^2) \leq \epsilon_n^2 \mathbb{E}\left[N^2\left((T + 1)^{\frac{1}{\epsilon_n}, s}\right)\right]$. By assumption 5.3.1 (more precisely, the regularity of the inhomogeneous Markov renewal process) we get:

$$\lim_{t \to \infty} \frac{\mathbb{E}(N^2(t))}{t^2} < \infty, \quad (5.122)$$

and therefore $\{Z_n\}$ is uniformly integrable. Then we show that $Z_n \overset{a.e.}{\to} Z := \frac{\eta}{\Pi m}$. Let:

$$\Omega^* := \left\{\lim_{\epsilon \to 0} \epsilon N\left((s + 1)^{\frac{1}{\epsilon}, s}\right) = \frac{1}{\Pi m}\right\}. \quad (5.123)$$

Let $\omega \in \Omega^*$ and $\delta > 0$. There exists some constant $r_2(\omega, \delta) > 0$ such that for $\epsilon < r_2$:

$$\left|\epsilon N\left((s + 1)^{\frac{1}{\epsilon}, s}\right) - \frac{1}{\Pi m}\right| < \frac{\delta}{T + \eta}, \quad (5.124)$$

and if $t \in [s + \eta, T + \eta]$:

$$\left|(t - s)\epsilon N\left((s + 1)^{\frac{1}{\epsilon}, s}\right) - \frac{t - s}{\Pi m}\right| < \frac{\delta(t - s)}{T + \eta} \leq \delta. \quad (5.125)$$

Let $\epsilon < \eta r_2$ (recall $\eta > 0$) and $\epsilon_2 := \frac{\epsilon}{t-s}$. Then $\epsilon_2 < \frac{\eta r_2}{\eta} = r_2$, and therefore:

$$\left|(t - s)\epsilon_2 N\left((s + 1)^{\frac{1}{\epsilon_2}, s}\right) - \frac{t - s}{\Pi m}\right| < \delta \quad (5.126)$$

$$\Rightarrow \left|\epsilon N\left(t^{\frac{1}{\epsilon}, s}\right) - \frac{t - s}{\Pi m}\right| < \delta. \quad (5.127)$$

And therefore for $\epsilon < \eta r_2$ and $t \in [s + \eta, T]$:

$$\left|\epsilon N\left((t + \eta)^{\frac{1}{\epsilon}, s}\right) - \epsilon N\left(t^{\frac{1}{\epsilon}, s}\right) - \frac{\eta}{\Pi m}\right| < 2\delta \quad (5.128)$$

$$\Rightarrow \left|\sup_{t \in [s+\eta, T]} \left[\epsilon N\left((t + \eta)^{\frac{1}{\epsilon}, s}\right) - \epsilon N\left(t^{\frac{1}{\epsilon}, s}\right)\right] - \frac{\eta}{\Pi m}\right| \leq 2\delta < 3\delta. \quad (5.129)$$

We have proved that $Z_n \overset{a.e.}{\to} Z$. By uniform integrability of $\{Z_n\}$, we get that $\lim_{n\to\infty} \mathbb{E}(Z_n) = \mathbb{E}(Z)$ and therefore since the sequence $\{\epsilon_n\}$ is arbitrary:

$$\lim_{\epsilon \to 0} \epsilon \mathbb{E} \left[\sup_{t \in [s+\eta, T]} \left[N\left((t+\eta)^{\frac{1}{\epsilon}, s}\right) - N\left(t^{\frac{1}{\epsilon}, s}\right) \right] \right] = \frac{\eta}{\Pi m}. \tag{5.130}$$

We now prove that the limit points of $\{V_\epsilon\}$ are continuous.

Lemma 5.3.12. *Let assumptions 5.3.1, 5.3.3 hold true. Then $\{V_\epsilon\}$ is $Y_1 - C$-relatively compact.*

Proof. *The proof is presented for the case $J = \mathbb{R}_+$. The proof for the case $J = [0, T_\infty]$ is exactly the same. By Lemma 5.3.11 it is relatively compact. By Theorem 5.3.6 it is sufficient to show that $j_s(V_\epsilon(s, \bullet)f) \overset{P}{\to} 0$. Let $\delta > 0$ and fix $T > 0$. For $u \in [s, T]$ we have:*

$$j_s(V_\epsilon(s, \bullet)f, u) \leq \sup_{t \in [s, T]} \| V_\epsilon(s, t)f - V_\epsilon(s, t^-)f \| \tag{5.131}$$

$$= \max_{k \in \left[1, N_s\left(T^{\frac{1}{\epsilon}, s} \right) \right]} \left\| V_\epsilon \left(s, T_k^{\epsilon, s}(s) \right) f - V_\epsilon \left(s, T_k^{\epsilon, s}(s)^- \right) f \right\| \tag{5.132}$$

(using Lemma 5.3.10) $= \max_{k \in \left[1, N_s\left(T^{\frac{1}{\epsilon}, s} \right) \right]}$

$$\| V_\epsilon \left(s, T_k^{\epsilon, s}(s)^- \right) \left(D^\epsilon(x_{k-1}(s), x_k(s))f - f \right) \| \tag{5.133}$$

$$\leq e^{\gamma(T-s)} \max_{(x,y) \in J^2} \| D^\epsilon(x, y)f - f \| \leq C_T \epsilon, \tag{5.134}$$

with $C_T := e^{\gamma(T-s)} \sup_{\epsilon, x, y} \| D_1^\epsilon(x, y)f \|$ (by assumption 5.3.1).

$$\tag{5.135}$$

Since:

$$j_s(V_\epsilon(s, \bullet)f) = \int_s^T e^{-u}(j_s(V_\epsilon(s, \bullet)f, u) \wedge 1)du + \int_T^\infty e^{-u}(j_s(V_\epsilon(s, \bullet)f, u) \wedge 1)du \tag{5.136}$$

$$\leq C_T \epsilon + e^{-T}, \tag{5.137}$$

we get $j_s(V_\epsilon(s, \bullet)f) \overset{a.e.}{\to} 0$ (choose T big enough, then ϵ small enough).

5.3.4 Martingale Characterization of the Inhomogeneous Random Evolution

In order to prove the weak law of large numbers for the random evolution, we use a martingale method similar to what is done in [23] (Section 4.2.1), but adapted rigorously to the inhomogeneous setting. We first introduce the quantity f_1, solution to a suitable "Poisson equation":

Definition 5.3.13. *Let assumption 5.3.1 hold true. For $f \in Y_1$, $x \in \mathbb{J}$, $t \in \mathbf{J}$, let $f^\epsilon(x,t) := f + \epsilon f_1(x,t)$, where f_1 is the unique solution of the equation:*

$$(P - I)f_1(\bullet, t)(x) = \Pi m[\widehat{A}(t) - a(x,t)]f \tag{5.138}$$

$$a(x,t) := \frac{1}{\Pi m}\left(m_1(x)A_x(t) + PD_1^0(x, \bullet)(x)\right) \tag{5.139}$$

$$\widehat{A}(t) := \Pi a(\bullet, t), \tag{5.140}$$

namely $f_1(x,t) = \Pi m R_0[\widehat{A}(t)f - a(\bullet, t)f](x)$, where $R_0 := (P - I + \Pi)^{-1}$ is the fundamental matrix associated to P.

Remark 5.3.14. *The existence of f_1 is guaranteed because $\Pi[\widehat{A}(t) - a(\bullet, t)]f = 0$ by construction (see [12], Proposition 4). In fact, in [12], the operators Π and P are defined on $B_{\mathbb{R}}^b(\mathbb{J})$ but the results hold true if we work on $B_E^b(\mathbb{J})$, where E is any Banach space such that $[\widehat{A}(t) - a(x,t)]f \in E$ (e.g. $E = Y_1$ if $f \in \widehat{D}$, $E = Y$ if $f \in Y_1$). To see that, first observe that P and Π can be defined the same way on $B_E^b(\mathbb{J})$ as they were on $B_{\mathbb{R}}^b(\mathbb{J})$. Then take $\ell \in E^*$ such that $||\ell|| = 1$ and $g \in B_E^b(\mathbb{J})$ such that $||g||_{B_E^b(\mathbb{J})} = \max_x ||g(x)||_E = 1$. We therefore have that: $||\ell \circ g||_{B_{\mathbb{R}}^b(\mathbb{J})} \leq 1$, and since we have the uniform ergodicity on $B_{\mathbb{R}}^b(\mathbb{J})$, we have that:*

$$\sup_{\substack{||\ell||=1 \\ ||g||_{B_E^b(\mathbb{J})}=1 \\ x \in \mathbb{J}}} |P^n(\ell \circ g)(x) - \Pi(\ell \circ g)(x)| \leq ||P^n - \Pi||_{\mathcal{B}(B_{\mathbb{R}}^b(\mathbb{J}))} \to 0. \tag{5.141}$$

By linearity of ℓ, P, Π we get that $|P^n(\ell \circ g)(x) - \Pi(\ell \circ g)(x)| = |\ell(P^n g(x) - \Pi g(x))|$. But because $||P^n g(x) - \Pi g(x)||_E = \sup_{||\ell||=1} |\ell(P^n g(x) - \Pi g(x))|$ and that this supremum is attained (see e.g. [1], Section III.6), then:

$$\sup_{\substack{||\ell||=1 \\ ||g||_{B_E^b(\mathbb{J})}=1 \\ x \in \mathbb{J}}} |\ell(P^n g(x) - \Pi g(x))| = \sup_{\substack{||g||_{B_E^b(\mathbb{J})}=1 \\ x \in \mathbb{J}}} ||P^n g(x) - \Pi g(x)||_E \tag{5.142}$$

$$= \sup_{||g||_{B_E^b(\mathbb{J})}=1} ||P^n g - \Pi g||_{B_E^b(\mathbb{J})} = ||P^n - \Pi||_{\mathcal{B}(B_E^b(\mathbb{J}))}, \tag{5.143}$$

and so we also have $||P^n - \Pi||_{\mathcal{B}(B_E^b(\mathbb{J}))} \to 0$, i.e., the uniform ergodicity in $B_E^b(\mathbb{J})$. Now, according to the proofs of Theorems 3.4, 3.5 Chapter VI of [17], $||P^n - \Pi||_{\mathcal{B}(B_E^b(\mathbb{J}))} \to 0$ is the only thing we need to prove that $P + \Pi - I$ is invertible on:

$$B_E^\Pi(\mathbb{J}) := \{f \in B_E^b(\mathbb{J}) : \Pi f = 0\}, \tag{5.144}$$

the space E plays no role. Further, $(P + \Pi - I)^{-1} \in \mathcal{B}(B_E^\Pi(\mathbb{J}))$ by the bounded inverse theorem.

We now introduce the martingale $(\widetilde{M}_t^\epsilon(s))_{t \geq 0}$ which will play a central role in the following.

Lemma 5.3.15. *Let assumption 5.3.1 hold true. Define recursively for $\epsilon \in (0,1]$, $s \in \mathbf{J}$:*

$$V_0^\epsilon(s) := I \tag{5.145}$$

$$V_{n+1}^\epsilon(s) := V_n^\epsilon(s)\Gamma_{x_n(s)}\left(T_n^{\epsilon,s}(s), T_{n+1}^{\epsilon,s}(s)\right) D^\epsilon(x_n(s), x_{n+1}(s)), \tag{5.146}$$

i.e. $V_n^\epsilon(s) = V_\epsilon(s, T_n^{\epsilon,s}(s))$; and for $f \in Y_1$ (we recall that $f^\epsilon(x,t) := f + \epsilon f_1(x,t)$):

$$M_n^\epsilon(s)f := V_n^\epsilon(s)f^\epsilon(x_n(s), T_n^{\epsilon,s}(s)) - f^\epsilon(x(s), s)$$

$$- \sum_{k=0}^{n-1} \mathbb{E}[V_{k+1}^\epsilon(s)f^\epsilon(x_{k+1}(s), T_{k+1}^{\epsilon,s}(s)) - V_k^\epsilon(s)f^\epsilon(x_k(s), T_k^{\epsilon,s}(s))|\mathcal{F}_k(s)],$$

$$\tag{5.147}$$

so that $(M_n^\epsilon(s)f)_{n\in\mathbb{N}}$ is a $\mathcal{F}_n(s)$-martingale by construction. Let for $t \in \mathbf{J}(s)$:

$$\widetilde{M}_t^\epsilon(s)f := M_{N_s\left(t^{\frac{1}{\epsilon},s}\right)+1}^\epsilon(s)f \tag{5.148}$$

$$\widetilde{\mathcal{F}}_t^\epsilon(s) := \mathcal{F}_{N_s\left(t^{\frac{1}{\epsilon},s}\right)+1}(s), \tag{5.149}$$

where $\mathcal{F}_n(s) := \sigma[x_k(s), T_k(s) : k \leq n] \vee \sigma(\mathbb{P}-null\ sets)$ and $\mathcal{F}_{N_s\left(t^{\frac{1}{\epsilon},s}\right)+1}(s)$ is defined the usual way (provided we have shown that $N_s\left(t^{\frac{1}{\epsilon},s}\right) + 1$ is a $\mathcal{F}_n(s)$-stopping time $\forall t \in \mathbf{J}(s)$). Then $\forall \ell \in Y^*$, $\forall s \in \mathbf{J}$, $\forall \epsilon \in (0,1]$, $\forall f \in Y_1$, $(\ell(\widetilde{M}_t^\epsilon(s)f), \widetilde{\mathcal{F}}_t^\epsilon(s))_{t\in\mathbf{J}(s)}$ is a real-valued square-integrable martingale.

Proof. *By construction $(\ell(M_n^\epsilon(s)f), \mathcal{F}_n(s))$ is a martingale. Let $\theta_s^\epsilon(t) := N_s\left(t^{\frac{1}{\epsilon},s}\right) + 1$. $\forall t \in \mathbf{J}(s)$, $\theta_s^\epsilon(t)$ is a $\mathcal{F}_n(s)$-stopping time, because:*

$$\{\theta_s^\epsilon(t) = n\} = \left\{N_s\left(t^{\frac{1}{\epsilon},s}\right) = n-1\right\} = \left\{T_{n-1}(s) \leq t^{\frac{1}{\epsilon},s}\right\} \cap \left\{T_n(s) > t^{\frac{1}{\epsilon},s}\right\} \in \mathcal{F}_n(s). \tag{5.150}$$

Let $t_1 \leq t_2 \in \mathbf{J}(s)$. We have that $(\ell(M_{\theta_s^\epsilon(t_2)\wedge n}^\epsilon(s)f), \mathcal{F}_n(s))$ is a martingale. Assume we have shown that it is uniformly integrable, then we can apply the optional sampling theorem for UI martingales to the stopping times $\theta_s^\epsilon(t_1) \leq \theta_s^\epsilon(t_2)$ a.e and get:

$$\mathbb{E}[\ell(M_{\theta_s^\epsilon(t_2)\wedge\theta_s^\epsilon(t_2)}^\epsilon(s)f)|\mathcal{F}_{\theta_s^\epsilon(t_1)}(s)] = \ell(M_{\theta_s^\epsilon(t_1)\wedge\theta_s^\epsilon(t_2)}^\epsilon(s)f)\ a.e. \tag{5.151}$$

$$\Rightarrow \mathbb{E}[\ell(M_{\theta_s^\epsilon(t_2)}^\epsilon(s)f)|\mathcal{F}_{\theta_s^\epsilon(t_1)}(s)] = \ell(M_{\theta_s^\epsilon(t_1)}^\epsilon(s)f)\ a.e. \tag{5.152}$$

$$\Rightarrow \mathbb{E}[\ell(\widetilde{M}_{t_2}^\epsilon(s)f)|\widetilde{\mathcal{F}}_{t_1}^\epsilon(s)] = \ell(\widetilde{M}_{t_1}^\epsilon(s)f)\ a.e., \tag{5.153}$$

which shows that $(\ell(\widetilde{M_t^\epsilon}(s)f), \widetilde{\mathcal{F}}_t^\epsilon(s))_{t \in \mathbf{J}(s)}$ is a martingale. Now to show the uniform integrability, by [3] it is sufficient to show that $\sup_n \mathbb{E}(\|M_{\theta_s^\epsilon(t_2) \wedge n}^\epsilon(s)f\|^2) < \infty$. But:

$$\|M_{\theta_s^\epsilon(t_2) \wedge n}^\epsilon(s)f\| \leq 2e^{\gamma(t_2 + \bar{\tau} - s)}(\|f\| + \|f_1\|) + 2e^{\gamma(t_2 + \bar{\tau} - s)}(\|f\| + \|f_1\|)(\theta_s^\epsilon(t_2) \wedge n) \tag{5.154}$$

$$\leq 2e^{\gamma(t_2 + \bar{\tau} - s)}(\|f\| + \|f_1\|)(1 + \theta_s^\epsilon(t_2)), \tag{5.155}$$

where $\|f_1\| := \sup_{\substack{x \in \mathbb{J} \\ u \in [0, t_2 + \bar{\tau}]}} \|f_1(x, u)\|$ ($\|f_1\| < \infty$ by assumption 5.3.1). The fact that $\mathbb{E}(\theta_s^\epsilon(t_2)^2) < \infty$ (assumption 5.3.1) concludes the proof.

Remark 5.3.16. *In the following we will make use of the fact that can be found in [1] (Theorem 3.1) that for sequences (X_n), (Y_n) of random variables with value in a separable metric space with metric d, if $X_n \Rightarrow X$ and $d(X_n, Y_n) \Rightarrow 0$, then $Y_n \Rightarrow X$. In our case, (X_n), (Y_n) take value in $D(\mathbf{J}(s), Y)$ and to show that $d(X_n, Y_n) \Rightarrow 0$, we will use the fact that:*

$$d(X_n, Y_n) \leq \sup_{t \in [s, T]} \|X_n(t) - Y_n(t)\| + e^{-T} \quad \forall T \in \mathbf{J}(s), \; if \; J = \mathbb{R}_+ \tag{5.156}$$

$$d(X_n, Y_n) \leq \sup_{t \in [s, T_\infty - \bar{\tau}]} \|X_n(t) - Y_n(t)\|, \; if \; J = [0, T_\infty], \tag{5.157}$$

and therefore that it is sufficient to have:

$$\sup_{t \in [s, T]} \|X_n(t) - Y_n(t)\| \stackrel{a.e.}{\to} 0 \; (resp. \; in \; probability), \quad \forall T \in \mathbf{J}(s) \tag{5.158}$$

to obtain $d(X_n, Y_n) \stackrel{a.e.}{\to} 0$ (resp. in probability).

Lemma 5.3.17. *Let assumption 5.3.1 hold true. For $f \in \widehat{\mathcal{D}}$ and $s \in \mathbf{J}$, $\widetilde{M_\bullet^\epsilon}(s)f$ has the asymptotic representation:*

$$\widetilde{M_\bullet^\epsilon}(s)f = V_\epsilon(s, \bullet)f - f - \epsilon \Pi m \sum_{k=1}^{N_s\left(t^{\frac{1}{\epsilon}, s}\right)} V_\epsilon(s, T_k^{\epsilon, s}(s)) \widehat{A}(T_k^{\epsilon, s}(s)) f + O(\epsilon) \; a.e., \tag{5.159}$$

where $O(\epsilon^p)$ is an element of the space $D(\mathbf{J}(s), Y)$ and is defined by the following property: $\forall r > 0$, $\forall T \in \mathbb{Q}^+ \cap \mathbf{J}(s)$, $\epsilon^{-p+r} \sup_{t \in [s, T]} \|O(\epsilon^p)\| \Rightarrow 0$ as $\epsilon \to 0$ (so that the Remark 5.3.16 will be satisfied).

Proof.
For sake of clarity let:

$$f_k^\epsilon := f^\epsilon(x_k(s), T_k^{\epsilon, s}(s)) \qquad f_{1,k} := f_1(x_k(s), T_k^{\epsilon, s}(s)) \tag{5.160}$$

Let $T \in \mathbf{J}(s)$. First we have that:

$$V^\epsilon_{N_s\left(t^{\frac{1}{\epsilon},s}\right)+1}(s)f^\epsilon_{N_s\left(t^{\frac{1}{\epsilon},s}\right)+1} = V^\epsilon_{N_s\left(t^{\frac{1}{\epsilon},s}\right)+1}(s)f + O(\epsilon) \qquad (5.161)$$

because $\sup_{t\in[s,T]} ||V^\epsilon_{N_s\left(t^{\frac{1}{\epsilon},s}\right)+1}(s)f_{1,N_s\left(t^{\frac{1}{\epsilon},s}\right)+1}|| \leq e^{\gamma(T+\bar{\tau}-s)}||f_1||.$ *Again and as in Lemma 5.3.15, we denote* $||f_1|| := \sup_{\substack{x\in\mathbb{J} \\ u\in[0,T+\tau]}} ||f_1(x,u)||,$ *and* $||f_1|| < \infty$ *by assumption 5.3.1. Now we have:*

$$V^\epsilon_{k+1}(s)f^\epsilon_{k+1} - V^\epsilon_k(s)f^\epsilon_k = V^\epsilon_k(s)(f^\epsilon_{k+1} - f^\epsilon_k) + V^\epsilon_{k+1}(s)f^\epsilon_{k+1} - V^\epsilon_k(s)f^\epsilon_{k+1}, \qquad (5.162)$$

and:

$$\mathbb{E}[V^\epsilon_k(s)(f^\epsilon_{k+1} - f^\epsilon_k)|\mathcal{F}_k(s)] = \epsilon V^\epsilon_k(s)\mathbb{E}[(f_{1,k+1} - f_{1,k})|\mathcal{F}_k(s)], \qquad (5.163)$$

as $V^\epsilon_k(s)$ is $\mathcal{F}_k(s) - Bor(\mathcal{B}(Y))$ measurable. Now, we know that every discrete time Markov process has the strong Markov property, so the Markov process (x_n, T_n) has it. For $k \geq 1$, the times $N(s) + k$ are $\mathcal{F}_n(0)-$stopping times. Therefore for $k \geq 1$:

$$\mathbb{E}[(f_{1,k+1} - f_{1,k})|\mathcal{F}_k(s)] = \mathbb{E}[(f_{1,k+1} - f_{1,k})|T_k(s), x_k(s)] \qquad (5.164)$$

$$= \sum_{y\in\mathbb{J}} \int_0^\infty f_1\left(y, T^{\epsilon,s}_k(s) + \epsilon u\right) Q_{T_k(s)}(x_k(s), y, du) - f_{1,k}. \qquad (5.165)$$

Let the following derivatives with respect to t:

$$a'(x,t) := \frac{1}{\Pi m} m_1(x) A'_x(t), \qquad (5.166)$$

$$\widehat{A}'(t) := \Pi a'(\bullet, t) \qquad (5.167)$$

$$f'_1(x,t) := \Pi m R_0[\widehat{A}'(t)f - a'(\bullet, t)f](x), \qquad (5.168)$$

which exist because $R_0 \in \mathcal{B}(B^\Pi_{Y_1}(\mathbb{J}))$ and $f \in \cap_{x\in\mathbb{J}} \mathcal{D}(A'_x)$. Using the fundamental theorem of calculus for the Bochner integral $(v \to f'_1(y,v) \in L^1_Y([a,b])$ $\forall [a,b]$ by assumption 5.3.1) we get:

$$\mathbb{E}[(f_{1,k+1} - f_{1,k})|T_k(s), x_k(s)] \qquad (5.169)$$

$$= \sum_{y\in\mathbb{J}} \int_0^\infty \left[f_1\left(y, T^{\epsilon,s}_k(s)\right) + \int_{T^{\epsilon,s}_k(s)}^{T^{\epsilon,s}_k(s)+\epsilon u} f'_1(y,v)dv \right] Q_{T_k(s)}(x_k(s), y, du)$$

$$- f_1\left(x_k(s), T^{\epsilon,s}_k(s)\right) \qquad (5.170)$$

$$= (P_{T_k(s)} - I)f_1\left(\bullet, T^{\epsilon,s}_k(s)\right)(x_k(s)) \qquad (5.171)$$

$$+ \sum_{y\in\mathbb{J}} \int_0^\infty \left[\int_0^{\epsilon u} f'_1(y, T^{\epsilon,s}_k(s) + v)dv \right] Q_{T_k(s)}(x_k(s), y, du) \qquad (5.172)$$

$$= (P_{T_k(s)} - I)f_1\left(\bullet, T^{\epsilon,s}_k(s)\right)(x_k(s)) + O(\epsilon). \qquad (5.173)$$

because by assumption 5.3.1:

$$\sum_{y \in \mathbb{J}} \int_0^\infty \left[\int_0^{\epsilon u} f_1'(y, T_k^{\epsilon,s}(s) + v) dv \right] Q_{T_k(s)}(x_k(s), y, du) \leq \epsilon \|f_1'\| \bar{\tau}. \quad (5.174)$$

We note that the contribution of the terms of order $O(\epsilon^2)$ inside the sum will make the sum of order $O(\epsilon)$, since for some constant C_T:

$$\sup_{t \in [s,T]} \sum_{k=1}^{N_s\left(t^{\frac{1}{\epsilon},s}\right)} \|O(\epsilon^2)\| \leq N_s\left(T^{\frac{1}{\epsilon},s}\right) C_T \epsilon^2 = O(\epsilon) \quad (5.175)$$

because $\epsilon N_s\left(T^{\frac{1}{\epsilon},s}\right) \overset{a.e.}{\to} \frac{T-s}{\Pi m}$. All put together we have for $k \geq 1$, using the definition of f_1:

$$\mathbb{E}[V_k^\epsilon(s)(f_{k+1}^\epsilon - f_k^\epsilon)|\mathcal{F}_k(s)] = \epsilon V_k^\epsilon(s)(P_{T_k(s)} - I) f_1\left(\bullet, T_k^{\epsilon,s}(s)\right)(x_k(s)) + O(\epsilon^2) \quad (5.176)$$

$$= \epsilon \Pi m V_k^\epsilon(s) \left[\widehat{A}(T_k^{\epsilon,s}(s)) - a(x_k(s), T_k^{\epsilon,s}(s)) \right] f \quad (5.177)$$

$$+ \epsilon V_k^\epsilon(s)(P_{T_k(s)} - P) f_1\left(\bullet, T_k^{\epsilon,s}(s)\right)(x_k(s)) + O(\epsilon^2). \quad (5.178)$$

The term involving $P_{T_k(s)} - P$ above will vanish as $k \to \infty$ by assumption 5.3.1, as $\|P_t - P\| \to 0$. We also have for the first term $(k = 0)$:

$$\mathbb{E}[V_0^\epsilon(s)(f_1^\epsilon - f_0^\epsilon)|\mathcal{F}_k(s)] = O(\epsilon) \quad (\leq 2\epsilon \|f_1\|). \quad (5.179)$$

Now we have to compute the terms corresponding to $V_{k+1}^\epsilon(s) f_{k+1}^\epsilon - V_k^\epsilon(s) f_{k+1}^\epsilon$. We will show that the term corresponding to $k = 0$ is $O(\epsilon)$ and that for $k \geq 1$:

$$\mathbb{E}[V_{k+1}^\epsilon(s) f_{k+1}^\epsilon - V_k^\epsilon(s) f_{k+1}^\epsilon | \mathcal{F}_k(s)] = \epsilon \Pi m V_k^\epsilon(s) a(x_k(s), T_k^{\epsilon,s}(s)) f$$
$$+ \text{ "negligible terms."} \quad (5.180)$$

For the term $k = 0$, we have, using assumption 5.3.1, the definition of V_k^ϵ and Theorem 5.1.8:

$$V_1^\epsilon(s) f_1^\epsilon - V_0^\epsilon(s) f_1^\epsilon = V_1^\epsilon(s) f - f + O(\epsilon) \quad (5.181)$$

$$= \Gamma_{x(s)}\left(s, T_1^{\epsilon,s}(s)\right) D^\epsilon(x(s), x_1(s)) f - f + O(\epsilon) \quad (5.182)$$

$$= \Gamma_{x(s)}\left(s, T_1^{\epsilon,s}(s)\right) \left(D^\epsilon(x(s), x_1(s)) f - f\right) + \Gamma_{x(s)}\left(s, T_1^{\epsilon,s}(s)\right) f - f + O(\epsilon) \quad (5.183)$$

$$\Rightarrow \mathbb{E}[V_1^\epsilon(s) f_1^\epsilon - V_0^\epsilon(s) f_1^\epsilon | \mathcal{F}_0(s)] \leq e^{\gamma \bar{\tau}} \max_{x,y} \|D^\epsilon(x,y) f - f\|$$

$$+ \int_0^\infty \epsilon u \sup_{x, t \in [s, s+\tau]} \|A_x(t)\|_{\mathcal{B}(Y_1, Y)} \|f\|_{Y_1} F_{T_0(s)}(x, du) \quad (5.184)$$

$$\leq \epsilon \left(e^{\gamma \bar{\tau}} \sup_{\epsilon, x, y} \|D_1^\epsilon(x,y) f\| + \|f\|_{Y_1} \bar{\tau} \|A_x(t)\|_{\mathcal{B}(Y_1, Y)} \right) = O(\epsilon). \quad (5.185)$$

Now we have for $k \geq 1$:

$$V_{k+1}^\epsilon(s) - V_k^\epsilon(s) = V_k^\epsilon(s) \left[\Gamma_{x_k(s)} \left(T_k^{\epsilon,s}(s), T_{k+1}^{\epsilon,s}(s) \right) D^\epsilon(x_k(s), x_{k+1}(s)) - I \right]. \tag{5.186}$$

By assumption 5.3.1 we have $\sup_{\epsilon,x,y} \|D_1^\epsilon(x,y)g\| < \infty$ for $g \in Y_1$. Therefore we get using Theorem 5.1.8, for $g \in Y_1$:

$$D^\epsilon(x_k(s), x_{k+1}(s))g = g + \int_0^\epsilon D_1^u(x_k(s), x_{k+1}(s))g\,du \tag{5.187}$$

and $\Gamma_{x_k(s)} \left(T_k^{\epsilon,s}(s), T_{k+1}^{\epsilon,s}(s) \right) g = g + \int_{T_k^{\epsilon,s}(s)}^{T_{k+1}^{\epsilon,s}(s)} \Gamma_{x_k(s)}(T_k^{\epsilon,s}(s), u) A_{x_k(s)}(u)g\,du.$

$$\tag{5.188}$$

Because $f \in \widehat{D}$, we get that $\widehat{A}(t)f \in Y_1$, $a(x,t)f \in Y_1 \; \forall t, x$. Since $R_0 \in \mathcal{B}(B_{Y_1}^{\mathrm{II}}(\mathbb{J}))$, we get that $f_{1,k+1} \in Y_1$ and therefore:

$$\Gamma_{x_k(s)} \left(T_k^{\epsilon,s}(s), T_{k+1}^{\epsilon,s}(s) \right) D^\epsilon(x_k(s), x_{k+1}(s))f_{1,k+1}$$

$$= \Gamma_{x_k(s)} \left(T_k^{\epsilon,s}(s), T_{k+1}^{\epsilon,s}(s) \right) f_{1,k+1} + O(\epsilon) \tag{5.189}$$

$$= f_{1,k+1} + \int_{T_k^{\epsilon,s}(s)}^{T_{k+1}^{\epsilon,s}(s)} \Gamma_{x_k(s)}(T_k^{\epsilon,s}(s), u) A_{x_k(s)}(u)f_{1,k+1}\,du + O(\epsilon). \tag{5.190}$$

Therefore taking the conditional expectation we get:

$$\mathbb{E}[V_{k+1}^\epsilon(s)f_{1,k+1} - V_k^\epsilon(s)f_{1,k+1}|\mathcal{F}_k(s)] \tag{5.191}$$

$$= V_k^\epsilon(s) \sum_{y \in \mathbb{J}} \int_0^\infty \left[\int_0^{\epsilon u} \Gamma_{x_k(s)}(T_k^{\epsilon,s}(s), T_k^{\epsilon,s}(s) + v) A_{x_k(s)}(T_k^{\epsilon,s}(s) + v)f_{1,k+1}dv \right]$$

$$\times Q_{T_k(s)}(x_k(s), y, du) + O(\epsilon) \tag{5.192}$$

$$= O(\epsilon) \quad (\leq \epsilon C \bar{\tau} \text{ for some constant } C \text{ by assumption 5.3.1)}, \tag{5.193}$$

and so:

$$\mathbb{E}[V_{k+1}^\epsilon(s)f_{k+1}^\epsilon - V_k^\epsilon(s)f_{k+1}^\epsilon|\mathcal{F}_k(s)] = \mathbb{E}[V_{k+1}^\epsilon(s)f - V_k^\epsilon(s)f|\mathcal{F}_k(s)] + O(\epsilon^2). \tag{5.194}$$

Now, because $f \in \widehat{D}$ and by assumption 5.3.1 (which ensures that the integral below exists):

$$D^\epsilon(x_k(s), x_{k+1}(s))f = f + \epsilon D_1^0(x_k(s), x_{k+1}(s))f + \int_0^\epsilon (\epsilon - u) D_2^u(x_k(s), x_{k+1}(s))f\,du. \tag{5.195}$$

And so, using boundedness of D_2^ϵ (again assumption 5.3.1):

$$\Gamma_{x_k(s)}\left(T_k^{\epsilon,s}(s), T_{k+1}^{\epsilon,s}(s)\right) D^\epsilon(x_k(s), x_{k+1}(s))f$$
$$= \Gamma_{x_k(s)}\left(T_k^{\epsilon,s}(s), T_{k+1}^{\epsilon,s}(s)\right) f$$
$$+ \epsilon\Gamma_{x_k(s)}\left(T_k^{\epsilon,s}(s), T_{k+1}^{\epsilon,s}(s)\right) D_1^0(x_k(s), x_{k+1}(s))f + O(\epsilon^2).$$

The first term above has the representation (by Theorem 5.1.12):

$$\Gamma_{x_k(s)}\left(T_k^{\epsilon,s}(s), T_{k+1}^{\epsilon,s}(s)\right) f = f + \int_{T_k^{\epsilon,s}(s)}^{T_{k+1}^{\epsilon,s}(s)} A_{x_k(s)}(u)f\,du$$

$$+ \int_{T_k^{\epsilon,s}(s)}^{T_{k+1}^{\epsilon,s}(s)} \int_{T_k^{\epsilon,s}(s)}^{u} \Gamma_{x_k(s)}(T_k^{\epsilon,s}(s), r)A_{x_k(s)}(r)A_{x_k(s)}(u)f\,dr\,du. \tag{5.196}$$

Taking the conditional expectation and using the fact that $\sup_{\substack{u\in[0,T+\bar\tau] \\ x\in\mathbb{J}}} \|A_x(u)f\|_{Y_1} < \infty$, *we can show as we did before that:*

$$\mathbb{E}\left[\int_{T_k^{\epsilon,s}(s)}^{T_{k+1}^{\epsilon,s}(s)} \int_{T_k^{\epsilon,s}(s)}^{u} \Gamma_{x_k(s)}(T_k^{\epsilon,s}(s), r)A_{x_k(s)}(r)A_{x_k(s)}(u)f\,dr\,du \,\middle|\, \mathcal{F}_k(s)\right] = O(\epsilon^2).$$
$$\tag{5.197}$$

The second term has the following representation, because $f \in \widehat{\mathcal{D}}$ (which ensures that $D_1^0(x,y)f \in Y_1$) and using Theorem 5.1.8:

$$\epsilon\Gamma_{x_k(s)}\left(T_k^{\epsilon,s}(s), T_{k+1}^{\epsilon,s}(s)\right) D_1^0(x_k(s), x_{k+1}(s))f \tag{5.198}$$

$$= \epsilon D_1^0(x_k(s), x_{k+1}(s))f$$

$$+ \epsilon\int_{T_k^{\epsilon,s}(s)}^{T_{k+1}^{\epsilon,s}(s)} \Gamma_{x_k(s)}\left(T_k^{\epsilon,s}(s), u\right) A_{x_k(s)}(u)D_1^0(x_k(s), x_{k+1}(s))f\,du \tag{5.199}$$

$$= \epsilon D_1^0(x_k(s), x_{k+1}(s))f + O(\epsilon^2). \tag{5.200}$$

And so we have overall:

$$\mathbb{E}[V_{k+1}^\epsilon(s)f - V_k^\epsilon(s)f|\mathcal{F}_k(s)]$$
$$= V_k^\epsilon(s)\mathbb{E}\left[\int_{T_k^{\epsilon,s}(s)}^{T_{k+1}^{\epsilon,s}(s)} A_{x_k(s)}(u)f\,du + \epsilon D_1^0(x_k(s), x_{k+1}(s))f \,\middle|\, \mathcal{F}_k(s)\right] + O(\epsilon^2).$$
$$\tag{5.201}$$

We have by the strong Markov property (because $k \geq 1$):

$$\mathbb{E}\left[D_1^0(x_k(s), x_{k+1}(s))f \,\middle|\, \mathcal{F}_k(s)\right] = P_{T_k(s)}D_1^0(x_k(s), \bullet)(x_k(s))f, \tag{5.202}$$

and:

$$\mathbb{E}\left[\int_{T_k^{\epsilon,s}(s)}^{T_{k+1}^{\epsilon,s}(s)} A_{x_k(s)}(u)f\,du\,\bigg|\,x_k(s),T_k(s)\right] \tag{5.203}$$

$$= \sum_{y\in\mathbb{J}}\int_0^\infty\left[\int_{T_k^{\epsilon,s}(s)}^{T_k^{\epsilon,s}(s)+\epsilon u} A_{x_k(s)}(v)f\,dv\right]Q_{T_k(s)}(x_k(s),y,du) \tag{5.204}$$

$$= \sum_{y\in\mathbb{J}}\int_0^\infty\left[\epsilon u A_x(T_k^{\epsilon,s}(s))f + \int_0^{\epsilon u}(\epsilon u - v)A'_{x_k(s)}(T_k^{\epsilon,s}(s)+v)f\,dv\right]$$

$$Q_{T_k(s)}(x_k(s),y,du) \tag{5.205}$$

$$= \epsilon m_1(x_k(s),T_k(s))A_{x_k(s)}(T_k^{\epsilon,s}(s))f + O(\epsilon^2), \tag{5.206}$$

as $\sup_{u\in[0,T+\bar\tau]}||A'_x(u)f|| < \infty$. *So finally we get:*

$$\mathbb{E}[V_{k+1}^\epsilon(s)f_{k+1}^\epsilon - V_k^\epsilon(s)f_{k+1}^\epsilon|\mathcal{F}_k(s)] = \epsilon\Pi m V_k^\epsilon(s)a(x_k(s),T_k^{\epsilon,s}(s))f$$
$$+ \epsilon V_k^\epsilon(s)\left[(m_1(x_k(s),T_k(s)) - m_1(x_k(s)))A_{x_k(s)}(T_k^{\epsilon,s}(s))f\right] \tag{5.207}$$
$$+ \epsilon V_k^\epsilon(s)\left[(P_{T_k(s)} - P)D_1^0(x_k(s),\bullet)(x_k(s))f\right] + O(\epsilon^2).$$

In the expression above, assumption 5.3.1 ensures that the terms containing $(P - P_{T_k(s)})$ *and* $(m_1(x_k(s),T_k(s)) - m_1(x_k(s)))$ *will vanish as* $k\to\infty$, *and therefore we get overall:*

$$\widetilde{M_t^\epsilon}(s)f = V_{N_s\left(t^{\frac{1}{\epsilon},s}\right)+1}^\epsilon(s)f - f - \epsilon\Pi m\sum_{k=1}^{N_s\left(t^{\frac{1}{\epsilon},s}\right)}V_\epsilon(s,T_k^{\epsilon,s}(s))\widehat{A}(T_k^{\epsilon,s}(s))f + O(\epsilon). \tag{5.208}$$

Now, let $\theta_s^\epsilon(t) := N_s\left(t^{\frac{1}{\epsilon},s}\right) + 1$. *Using assumption 5.3.1 (in particular uniform boundedness of sojourn times):*

$$V_{\theta_s^\epsilon(t)}^\epsilon(s)f = V_\epsilon(s,t)\Gamma_{x_{\theta_s^\epsilon(t)-1}(s)}(t,T_{\theta_s^\epsilon(t)}^{\epsilon,s}(s))D^\epsilon(x_{\theta_s^\epsilon(t)-1}(s),x_{\theta_s^\epsilon(t)}(s))f \tag{5.209}$$

$$\Rightarrow ||V_{\theta_s^\epsilon(t)}^\epsilon(s)f - V_\epsilon(s,t)f|| \le e^{\gamma(t-s)}$$

$$||\Gamma_{x_{\theta_s^\epsilon(t)-1}(s)}(t,T_{\theta_s^\epsilon(t)}^{\epsilon,s}(s))D^\epsilon(x_{\theta_s^\epsilon(t)-1}(s),x_{\theta_s^\epsilon(t)}(s))f - f|| \tag{5.210}$$

$$\le e^{\gamma(t+\bar\tau-s)}||D^\epsilon(x_{\theta_s^\epsilon(t)-1}(s),x_{\theta_s^\epsilon(t)}(s))f - f|| + e^{\gamma(t-s)}$$

$$||\Gamma_{x_{\theta_s^\epsilon(t)-1}(s)}(t,T_{\theta_s^\epsilon(t)}^{\epsilon,s}(s))f - f|| \tag{5.211}$$

$$\le \epsilon e^{\gamma(t+\bar\tau-s)}\sup_{\substack{(x,y)\in\mathbb{J}^2\\\epsilon\in[0,1]}}||D_1^\epsilon(x,y)f|| + e^{\gamma(t-s)}(T_{\theta_s^\epsilon(t)}^{\epsilon,s}(s) - t)$$

$$\sup_{\substack{x\in\mathbb{J}\\t\in[0,T+\bar\tau]}}||A_x(t)||_{\mathcal{B}(Y_1,Y)}||f||_{Y_1} \tag{5.212}$$

$$\le \epsilon e^{\gamma(t+\bar\tau-s)}\sup_{\substack{(x,y)\in\mathbb{J}^2\\\epsilon\in[0,1]}}||D_1^\epsilon(x,y)f|| + \epsilon e^{\gamma(t-s)}\bar\tau\sup_{\substack{x\in\mathbb{J}\\t\in[0,T+\bar\tau]}}||A_x(t)||_{\mathcal{B}(Y_1,Y)}||f||_{Y_1}. \tag{5.213}$$

And therefore:

$$V^\epsilon_{N_s\left(t^{\frac{1}{\epsilon},s}\right)+1}(s)f = V_\epsilon(s,t)f + O(\epsilon). \tag{5.214}$$

We now show that the martingale $\widetilde{M}^\epsilon_\bullet(s)f$ converges to 0 as $\epsilon \to 0$.

Lemma 5.3.18. *Let assumption 5.3.1 hold true. Let $f \in Y_1$, $s \in \mathbf{J}$. For every $\ell \in Y^*$, $\ell(\widetilde{M}^\epsilon_\bullet(s)f) \Rightarrow 0$. Further, if $\{\widetilde{M}^\epsilon_\bullet(s)f\}$ is relatively compact (in $D(\mathbf{J}(s), Y)$), then $\widetilde{M}^\epsilon_\bullet(s)f \Rightarrow 0$.*

Proof.

Take any sequence $\epsilon_n \to 0$. Weak convergence in $D(\mathbf{J}(s), \mathbb{R})$ is equivalent to weak convergence in $D([s,T), \mathbb{R})$ for every $T \in \mathbf{J}(s)$. So let's fix $T \in \mathbf{J}(s)$. Let $\ell \in Y^*$. Then $\ell(\widetilde{M}^{\epsilon_n}_\bullet(s)f)$ is a real-valued martingale by Lemma 5.3.15. To show that $\ell(\widetilde{M}^{\epsilon_n}_\bullet(s)f) \Rightarrow 0$, we are going to apply Theorem 3.11 of [10], Chapter VIII. First, $\ell(\widetilde{M}^{\epsilon_n}_\bullet(s)f)$ is square-integrable by Lemma 5.3.15. From the proof of Lemma 5.3.17, we have that:

$$M^{\epsilon_n}_{k+1}(s)f - M^{\epsilon_n}_k(s)f = \Delta V^{\epsilon_n}_{k+1}(s)f^{\epsilon_n}_{k+1} - \mathbb{E}[\Delta V^{\epsilon_n}_{k+1}(s)f^{\epsilon_n}_{k+1}|\mathcal{F}_k(s)] \tag{5.215}$$

$$\text{where } \Delta V^{\epsilon_n}_{k+1}(s)f^{\epsilon_n}_{k+1} := V^{\epsilon_n}_{k+1}(s)f^{\epsilon_n}_{k+1} - V^{\epsilon_n}_k(s)f^{\epsilon_n}_k, \tag{5.216}$$

and that:

$$\sup_{k \le N_s\left(t^{\frac{1}{\epsilon},s}\right)} \|\Delta V^{\epsilon_n}_{k+1}(s)f^{\epsilon_n}_{k+1}\| \le C_T \epsilon_n \text{ a.e.} \tag{5.217}$$

for some uniform constant C_T dependent on T, so that:

$$\left|\Delta\ell(\widetilde{M}^{\epsilon_n}_t(s)f)\right| = \left|\ell\left(\Delta\widetilde{M}^{\epsilon_n}_t(s)f\right)\right| \le 2\|\ell\| \sup_{k \le N_s\left(t^{\frac{1}{\epsilon},s}\right)} \|\Delta V^{\epsilon_n}_{k+1}(s)f^{\epsilon_n}_{k+1}\| \text{ a.e.}$$
$$\tag{5.218}$$

$$\le 2\|\ell\|C_T\epsilon_n \text{ a.e.} \tag{5.219}$$

The latter shows that the jumps are uniformly bounded by $2\|\ell\|C_T$ and that

$$\sup_{t<T}\left|\Delta\ell(\widetilde{M}^{\epsilon_n}_t(s)f)\right| \xrightarrow{P} 0. \tag{5.220}$$

To show that $\ell(\widetilde{M}^\epsilon_\bullet(s)f) \Rightarrow 0$, it remains to show that the following predictable quadratic variation goes to 0 in probability:

$$\left\langle \ell(\widetilde{M}^{\epsilon_n}_\bullet(s)f) \right\rangle_t \xrightarrow{P} 0 \text{ for all } t \in [s,T). \tag{5.221}$$

Using the definition of $\widetilde{M}^{\epsilon_n}_\bullet(s)f$ in Lemma 5.3.15, we get that:

$$\left\langle \ell(\widetilde{M}^{\epsilon_n}_\bullet(s)f) \right\rangle_t = \sum_{k=0}^{N_s\left(t^{\frac{1}{\epsilon_n},s}\right)} \mathbb{E}\left[\ell^2(M^{\epsilon_n}_{k+1}(s)f - M^{\epsilon_n}_k(s)f)|\mathcal{F}_k(s)\right]. \tag{5.222}$$

Therefore:

$$\left| \left\langle \ell(\widetilde{M}^{\epsilon_n}_\bullet(s)f) \right\rangle_t \right| \leq 4C_T^2 \|\ell\|^2 \epsilon_n^2 N_s \left(t^{\frac{1}{\epsilon_n},s} \right) \overset{a.e.}{\to} 0, \tag{5.223}$$

because $\epsilon_n N_s \left(t^{\frac{1}{\epsilon_n},s} \right) \overset{a.e.}{\to} \frac{t-s}{\Pi m}$. *Now to show the second part of the lemma, we observe that if* $\widetilde{M}^{\epsilon_n}_\bullet(s)f \Rightarrow \widetilde{M}^0_\bullet(s,f)$ *for some converging sequence* $\epsilon_n \to 0$, *we get that* $\ell(\widetilde{M}^{\epsilon_n}_\bullet(s)f) \Rightarrow \ell(\widetilde{M}^0_\bullet(s,f))$ *by problem 13, Section 3.11 of [3] together with the continuous mapping theorem. Therefore* $\ell(\widetilde{M}^0_\bullet(s,f))$ *is equal to the zero process, up to indistinguishability. In particular, it yields that* $\forall t \in \mathbf{J}(s)$, $\forall \ell \in Y^*$: $\ell(\widetilde{M}^0_t(s,f)) = 0$ *a.e.. Now, by [13], we know that the dual space of every separable Banach space has a countable total subset, more precisely there exists a countable subset* $S \subseteq Y^*$ *such that* $\forall g \in Y$:

$$(\ell(g) = 0 \quad \forall \ell \in S) \Rightarrow g = 0. \tag{5.224}$$

Since $\mathbb{P}[\ell(\widetilde{M}^0_t(s,f)) = 0, \forall \ell \in S] = 1$, *we get* $M^0_t(s,f) = 0$ *a.e., i.e.* $\widetilde{M}^0_\bullet(s,f)$ *is a modification of the zero process. Since both processes have a.e. right-continuous paths, they are in fact indistinguishable (see e.g. [10]). And so* $\widetilde{M}^0_\bullet(s,f) = 0$ *a.e..*

5.3.5   Weak Law of Large Numbers (WLLN)

To prove our next WLLN theorem - the main result of this section - we need the following assumption:

$$\widehat{A} \text{ is the generator of a regular } Y-\text{propagator } \widehat{\Gamma}. \tag{5.225}$$

In the applications we will consider in Chapter 6, the above assumption will be very easily checked as we will be able to exhibit $\widehat{\Gamma}$ almost immediately after looking at $\widehat{A}$. This assumption will be used in order to be able to use Theorem 5.1.10 which gives uniqueness of solutions to the Cauchy problem:

$$\begin{cases} \frac{d}{dt}G(t)f = G(t)\widehat{A}(t)f & \forall t \in J(s), f \in Y_1. \\ G(s) = G_s \end{cases} \tag{5.226}$$

For the cases where $\widehat{\Gamma}$ cannot be exhibited, we refer to [16], Chapter 5, Theorems 3.1 and 4.3 for requirements on $\widehat{A}$ to obtain the well-posedness of the Cauchy problem.

Theorem 5.3.19. *Under assumptions 5.3.1, 5.3.3, 5.3.5, we get that* $\widehat{\Gamma}$ *is* $\mathcal{B}(Y)-$*exponentially bounded (with constant* γ). *Further, for every* $s \in \mathbf{J}$ *we have the weak convergence in the Skorokhod topology* $(\mathbb{D}_s, d)$:

$$V_\epsilon(s, \bullet) \Rightarrow \widehat{\Gamma}(s, \bullet), \tag{5.227}$$

where $(\mathbb{D}_s, d)$ *is defined similarly to* $(\mathbb{D}, d)$ *in [27], with* $\mathbb{R}_+$ *replaced by* $\mathbf{J}(s)$.

Proof. *By Lemma 5.3.12, we know that $\{V_\epsilon(s, \bullet)\}$ is $Y_1 - C$-relatively compact. Take $\{f_p\}$ a countable family of $\widehat{D}$ that is dense in both Y_1 and Y. Marginal relative compactness implies countable relative compactness and therefore, $\{V_\epsilon(s, \bullet)f_p\}_{p\in\mathbb{N}}$ is C-relatively compact in $D(\mathbf{J}(s), Y)^\infty$, and actually in $D(\mathbf{J}(s), Y^\infty)$ since the limit points are continuous. Take a weakly converging sequence ϵ_n. By the proof of Theorem 3.4.1, we can find a probability space $(\Omega', \mathcal{F}', \mathbb{P}')$ and $\mathbb{D}_s$-valued random variables $V'_{\epsilon_n}(s, \bullet)$, $V'_0(s, \bullet)$ on it, such that:*

$$V'_{\epsilon_n}(s, \bullet) \to V'_0(s, \bullet) \text{ in } (\mathbb{D}_s, d), \text{ on } \Omega', \tag{5.228}$$

$$\|V'_0(s, t)\|_{\mathcal{B}(Y)} \le e^{\gamma(t-s)} \quad \forall t \in \mathbf{J}(s), \text{ on } \Omega', \tag{5.229}$$

$$V'_{\epsilon_n}(s, \bullet) \stackrel{d}{=} V_{\epsilon_n}(s, \bullet) \text{ in } (\mathbb{D}_s, d), \forall n. \tag{5.230}$$

We can extend the probability space $(\Omega', \mathcal{F}', \mathbb{P}')$ so that it supports a Markov renewal process $(x'_k, T'_k)_{k\ge 0}$ such that:

$$(V'_{\epsilon_n}(s, \bullet), (x'_k, T'_k)_{k\ge 0}) \stackrel{d}{=} (V_{\epsilon_n}(s, \bullet), (x_k, T_k)_{k\ge 0}) \text{ in } (\mathbb{D}_s, d) \times (\mathbb{J} \times \mathbb{R}_+)^\infty, \forall n. \tag{5.231}$$

Now let's take some f_p and show that on Ω' we have the following convergence in the Skorokhod topology:

$$\epsilon_n \sum_{k=1}^{N'_s\left(t^{\frac{1}{\epsilon_n},s}\right)} V'_{\epsilon_n}(s, T_k^{\epsilon_n, s}(s)')\widehat{A}(T_k^{\epsilon_n, s}(s)')f_p \to \frac{1}{\Pi m} \int_s^\bullet V'_0(s, u)\widehat{A}(u)f_p du. \tag{5.232}$$

We have on Ω':

$$d\left(\epsilon_n \sum_{k=1}^{N'_s\left(t^{\frac{1}{\epsilon_n},s}\right)} V'_{\epsilon_n}(s, T_k^{\epsilon_n, s}(s)')\widehat{A}(T_k^{\epsilon_n, s}(s)')f_p, \frac{1}{\Pi m} \int_s^\bullet V'_0(s, u)\widehat{A}(u)f_p du \right) \tag{5.233}$$

$$\le d\underbrace{\left(\epsilon_n \sum_{k=1}^{N'_s\left(t^{\frac{1}{\epsilon_n},s}\right)} V'_{\epsilon_n}(s, T_k^{\epsilon_n, s}(s)')\widehat{A}(T_k^{\epsilon_n, s}(s)')f_p, \epsilon_n \sum_{k=1}^{N'_s\left(t^{\frac{1}{\epsilon_n},s}\right)} V'_0(s, T_k^{\epsilon_n, s}(s)')\widehat{A}(T_k^{\epsilon_n, s}(s)')f_p \right)}_{(i)} + \tag{5.234}$$

$$d\underbrace{\left(\epsilon_n \sum_{k=1}^{N'_s\left(t^{\frac{1}{\epsilon_n},s}\right)} V'_0(s, T_k^{\epsilon_n, s}(s)')\widehat{A}(T_k^{\epsilon_n, s}(s)')f_p, \frac{1}{\Pi m} \int_s^\bullet V'_0(s, u)\widehat{A}(u)f_p du \right)}_{(ii)}. \tag{5.235}$$

Let $T \in \mathbf{J}(s)$. By continuity of $\widehat{A}$, we get that $V'_{\epsilon_n}(s, \bullet)\widehat{A}(\bullet)f_p \to V'_0(s, \bullet)\widehat{A}(\bullet)f_p$. Since the latter limit is continuous, the convergence in the Skorokhod topology is equivalent to convergence in the local uniform topology, that is:

$$\sup_{t \in [s,T]} \|V'_{\epsilon_n}(s,t)\widehat{A}(t)f_p - V'_0(s,t)\widehat{A}(t)f_p\| \to 0. \tag{5.236}$$

Therefore we get on Ω':

$$\sup_{t \in [s,T]} \left\| \epsilon_n \sum_{k=1}^{N'_s\left(t^{\frac{1}{\epsilon_n},s}\right)} V'_{\epsilon_n}(s, T_k^{\epsilon_n,s}(s)')\widehat{A}(T_k^{\epsilon_n,s}(s)')f_p - \epsilon_n \sum_{k=1}^{N'_s\left(t^{\frac{1}{\epsilon_n},s}\right)} V'_0(s, T_k^{\epsilon_n,s}(s)')\widehat{A}(T_k^{\epsilon_n,s}(s)')f_p \right\| \tag{5.237}$$

$$\leq \epsilon_n N'_s\left(t^{\frac{1}{\epsilon_n},s}\right) \sup_{t \in [s,T]} \|V'_{\epsilon_n}(s,t)\widehat{A}(t)f_p - V'_0(s,t)\widehat{A}(t)f_p\| \to 0, \tag{5.238}$$

since $\epsilon_n N'_s\left(t^{\frac{1}{\epsilon_n},s}\right) \to \frac{t-s}{\Pi m} \leq \frac{T-s}{\Pi m}$. By Remark 5.3.16 we get $(i) \to 0$. For (ii), we observe that we can refine as we wish the partition $\{T_k^{\epsilon_n,s}(s)'\}$ uniformly in k, since $T_{k+1}^{\epsilon_n,s}(s)' - T_k^{\epsilon_n,s}(s)' \leq \epsilon_n \bar{\tau}$ almost surely. Therefore we get the convergence of the Riemann sum to the Riemann integral uniformly on $[s,T]$, i.e.,

$$\sup_{t \in [s,T]} \left\| \epsilon_n \sum_{k=1}^{N'_s\left(t^{\frac{1}{\epsilon_n},s}\right)} V'_0(s, T_k^{\epsilon_n,s}(s)')\widehat{A}(T_k^{\epsilon_n,s}(s)')f_p - \frac{1}{\Pi m}\int_s^t V'_0(s,u)\widehat{A}(u)f_p du \right\| \overset{n \to \infty}{\to} 0. \tag{5.239}$$

Finally we get on Ω', by continuity of the limit points:

$$V'_{\epsilon_n}(s, \bullet)f_p - f_p - \Pi m \epsilon_n \sum_{k=1}^{N'_s\left(t^{\frac{1}{\epsilon_n},s}\right)} V'_{\epsilon_n}(s, T_k^{\epsilon_n,s}(s)')\widehat{A}(T_k^{\epsilon_n,s}(s)')f_p \tag{5.240}$$

$$\to V'_0(s, \bullet)f_p - f_p - \int_s^{\bullet} V'_0(s,u)\widehat{A}(u)f_p du. \tag{5.241}$$

Since we have:

$$(V'_{\epsilon_n}(s, \bullet), (x'_k, T'_k)_{k \geq 0}) \overset{d}{=} (V_{\epsilon_n}(s, \bullet), (x_k, T_k)_{k \geq 0}) \text{ in } (\mathbb{D}_s, d) \times (\mathbf{J} \times \mathbb{R}_+)^{\infty}, \forall n, \tag{5.242}$$

by Lemma 5.3.17, 5.3.18 we get that:

$$V'_0(s, \bullet)f_p - f_p - \int_s^{\bullet} V'_0(s,u)\widehat{A}(u)f_p du = 0 \tag{5.243}$$

for all p on some set of probability one $\Omega'_0 \subseteq \Omega'$. Let $g \in Y_1$ and $t \in \mathbf{J}(s)$. Since $\{f_p\}$ is dense in Y_1, there exists a sequence $g_p \subseteq \{f_p\} \xrightarrow{Y_1} g$. We have on Ω'_0:

$$\left\| V'_0(s,t)g - g - \int_s^t V'_0(s,u)\widehat{A}(u)g \right\| \tag{5.244}$$

$$\leq \|V'_0(s,t)g - V'_0(s,t)g_p\| + \left\| V'_0(s,t)g_p - g - \int_s^t V'_0(s,u)\widehat{A}(u)g \right\| \tag{5.245}$$

$$\leq 2e^{\gamma(t-s)}\|g_p - g\| + \int_s^t \|V'_0(s,u)\widehat{A}(u)g_p - V'_0(s,u)\widehat{A}(u)g\|du \tag{5.246}$$

$$\leq 2e^{\gamma(t-s)}\|g_p - g\| + e^{\gamma(t-s)}\|g_p - g\|_{Y_1} \int_s^t \|\widehat{A}(u)\|_{\mathcal{B}(Y_1,Y)}du \to 0, \tag{5.247}$$

using assumption 5.3.1. Therefore on Ω'_0 we have $\forall g \in Y_1$:

$$V'_0(s,\bullet)g = g + \int_s^\bullet V'_0(s,u)\widehat{A}(u)gdu. \tag{5.248}$$

By continuity of $\widehat{A}$ on Y_1 (assumption 5.3.1), and exponential boundedness + continuity of V'_0: $t \to V'_0(s,t)\widehat{A}(t)g \in C(\mathbf{J}(s),Y)$ on Ω'_0 and therefore we have on Ω'_0:

$$\frac{\partial}{\partial t}V'_0(s,t)g = V'_0(s,t)\widehat{A}(t)g \quad \forall t \in \mathbf{J}(s), \forall g \in Y_1 \tag{5.249}$$

$$V'_0(s,s) = I. \tag{5.250}$$

By assumption 5.3.5 and Theorem 5.1.10, $V'_0(s,t)(\omega')g = \widehat{\Gamma}(s,t)g$, $\forall \omega' \in \Omega'_0$, $\forall g \in Y_1$, $\forall t \in \mathbf{J}(s)$. By density of Y_1 in Y, the previous equality is true in Y.

5.4 Central Limit Theorem (CLT)

In the previous section, we obtained in Theorem 5.3.19 a WLLN for the random evolution. The goal of this section is to obtain a CLT. The main result is Theorem 5.4.7 which states that - in a suitable operator Skorokhod topology - we have the weak convergence:

$$\epsilon^{-1/2}(V_\epsilon(s,\bullet) - \widehat{\Gamma}(s,\bullet)) \Rightarrow I_\sigma(s,\bullet), \tag{5.251}$$

where $I_\sigma(s,\bullet)$ is a "Gaussian operator" defined in Theorem 5.4.7. In the WLLN result Theorem 5.3.19, it was relatively obvious that the convergence would hold on the space $\mathbb{D}_s$. Here, because of the rescaling by $\epsilon^{-1/2}$, it is not *a priori* obvious. In fact, the above weak convergence will prove to hold in the space $\mathbb{D}_s^{Y_2 \to Y}$, which is defined very similarly to $\mathbb{D}_s$:

$$\mathbb{D}_s^{Y_2 \to Y} := \{S \in \mathcal{B}(Y_2, Y)^{\mathbf{J}(s)} : Sf \in D(\mathbf{J}(s), Y) \ \forall f \in Y_2\}, \qquad (5.252)$$

where $(Y_2, || \cdot ||_{Y_2})$ is a Banach space continuously embedded in $(Y_1, || \cdot ||_{Y_1})$. We need the Banach space $Y_2 \subseteq Y_1$ because the limiting Gaussian operator I_σ takes value in $\mathbb{C}_s^{Y_2 \to Y}$, but not in $\mathbb{C}_s^{Y_1 \to Y}$, as shown in Lemma 5.4.5. It will be proved that the volatility vector σ defined in Proposition 5.4.2 (which governs the convergence of V_ϵ to $\widehat{\Gamma}$ as it will be shown in Theorem 5.4.7) comes from three *independent sources*, which is what we expected intuitively:

- $\sigma^{(1)}$ which comes from the variance of the sojourn times $\{T_{n+1} - T_n\}_{n \geq 0}$.

- $\sigma^{(2)}$ which comes from the jump operators D.

- $\sigma^{(3)}$ which comes from the "state process" $\{x_n\}_{n \geq 0}$.

For example, if there is only 1 state in $\mathbf{J}$, then $\sigma^{(2)} = \sigma^{(3)} = 0$ and σ only comes from the randomness of the times $\{T_n\}$. If on the other hand the times T_n are assumed to be deterministic, then $\sigma^{(1)} = 0$. If the random evolution is continuous ($D = I$), then $\sigma^{(2)} = 0$. This decomposition of the volatility σ in three independent sources is obtained using the concept of so-called orthogonal martingale measures (see e.g. [2], Definition I-1), which is the heart of this section.

In order to get the main result of Theorem 5.4.7, we need a series of preliminary results: Theorem 5.4.1 establishes the weak convergence of the martingale $\epsilon^{-\frac{1}{2}} \ell(\widetilde{M}_\bullet^\epsilon(s)f)$ as $\epsilon \to 0$ for $\ell \in Y^*$, Proposition 5.4.2 and Theorem 5.4.4 are allowed to express the limit of the previous martingale in a neat way using so-called orthogonal martingale measures and weak Banach-valued stochastic integrals introduced in Definition 5.4.3 (which are an extension of the weak Banach-valued stochastic integrals of [18]). Lemmata 5.4.5, 5.4.6 give well-posedness of a Cauchy problem which will be used in the proof of Theorem 5.4.7, and show that the correct space to establish the convergence of $\epsilon^{-1/2}(V_\epsilon(s, \bullet) - \widehat{\Gamma}(s, \bullet))$ is indeed $\mathbb{D}_s^{Y_2 \to Y}$.

Theorem 5.4.1. *Let $\ell \in Y^*$, $s \in \mathbf{J}$ and $f \in \widehat{D}$. Under assumptions 5.3.1, 5.3.3, 5.3.5, we have the following weak convergence in the space $D(\mathbf{J}(s), \mathbb{R})$:*

$$\epsilon^{-\frac{1}{2}} \ell(\widetilde{M}_\bullet^\epsilon(s)f) \Rightarrow N_\bullet^\ell(s, f), \qquad (5.253)$$

where $N_\bullet^\ell(s, f)$ is a continuous Gaussian martingale with characteristics $(0, \widehat{\sigma}^\ell(s, \bullet, f)^2, 0)$ given by:

$$\widehat{\sigma}^\ell(s, t, f)^2 = \int_s^t \Pi[\widehat{\sigma}^\ell(\bullet, s, u, f)^2] du \qquad (5.254)$$

$$\widehat{\sigma}^\ell(x, s, t, f)^2 := \qquad (5.255)$$

$$\frac{1}{\Pi m}\left(\sigma^2(x)\ell^2(\widehat{\Gamma}(s,t)A_x(t)f) + P^{var}[\ell(\widehat{\Gamma}(s,t)D_1^0(x,\bullet)f)](x) + P^{var}[\ell(\widehat{\Gamma}(s,t)f_1(\bullet,t))](x)\right),$$

$$(5.256)$$

where $\sigma^2(x) := m_2(x) - m_1^2(x)$ and P^{var} is the following non-linear variance operator:

$$P^{var}g(x) := Pg^2(x) - (Pg(x))^2 \quad \text{for } x \in \mathbb{J}, g \in B_\mathbb{R}^b(\mathbb{J}). \qquad (5.257)$$

Proof. *Because the proof of this theorem is long, we explain the main parts of the proof before actually going into proving. The main idea is to prove that the quadratic variation of $\left\langle \epsilon^{-\frac{1}{2}}\ell(\widetilde{M}_\bullet^\epsilon(s)f)\right\rangle_t$ converges to $\widehat{\sigma}^\ell(s,t,f)^2$ in probability. We express the martingale-difference as the sum of two terms in (5.10) and (5.11). Then we calculate the expected values of the square of the functional of each term (see (5.12)). It was done in (5.26) and ((5.32)-(5.33)). Thus, our martingale-difference can be expressed in the form of (5.43)-(5.44), or (5.45)-(5.46) after $k \to +\infty$. Therefore, the quadratic variation $\left\langle \epsilon^{-\frac{1}{2}}\ell(\widetilde{M}_\bullet^\epsilon(s)f)\right\rangle_t$ is expressed as shown in (5.48)-(5.50). To show the last convergence in (5.53), we express the difference between the lefthand side and the righthand side in (5.5) by two terms, (i) and (ii), as in (5.55) and (5.56), respectively. Finally, we show that both terms (i) and (ii) converge to 0 in probability.*

Now, let us go into details of the proof. Weak convergence in $D(\mathbf{J}(s),\mathbb{R})$ is equivalent to weak convergence in $D([s,T],\mathbb{R})$ for every $T \in \mathbf{J}(s)$. So let's fix $T \in \mathbf{J}(s)$. In order to apply Theorem 3.11 of [10], Chapter VIII, we first need to show that the jumps of the martingale $\epsilon^{-\frac{1}{2}}\ell(\widetilde{M}_\bullet^\epsilon(s)f)$ are uniformly bounded. By the proof of Lemma 5.3.18, we know that these jumps are uniformly bounded by $C_T\epsilon^{\frac{1}{2}}$, where C_T is some constant dependent on T. The theorem we want to apply also requires the two following conditions, for all t in some dense subset of $[s,T]$:

1. $\left\langle \epsilon^{-\frac{1}{2}}\ell(\widetilde{M}_\bullet^\epsilon(s)f)\right\rangle_t \xrightarrow{P} \widehat{\sigma}^\ell(s,t,f)^2$

2. *condition "$[\widehat{\delta}_5 - D]$", which is equivalent by equation (3.5) of [10], chapter VIII to:*

$$\epsilon^{-\frac{1}{2}}\sup_{u \leq t}\left|\Delta\ell(\widetilde{M}_u^\epsilon(s)f)\right| \xrightarrow{P} 0. \qquad (5.258)$$

Condition 2) is true since as we mentioned earlier, the jumps of the martingale $\epsilon^{-\frac{1}{2}}\ell(\widetilde{M}_\bullet^\epsilon(s)f)$ are uniformly bounded by $C_T\epsilon^{\frac{1}{2}}$. Now we need to show 1). The same way as in the proof of Lemma 5.3.18, we use the definition of $\widetilde{M}_\bullet^\epsilon(s)f$ in Lemma 5.3.15 to get:

$$\left\langle \epsilon^{-\frac{1}{2}}\ell(\widetilde{M}_\bullet^\epsilon(s)f)\right\rangle_t = \epsilon^{-1}\sum_{k=0}^{N_s\left(t^{\frac{1}{\epsilon},s}\right)}\mathbb{E}\left[\ell^2(M_{k+1}^\epsilon(s)f - M_k^\epsilon(s)f)\big|\mathcal{F}_k(s)\right].$$

$$(5.259)$$

We have by definition, using the same notations as in the proof of Lemma 5.3.18:

$$M_{k+1}^\epsilon(s)f - M_k^\epsilon(s)f = \Delta V_{k+1}^\epsilon(s)f_{k+1}^\epsilon - \mathbb{E}[\Delta V_{k+1}^\epsilon(s)f_{k+1}^\epsilon|\mathcal{F}_k(s)] \quad (5.260)$$

$$\text{where } \Delta V_{k+1}^\epsilon(s)f_{k+1}^\epsilon := V_{k+1}^\epsilon(s)f_{k+1}^\epsilon - V_k^\epsilon(s)f_k^\epsilon. \quad (5.261)$$

It follows that, denoting $\mathbb{E}_k$ the conditional expectation w.r.t. $\mathcal{F}_k(s)$:

$$\mathbb{E}_k\left[\ell^2(M_{k+1}^\epsilon(s)f - M_k^\epsilon(s)f)\right] = \mathbb{E}_k[\ell^2(\Delta V_{k+1}^\epsilon(s)f_{k+1}^\epsilon)] - \ell^2\left(\mathbb{E}_k[\Delta V_{k+1}^\epsilon(s)f_{k+1}^\epsilon]\right). \quad (5.262)$$

In order to make the following lines of computation more readable, we let:

$$P_k f_{1,k} := P_{T_k(s)}f_1\left(\bullet, T_k^{\epsilon,s}\right)(x_k(s)) \quad P_k D_{1,k}^0 := P_{T_k(s)}D_1^0(x_k(s),\bullet)(x_k(s)) \quad (5.263)$$

$$m_{n,k}' := m_n(x_k(s), T_k(s)) \qquad\qquad m_{n,k} := m_n(x_k(s)) \quad (5.264)$$

$$A_k := A_{x_k(s)}(T_k^{\epsilon,s}(s)) \qquad\qquad a_k' := \frac{1}{\Pi m}\left(m_{1,k}'A_k + P_k D_{1,k}^0\right). \quad (5.265)$$

Using the proof of Lemma 5.3.17, we get:

$$\ell^2\left(\mathbb{E}_k[\Delta V_{k+1}^\epsilon(s)f_{k+1}^\epsilon]\right) = \epsilon^2\ell^2\left(V_k^\epsilon(s)P_k f_{1,k} - V_k^\epsilon(s)f_{1,k} + \Pi m V_k^\epsilon(s)a_k'f\right) + O(\epsilon^4) \quad (5.266)$$

$$= \epsilon^2\ell^2(V_k^\epsilon(s)P_k f_{1,k}) + \underbrace{\epsilon^2\ell^2(V_k^\epsilon(s)f_{1,k})}_{(A)} - \underbrace{2\epsilon^2\ell(V_k^\epsilon(s)P_k f_{1,k})\ell(V_k^\epsilon(s)f_{1,k})}_{(B)} \quad (5.267)$$

$$+ \epsilon^2\ell^2(V_k^\epsilon(s)m_{1,k}'A_kf) + \epsilon^2\ell^2(V_k^\epsilon(s)P_k D_{1,k}^0f) + \underbrace{2\epsilon^2\ell(V_k^\epsilon(s)m_{1,k}'A_kf)\ell(V_k^\epsilon(s)P_k D_{1,k}^0f)}_{(C)} \quad (5.268)$$

$$+ \underbrace{2\epsilon^2\ell(\Pi m V_k^\epsilon(s)a_k'f)\ell(V_k^\epsilon(s)P_k f_{1,k} - V_k^\epsilon(s)f_{1,k})}_{(D)} + O(\epsilon^4). \quad (5.269)$$

On the other hand, we have:

$$\mathbb{E}_k[\ell^2(\Delta V_{k+1}^\epsilon(s)f_{k+1}^\epsilon)] = \mathbb{E}_k[\ell^2((V_{k+1}^\epsilon(s) - V_k^\epsilon(s))f_{k+1}^\epsilon + V_k^\epsilon(s)(f_{k+1}^\epsilon - f_k^\epsilon))] \quad (5.270)$$

$$= \mathbb{E}_k[\ell^2((V_{k+1}^\epsilon(s) - V_k^\epsilon(s))f_{k+1}^\epsilon)] + \mathbb{E}_k[\ell^2(V_k^\epsilon(s)(f_{k+1}^\epsilon - f_k^\epsilon))] \quad (5.271)$$

$$+ \underbrace{2\mathbb{E}_k[\ell((V_{k+1}^\epsilon(s) - V_k^\epsilon(s))f_{k+1}^\epsilon)]\mathbb{E}_k[\ell(V_k^\epsilon(s)(f_{k+1}^\epsilon - f_k^\epsilon))]}_{=(D)+O(\epsilon^4)}. \quad (5.272)$$

The fact that the last term is equal to $(D) + O(\epsilon^4)$ is what we have shown in the proof of Lemma 5.3.17. Now:

$$\mathbb{E}_k[\ell^2(V_k^\epsilon(s)(f_{k+1}^\epsilon - f_k^\epsilon))] = \epsilon^2 \mathbb{E}_k[\ell^2(V_k^\epsilon(s)f_{1,k+1})] + \underbrace{\epsilon^2 \ell^2(V_k^\epsilon(s)f_{1,k})}_{(A)} \quad (5.273)$$

$$- \underbrace{2\epsilon^2 \mathbb{E}_k[\ell(V_k^\epsilon(s)f_{1,k+1})]\ell(V_k^\epsilon(s)f_{1,k})}_{=(B)+O(\epsilon^4)}. \quad (5.274)$$

The fact that the last term is equal to $(B) + O(\epsilon^4)$ comes from the fact that we proved in the proof of Lemma 5.3.17 that $\mathbb{E}_k[\ell(V_k^\epsilon(s)f_{1,k+1})] = \ell(V_k^\epsilon(s)P_k f_{1,k}) + O(\epsilon)$. The same way, we can prove that:

$$\epsilon^2 \mathbb{E}_k[\ell^2(V_k^\epsilon(s)f_{1,k+1})] = \epsilon^2 P_k \ell^2(V_k^\epsilon(s)f_1(\bullet, T_k^{\epsilon,s}(s)))(x_k(s)) + O(\epsilon^4). \quad (5.275)$$

So overall:

$$\mathbb{E}_k[\ell^2(V_k^\epsilon(s)(f_{k+1}^\epsilon - f_k^\epsilon))] = \epsilon^2 P_k \ell^2(V_k^\epsilon(s)f_1(\bullet, T_k^{\epsilon,s}(s)))(x_k(s)) + (A) + (B) + O(\epsilon^4). \quad (5.276)$$

It remains to compute the term $\mathbb{E}_k[\ell^2((V_{k+1}^\epsilon(s) - V_k^\epsilon(s))f_{k+1}^\epsilon)]$. We can show the same way as in the proof of Lemma 5.3.17 that:

$$\mathbb{E}_k[\ell^2((V_{k+1}^\epsilon(s) - V_k^\epsilon(s))f_{k+1}^\epsilon)] = \mathbb{E}_k[\ell^2((V_{k+1}^\epsilon(s) - V_k^\epsilon(s))f)] + O(\epsilon^4). \quad (5.277)$$

We get, using what we have computed in the proof of Lemma 5.3.17, that:

$$\mathbb{E}_k[\ell^2((V_{k+1}^\epsilon(s) - V_k^\epsilon(s))f)] \quad (5.278)$$

$$= \mathbb{E}_k\left[\ell^2\left(V_k^\epsilon(s)\int_{T_k^{\epsilon,s}(s)}^{T_{k+1}^{\epsilon,s}(s)} A_{x_k(s)}(u)f du + \epsilon V_k^\epsilon(s)D_1^0(x_k(s), x_{k+1}(s))f\right)\right] + O(\epsilon^4) \quad (5.279)$$

$$= \mathbb{E}_k\left[\ell^2\left(V_k^\epsilon(s)\int_{T_k^{\epsilon,s}(s)}^{T_{k+1}^{\epsilon,s}(s)} A_{x_k(s)}(u)f du\right)\right] + \epsilon^2 \mathbb{E}_k\left[\ell^2\left(V_k^\epsilon(s)D_1^0(x_k(s), x_{k+1}(s))f\right)\right] \quad (5.280)$$

$$+ 2\epsilon \mathbb{E}_k\left[\ell\left(V_k^\epsilon(s)\int_{T_k^{\epsilon,s}(s)}^{T_{k+1}^{\epsilon,s}(s)} A_{x_k(s)}(u)f du\right)\right]\mathbb{E}_k\left[\ell\left(V_k^\epsilon(s)D_1^0(x_k(s), x_{k+1}(s))f\right)\right]$$

$$+ O(\epsilon^4) \quad (5.281)$$

$$= \mathbb{E}_k\left[\ell^2\left(V_k^\epsilon(s)\int_{T_k^{\epsilon,s}(s)}^{T_{k+1}^{\epsilon,s}(s)} A_{x_k(s)}(u)f du\right)\right] + \epsilon^2 \mathbb{E}_k\left[\ell^2\left(V_k^\epsilon(s)D_1^0(x_k(s), x_{k+1}(s))f\right)\right]$$

$$+ (C) + O(\epsilon^4). \quad (5.282)$$

Again, using the same method as in the proof of Lemma 5.3.17, we have:

$$\mathbb{E}_k \left[\ell^2 \left(V_k^\epsilon(s) \int_{T_k^{\epsilon,s}(s)}^{T_{k+1}^{\epsilon,s}(s)} A_{x_k(s)}(u) f\, du \right) \right]$$

$$= \mathbb{E}_k \left[\ell^2 \left(V_k^\epsilon(s)(T_{k+1}^{\epsilon,s}(s) - T_k^{\epsilon,s}(s)) A_k f \right) \right] + o_k(\epsilon^2) \tag{5.283}$$

$$= \mathbb{E}_k \left[(T_{k+1}^{\epsilon,s}(s) - T_k^{\epsilon,s}(s))^2 \ell^2 \left(V_k^\epsilon(s) A_k f \right) \right] + O(\epsilon^4) \tag{5.284}$$

$$= \epsilon^2 m'_{2,k} \ell^2 (V_k^\epsilon(s) A_k f) + O(\epsilon^4). \tag{5.285}$$

And:

$$\mathbb{E}_k[\ell^2(V_k^\epsilon(s) D_1^0(x_k(s), x_{k+1}(s)) f)] = P_k \ell^2 (V_k^\epsilon(s) D_1^0(x_k(s), \bullet) f)(x_k(s)), \tag{5.286}$$

So that:

$$\mathbb{E}_k[\ell^2((V_{k+1}^\epsilon(s) - V_k^\epsilon(s)) f_{k+1}^\epsilon)] = \epsilon^2 m'_{2,k} \ell^2 (V_k^\epsilon(s) A_k f) \tag{5.287}$$

$$+ P_k \ell^2 (V_k^\epsilon(s) D_1^0(x_k(s), \bullet) f)(x_k(s)) + (C) + O(\epsilon^4). \tag{5.288}$$

Now we gather all the terms (so that (A), (B), (C), (D) cancel each other):

$$\mathbb{E}_k \left[\ell^2 (M_{k+1}^\epsilon(s) f - M_k^\epsilon(s) f) \right] = \epsilon^2 \sigma_k'^2 \ell^2 (V_k^\epsilon(s) A_k f) \tag{5.289}$$

$$+ \epsilon^2 P_k \ell^2 (V_k^\epsilon(s) D_1^0(x_k(s), \bullet) f)(x_k(s)) - \epsilon^2 \ell^2 (V_k^\epsilon(s) P_k D_{1,k}^0 f) \tag{5.290}$$

$$+ \epsilon^2 P_k \ell^2 (V_k^\epsilon(s) f_1(\bullet, T_k^{\epsilon,s}(s)))(x_k(s)) - \epsilon^2 \ell^2 (V_k^\epsilon(s) P_k f_{1,k}) + O(\epsilon^4), \tag{5.291}$$

where $\sigma_k'^2 := m'_{2,k} - m'^2_{1,k}$. We can also write the last result in terms of the nonlinear variance operator P_k^{var} associated to the kernel $P_{T_k(s)}$:

$$\mathbb{E}_k \left[\ell^2 (M_{k+1}^\epsilon(s) f - M_k^\epsilon(s) f) \right] = \epsilon^2 \sigma_k^2 \ell^2 (V_k^\epsilon(s) A_k f)$$

$$+ \epsilon^2 P_k^{var}[\ell(V_k^\epsilon(s) D_1^0(x_k(s), \bullet) f)](x_k(s)) \tag{5.292}$$

$$+ \epsilon^2 P_k^{var}[\ell(V_k^\epsilon(s) f_1(\bullet, T_k^{\epsilon,s}(s)))](x_k(s)) + O(\epsilon^4). \tag{5.293}$$

As $k \to \infty$ and by assumption, $P_{T_k(s)}$ will tend to P, $\sigma_k'^2$ to $\sigma_k^2 := m_{2,k} - m_{1,k}^2$ and therefore we get:

$$\left\langle \epsilon^{-\frac{1}{2}} \ell(\widetilde{M}_\bullet^\epsilon(s) f) \right\rangle_t = \epsilon \sum_{k=0}^{N_s\left(\frac{1}{\epsilon} \cdot s\right)} (\sigma_k^2 \ell^2 (V_k^\epsilon(s) A_k f) + P^{var}[\ell(V_k^\epsilon(s) D_1^0(x_k(s), \bullet) f)](x_k(s))$$

$$\tag{5.294}$$

$$+ P^{var}[\ell(V_k^\epsilon(s) f_1(\bullet, T_k^{\epsilon,s}(s)))](x_k(s))) + O(\epsilon). \tag{5.295}$$

To obtain the convergence of the latter predictable quadratic variation process, we first "replace" - in the previous Riemann sum - the random *times $T_k^{\epsilon,s}(s)$ and counting process $N_s\left(t^{\frac{1}{\epsilon},s}\right)$ by their deterministic equivalents $t_k^\epsilon(s,t)$ and $n_\epsilon(s,t)$ defined by:*

$$n_\epsilon(s,t) := \left\lfloor \frac{t-s}{\epsilon \Pi m} \right\rfloor \qquad t_k^\epsilon(s,t) := \left(s + k\frac{t-s}{n_\epsilon(s,t)} \right) \wedge t. \qquad (5.296)$$

It is possible to do so using assumption 5.3.1 and up to some technicalities. We do so for technical reasons in order to be able to use a SLLN for weakly correlated random variables [14] as specified later on. We get:

$$\left\langle \epsilon^{-\frac{1}{2}}\ell(\widetilde{M}_\bullet^\epsilon(s)f) \right\rangle_t = \epsilon \sum_{k=0}^{n_\epsilon(s,t)} \left(\sigma_k^2 \ell^2(V_\epsilon(s,t_k^\epsilon(s,t)) A_{x_k(s)}(t_k^\epsilon(s,t))f) \right. \qquad (5.297)$$

$$+ P^{var}[\ell(V_\epsilon(s,t_k^\epsilon(s,t)) D_1^0(x_k(s),\bullet)f)](x_k(s)) \qquad (5.298)$$

$$+ P^{var}[\ell(V_\epsilon(s,t_k^\epsilon(s,t)) f_1(\bullet,t_k^\epsilon(s,t)))](x_k(s))) + o(1). \qquad (5.299)$$

The second step is the following: denote for sake of clarity the inside of the Riemann sum by ϕ:

$$\phi(V_\epsilon,t_k^\epsilon(s,t),x_k(s)) := \sigma_k^2 \ell^2(V_\epsilon(s,t_k^\epsilon(s,t)) A_{x_k(s)}(t_k^\epsilon(s,t))f) \qquad (5.300)$$

$$+ P^{var}[\ell(V_\epsilon(s,t_k^\epsilon(s,t)) D_1^0(x_k(s),\bullet)f)](x_k(s))$$

$$+ P^{var}[\ell(V_\epsilon(s,t_k^\epsilon(s,t)) f_1(\bullet,t_k^\epsilon(s,t)))](x_k(s)). \qquad (5.301)$$

We want to show that:

$$\epsilon \sum_{k=0}^{n_\epsilon(s,t)} \phi(V_\epsilon,t_k^\epsilon(s,t),x_k(s)) \xrightarrow{P} \frac{1}{\Pi m} \int_s^t \Pi\phi(\widehat{\Gamma},u,\bullet)du. \qquad (5.302)$$

To do so, we write the decomposition:

$$\epsilon \sum_{k=0}^{n_\epsilon(s,t)} \phi(V_\epsilon,t_k^\epsilon(s,t),x_k(s)) - \frac{1}{\Pi m} \int_s^t \Pi\phi(\widehat{\Gamma},u,\bullet)du = \qquad (5.303)$$

$$\underbrace{\epsilon \sum_{k=0}^{n_\epsilon(s,t)} \phi(V_\epsilon,t_k^\epsilon(s,t),x_k(s)) - \epsilon \sum_{k=0}^{n_\epsilon(s,t)} \phi(\widehat{\Gamma},t_k^\epsilon(s,t),x_k(s))}_{(i)} \qquad (5.304)$$

$$\underbrace{+\epsilon \sum_{k=0}^{n_\epsilon(s,t)} \phi(\widehat{\Gamma},t_k^\epsilon(s,t),x_k(s)) - \frac{1}{\Pi m} \int_s^t \Pi\phi(\widehat{\Gamma},u,\bullet)du}_{(ii)}. \qquad (5.305)$$

First let's show that the term (i) *converges in probability to 0. We have:*

$$\left| \epsilon \sum_{k=0}^{n_\epsilon(s,t)} \phi(V_\epsilon, t_k^\epsilon(s,t), x_k(s)) - \epsilon \sum_{k=0}^{n_\epsilon(s,t)} \phi(\widehat{\Gamma}, t_k^\epsilon(s,t), x_k(s)) \right| \tag{5.306}$$

$$\leq \epsilon n_\epsilon(s,t) \sup_{\substack{u \in [s,t] \\ x \in \mathbb{J}}} \left| \phi(V_\epsilon, u, x) - \phi(\widehat{\Gamma}, u, x) \right|. \tag{5.307}$$

In the proof of Theorem 5.3.19, we got that $\{V_\epsilon(s, \bullet) \widehat{A}(\bullet) f\}$ *is C-relatively compact. Similarly, we also have that* $\forall (x, y) \in \mathbb{J} \times \mathbb{J}$, $\{V_\epsilon(s, \bullet) A_x(\bullet) f\}$, $\{V_\epsilon(s, \bullet) D_1^0(x, y) f\}$ *and* $\{V_\epsilon(s, \bullet) f_1(x, \bullet)\}$ *are C-relatively compact, and therefore they converge jointly, weakly in* $D(\mathbf{J}(s), Y)$ *to:*

$$(\widehat{\Gamma}(s, \bullet) A_x(\bullet) f, \widehat{\Gamma}(s, \bullet) D_1^0(x, y) f, \widehat{\Gamma}(s, \bullet) f_1(x, \bullet)). \tag{5.308}$$

Using the continuous mapping theorem and the fact that the limit points belong to $C(\mathbf{J}(s), Y)$ *(and so that the limit of the sum is equal to the sum of the limits), we get that:*

$$\phi(V_\epsilon, \bullet, x) \Rightarrow \phi(\widehat{\Gamma}, \bullet, x). \tag{5.309}$$

Since $\phi(\widehat{\Gamma}, \bullet, x)$ *is continuous, convergence in the Skorokhod topology is equivalent to convergence in the local uniform topology, and therefore we get that* $\sup_{u \in [s,t]} |\phi(V_\epsilon, u, x) - \phi(\widehat{\Gamma}, u, x)| \Rightarrow 0$. *The previous weak convergence to 0 also holds in probability and that completes the proof that* (i) *converges in probability to 0, since* $\epsilon n_\epsilon(s,t)$ *is bounded uniformly in* ϵ. *To show that the term* (ii) *converges a.e. to 0 we write:*

$$\epsilon \sum_{k=0}^{n_\epsilon(s,t)} \phi(\widehat{\Gamma}, t_k^\epsilon(s,t), x_k(s)) = \sum_{x \in \mathbb{J}} \epsilon \sum_{k=0}^{n_\epsilon(s,t)} \phi(\widehat{\Gamma}, t_k^\epsilon(s,t), x) 1_{\{x_k(s)=x\}}. \tag{5.310}$$

We have to prove that :

$$\epsilon \sum_{k=0}^{n_\epsilon(s,t)} \phi(\widehat{\Gamma}, t_k^\epsilon(s,t), x) 1_{\{x_k(s)=x\}} \overset{a.e.}{\to} \frac{\pi(x)}{\Pi m} \int_s^t \phi(\widehat{\Gamma}, u, x) du. \tag{5.311}$$

To do so, we shall use corollary 11 of [14]. *It is a SLLN for weakly correlated random variables* Y_k. *It tells us that if* $\sup_k var(Y_k) < \infty$ *and if* $cov(Y_m, Y_n) \leq \alpha(|m - n|)$ *with* $\sum_{n \geq 1} \frac{\alpha(n)}{n} < \infty$, *then we have:*

$$\lim_{n \to \infty} \frac{1}{n} \sum_{k=1}^{n} (Y_k - \mathbb{E}(Y_k)) = 0 \text{ a.e.} \tag{5.312}$$

In fact, it turns out that we need a generalized version of this corollary, that can be obtained exactly the same way as in the original proof. We need:

$$\lim_{n \to \infty} \frac{1}{n} \sum_{k=1}^{n} a_{nk}(Y_k - \mathbb{E}(Y_k)) = 0 \quad a.e. \tag{5.313}$$

for deterministic, nonnegative and uniformly bounded coefficients a_{nk}. Note that we can drop the nonnegativity assumption on the a_{nk}'s by requiring that $|cov(Y_m, Y_n)| \leq \alpha(|m-n|)$, and not only $cov(Y_m, Y_n) \leq \alpha(|m-n|)$. Here we let $Y_k := 1_{\{x_k(s)=x\}}$ so that $\sup_k var(Y_k) < \infty$ and $a_{\epsilon k} := \phi(\widehat{\Gamma}, t_k^\epsilon(s,t), x)$, which are indeed deterministic, nonnegative and uniformly bounded. Before showing $cov(Y_m, Y_n) \leq \alpha(|m-n|)$ for some suitable α, let's show that $\lim_{n \to \infty} \mathbb{E}(Y_n) = \pi(x)$ and so, by convergence of the Riemann sum:

$$\lim_{\epsilon \to 0} \epsilon \sum_{k=1}^{n_\epsilon(s,t)} a_{\epsilon k} \mathbb{E}(Y_k) = \frac{\pi(x)}{\Pi m} \int_s^t \phi(\widehat{\Gamma}, u, x) du. \tag{5.314}$$

As mentioned in the proof of Lemma 5.3.17, the process $(x_n, T_n)_{n \in \mathbb{N}}$ has the strong Markov property and therefore, since $x_k(s) = x_{N(s)+k}$ and that for $k \geq 1$, $N(s) + k$ is a stopping time of the augmented filtration $\mathcal{F}_n(0)$, where:

$$\mathcal{F}_n(s) := \sigma\left[x_k(s), T_k(s) : k \leq n\right] \vee \sigma(\mathbb{P} - null \; sets), \tag{5.315}$$

then we have for $k, m \geq 1$, noting $h_x(y) := 1_{\{x=y\}}$:

$$\mathbb{E}(Y_{k+m}) = \mathbb{E}(\mathbb{E}(h_x(x_{k+m}(s))|\mathcal{F}_k(s))) = \mathbb{E}(P_{T_k(s)}^{k,k+m} h_x(x_k(s))), \tag{5.316}$$

which by assumption 5.3.1, converges as $m \to \infty$ to $\Pi h_x = \pi(x)$. Now, it remains to study $cov(Y_m, Y_n)$. We have for $m > n$, again using the strong Markov property of the process (x_n, T_n):

$$\mathbb{E}(Y_m Y_n) = \mathbb{E}(1_{\{x_m(s)=x\}} 1_{\{x_n(s)=x\}}) = \mathbb{E}(1_{\{x_n(s)=x\}} \mathbb{E}(1_{\{x_m(s)=x\}}|x_n(s), T_n(s))) \tag{5.317}$$

$$= \mathbb{E}(1_{\{x_n(s)=x\}} P_{T_n(s)}^{m-n} h_x(x)) \tag{5.318}$$

$$= \mathbb{E}(1_{\{x_n(s)=x\}}(P_{T_n(s)}^{m-n} h_x(x) - \Pi h_x(x))) + \mathbb{E}(1_{\{x_n(s)=x\}} \Pi h_x(x)). \tag{5.319}$$

The proof of Lemma 5.3.18 then shows that $cov(Y_m, Y_n) \leq 2\psi(m - n)$, which completes the proof that the term (ii) converges a.e. to 0.

The following result allows us to represent the limiting process $N_\bullet^\ell(s, f)$ obtained above in a very neat fashion, using orthogonal martingale measures (see [2], Definition I-1).

Proposition 5.4.2. *Let $\ell \in Y^*$, $s \in \mathbf{J}$ and $f \in \widehat{\mathcal{D}}$. Under assumptions 5.3.1, 5.3.5 we have the following equality in distribution in $D(\mathbf{J}(s), \mathbb{R})$ (using the notations of Theorem 5.4.1):*

$$N_\bullet^\ell(s, f) \stackrel{d}{=} \sum_{x,y \in \mathbf{J}} \int_s^\bullet \ell \left(\sigma_s(x, y, u, f)\right)^T W(x, y, du) \tag{5.320}$$

$$= \sum_{k=1}^3 \sum_{x,y \in \mathbf{J}} \int_s^\bullet \ell \left(\sigma_s^{(k)}(x, y, u, f)\right) W^{(k)}(x, y, du), \tag{5.321}$$

where $W = \{W^{(k)} : k = 1, 2, 3\}$ is a vector of orthogonal martingale measures on $\mathbf{J} \times \mathbf{J} \times \mathbf{J}(s)$ with intensities $\frac{1}{\Pi m} P(x, y)\pi(x) dt$, $\ell(\sigma_s(x, y, t, f))$ is the vector with components $\ell(\sigma_s^{(k)}(x, y, t, f))$ $(k = 1, 2, 3)$ given by:

$$\sigma_s^{(1)}(x, y, t, f) = \sigma(x)\widehat{\Gamma}(s, t)A_x(t)f, \tag{5.322}$$

$$\sigma_s^{(2)}(x, y, t, f) = \widehat{\Gamma}(s, t)D_1^0(x, y)f - \sum_{y \in \mathbf{J}} P(x, y)\widehat{\Gamma}(s, t)D_1^0(x, y)f, \tag{5.323}$$

$$\sigma_s^{(3)}(x, y, t, f) = \widehat{\Gamma}(s, t)f_1(y, t) - \sum_{y \in \mathbf{J}} P(x, y)\widehat{\Gamma}(s, t)f_1(y, t). \tag{5.324}$$

Proof. *Take any probability space supporting three independent real-valued Brownian motions $\{B^{(k)} : k = 1, 2, 3\}$. Define for $t \in \mathbf{J}(s)$:*

$$m_t^{(k)} := \int_s^t \sqrt{\sum_{x,y \in \mathbf{J}} \ell^2(\sigma_s^{(k)}(x, y, u, f)) \frac{1}{\Pi m} P(x, y)\pi(x)} dB_u^{(k)}, \quad k = 1, 2, 3.$$
$$\tag{5.325}$$

We therefore have:

$$\left\langle m^{(i)}, m^{(j)} \right\rangle_t = \delta_{ij} \int_s^t \sum_{x,y \in \mathbf{J}} \ell^2(\sigma_s^{(i)}(x, y, u, f)) \frac{1}{\Pi m} P(x, y)\pi(x) du, \quad i, j = 1, 2, 3.$$
$$\tag{5.326}$$

By [2] (Theorem III-10), there exists an extension of the original space supporting $\{W^{(k)} : k = 1, 2, 3\}$ orthogonal martingale measures on $\mathbf{J} \times \mathbf{J} \times \mathbf{J}(s)$ with intensity $\frac{1}{\Pi m} P(x, y)\pi(x) dt$ such that:

$$m_t^{(k)} = \int_s^t \sum_{x,y \in \mathbf{J}} \ell(\sigma_s^{(k)}(x, y, u, f)) W^{(k)}(x, y, du), \quad k = 1, 2, 3. \tag{5.327}$$

$\sum_{k=1}^3 m_\bullet^{(k)}$ is a continuous Gaussian martingale with characteristics $(0, \widehat{\sigma}^\ell(s, \bullet, f)^2, 0)$ (using the notations of Theorem 5.4.1). Therefore $N_\bullet^\ell(s, f) \stackrel{d}{=} \sum_{k=1}^3 m_\bullet^{(k)}$ in $D(\mathbf{J}(s), \mathbb{R})$, wich completes the proof.

The above characterization of $N_\bullet^\ell(s, f)$ is relevant especially because it allows us to express the limiting process of $\ell(\epsilon^{-\frac{1}{2}}\widetilde{M}_\bullet^\epsilon(s)f)$ in the form $\ell(\cdot)$, which will allow us to express the limiting process of $\epsilon^{-\frac{1}{2}}\widetilde{M}_\bullet^\epsilon(s)f$ in terms of the weak random integral defined in [18] (Definition 3.1). In fact, we need a slight generalization of the latter integral to vectors, which is given in the following Definition 5.4.3 (the case $n = 1$ in the following definition coincides precisely with [18], Definition 3.1). We also note that the above representation of the Gaussian martingale $N_\bullet^\ell(s, f)$ allows us to see that the volatility vector σ (which governs the convergence of V_ϵ to $\widehat{\Gamma}$ as it will be shown in the main result Theorem 5.4.7) comes from three *independent sources*, which is what we expected intuitively:

- $\sigma^{(1)}$ which comes from the variance of the sojourn times $\{T_{n+1} - T_n\}_{n \geq 0}$ (component $\sigma(x)$).

- $\sigma^{(2)}$ which comes from the jump operators D.

- $\sigma^{(3)}$ which comes from the "state process" $\{x_n\}_{n \geq 0}$.

Definition 5.4.3. *Let $M = \{M^{(k)} : k = 1..n\}$ a vector of martingale measures on $(E, \mathcal{R})$, where E is a Lusin space and $\mathcal{R}$ a ring of subsets of E (see [2], Definition I-1). Let $\sigma : E \times \mathbb{R}_+ \to Y^n$ the vector $\sigma(x, t) := \{\sigma^{(k)}(x, t) : k = 1..n\}$ ($x \in E, t \in \mathbb{R}_+$), and the square-mean measures $\rho^{(k)}(A \times [0, t]) := \mathbb{E}[M^{(k)}(A, t)^2]$ ($A \in \mathcal{R}, t \in \mathbb{R}_+$). We say that σ is stochastically integrable with respect to M on $A \times [s, t]$ ($A \in \mathcal{R}, s \leq t \in \mathbb{R}_+$) if $\forall \ell \in Y^*$:*

$$\sum_{k=1}^{n} \int_{A \times [s,t]} \ell^2(\sigma^{(k)}(x, u))\rho^{(k)}(dx, du) < \infty, \tag{5.328}$$

and there exists a Y-valued random variable α such that $\forall \ell \in Y^*$:

$$\ell(\alpha) = \int_{A \times [s,t]} \ell(\sigma(x, u))^T M(dx, du) = \sum_{k=1}^{n} \int_{A \times [s,t]} \ell(\sigma^{(k)}(x, u)) M^{(k)}(dx, du) \text{ a.e.} \tag{5.329}$$

In this case we write:

$$\alpha = \int_{A \times [s,t]} \sigma(x, u)^T M(dx, du). \tag{5.330}$$

The previous definition will now allow us to express the limiting process of $\epsilon^{-\frac{1}{2}}\widetilde{M}_\bullet^\epsilon(s)f$ in $D(\mathbf{J}(s), Y)$ for all $f \in \widehat{\mathcal{D}}$.

Theorem 5.4.4. *Let $s \in \mathbf{J}$ and $\ell \in Y^*$. Under assumptions 5.3.1, 5.3.3, 5.3.5, we have the following weak convergence in $D(\mathbf{J}(s), \mathbb{R}^\infty)$ for each countable family $\{f_p\}_{p \in \mathbb{N}} \subseteq \widehat{\mathcal{D}}$:*

$$\left(\epsilon^{-\frac{1}{2}} \ell(\widetilde{M}_\bullet^\epsilon(s) f_p) \right)_{p \in \mathbb{N}} \Rightarrow \left(\sum_{x,y \in \mathbb{J}} \int_s^\bullet \ell(\sigma_s(x,y,u,f_p))^T W(x,y,du) \right)_{p \in \mathbb{N}},$$

$$(5.331)$$

where $\sigma_s(x, y, t, f)$ is the vector with components $\sigma_s^{(k)}(x, y, t, f)$ given in Proposition 5.4.2, $W = \{W^{(k)} : k = 1, 2, 3\}$ is a vector of orthogonal martingale measures on $\mathbb{J} \times \mathbb{J} \times \mathbf{J}(s)$ with intensities $\frac{1}{\Pi m} P(x, y)\pi(x)dt$. Further, if in addition $\epsilon^{-\frac{1}{2}} \widetilde{M}_\bullet^\epsilon(s)$ is $\widehat{\mathcal{D}}$–relatively compact, we have the following weak convergence in $D(\mathbf{J}(s), Y^\infty)$ for each countable family $\{f_p\}_{p \in \mathbb{N}} \subseteq \widehat{\mathcal{D}}$:

$$\left(\epsilon^{-\frac{1}{2}} \widetilde{M}_\bullet^\epsilon(s) f_p \right)_{p \in \mathbb{N}} \Rightarrow \left(\sum_{x,y \in \mathbb{J}} \int_s^\bullet \sigma_s(x,y,u,f_p)^T W(x,y,du) \right)_{p \in \mathbb{N}}, \quad (5.332)$$

where each integral on the righthand side of the above is a weak random integral in the sense of Definition 5.4.3.

Proof.

We will split the proof in two steps. For the first step, we fix $f \in \widehat{\mathcal{D}}$ and prove the weak convergence in $D(\mathbf{J}(s), Y)$:

$$\epsilon^{-\frac{1}{2}} \widetilde{M}_\bullet^\epsilon(s) f \Rightarrow \sum_{x,y \in \mathbb{J}} \int_s^\bullet \sigma_s(x,y,u,f)^T W(x,y,du). \quad (5.333)$$

In the second step, we prove the joint convergence in $D(\mathbf{J}(s), Y^\infty)$.

First step. Let a sequence $\epsilon_n \to 0$ such that $\epsilon_n^{-\frac{1}{2}} \widetilde{M}_\bullet^{\epsilon_n}(s) f \Rightarrow \widetilde{M}_\bullet(s, f)$ for some $D(\mathbf{J}(s), Y)-$ valued random variable $\widetilde{M}_\bullet(s, f)$. By the Skorokhod representation theorem, there exists a probability space $(\Omega', \mathcal{F}', \mathbb{P}')$ and random variables with the same distribution as the original ones, denoted by the subscript ', such that the previous convergence holds a.e. By Theorem 5.4.1, for all $\ell \in Y^*$, $\ell(\widetilde{M'}_\bullet(s, f))$ is a continuous Gaussian martingale with characteristics $(0, \widehat{\sigma}^\ell(s, \bullet, f)^2, 0)$. By the proof of Proposition 5.4.2, we can consider on an extension $(\Omega'', \mathcal{F}'', \mathbb{P}'')$ of $(\Omega', \mathcal{F}', \mathbb{P}')$, $\{W'^{(k)} : k = 1, 2, 3\}$ orthogonal martingale measures on $\mathbb{J} \times \mathbb{J} \times \mathbf{J}(s)$ with intensity $\frac{1}{\Pi m} P(x, y)\pi(x)dt$ such that $\forall \ell \in Y^*$:

$$\ell(\widetilde{M'}_\bullet(s, f)) = \int_s^\bullet \sum_{x,y \in \mathbb{J}} \ell(\sigma_s(x,y,t,f))^T W'(x,y,du) \quad a.e. \quad (5.334)$$

By definition of the weak random integrals in the sense of Definition 5.4.3, we get:

$$\widetilde{M'}_\bullet(s,f) = \sum_{x,y \in \mathbb{J}} \int_s^\bullet \sigma_s(x,y,u,f)^T W'(x,y,du). \tag{5.335}$$

Actually there is a subtlety that we voluntarily skipped: a priori, it is not obvious that the orthogonal martingale measures $\{W'^{(k)} : k = 1, 2, 3\}$ do not depend on ℓ, i.e. that we can represent all Gaussian martingales $\{\ell(\widetilde{M'}_\bullet(s,f)) : \ell \in Y^*\}$ according to the same vector W'. To see that it is the case, observe that the joint distribution of the Gaussian martingales $\{\ell(\widetilde{M'}_\bullet(s,f)) : \ell \in Y^*\}$ is completely determined by their pairwise predictable quadratic co-variations. Take $\ell_1, \ell_2 \in Y^*$. Then $\ell_1 + \ell_2 \in Y^*$ and $\ell_1 - \ell_2 \in Y^*$, therefore we can compute by Theorem 5.4.1 the quantity $\left\langle \ell_1(\widetilde{M'}_\bullet(s,f)), \ell_2(\widetilde{M'}_\bullet(s,f)) \right\rangle_t$ according to:

$$\left\langle \ell_1(\widetilde{M'}_\bullet(s,f)), \ell_2(\widetilde{M'}_\bullet(s,f)) \right\rangle_t = \frac{1}{4} \left(\widehat{\sigma}^{\ell_1 + \ell_2}(s,t,f)^2 - \widehat{\sigma}^{\ell_1 - \ell_2}(s,t,f)^2 \right)$$

$$(5.336)$$

$$= \sum_{k=1}^3 \sum_{x,y \in \mathbb{J}} \int_s^t \ell_1 \left(\sigma_s^{(k)}(x,y,u,f) \right) \ell_2 \left(\sigma_s^{(k)}(x,y,u,f) \right) \frac{1}{\Pi m} P(x,y)\pi(x)du.$$

$$(5.337)$$

The latter gives us that we can indeed represent all Gaussian martingales $\{\ell(\widetilde{M'}_\bullet(s,f)) : \ell \in Y^*\}$ according to the same vector W'.

Second step. The limit points are continuous, so weak convergence in $D(\mathbf{J}(s), Y^\infty)$ is equivalent to weak convergence in $D(\mathbf{J}(s), Y)^\infty$. Similarly to what we did for ℓ_1, ℓ_2, the joint distribution of the Gaussian martingales $\{\ell(\widetilde{M'}_\bullet(s,f_p)) : p \in \mathbb{N}\}$ is completely determined by their pairwise predictable quadratic co-variations. We can compute by Theorem 5.4.1 the quantity $\left\langle \ell(\widetilde{M'}_\bullet(s,f_i)), \ell(\widetilde{M'}_\bullet(s,f_j)) \right\rangle_t$ according to:

$$\left\langle \ell(\widetilde{M'}_\bullet(s,f_i)), \ell(\widetilde{M'}_\bullet(s,f_j)) \right\rangle_t = \frac{1}{4} \left(\widehat{\sigma}^\ell(s,t,f_i + f_j)^2 - \widehat{\sigma}^\ell(s,t,f_i - f_j)^2 \right)$$

$$(5.338)$$

$$= \sum_{k=1}^3 \sum_{x,y \in \mathbb{J}} \int_s^t \ell \left(\sigma_s^{(k)}(x,y,u,f_i) \right) \ell \left(\sigma_s^{(k)}(x,y,u,f_j) \right) \frac{1}{\Pi m} P(x,y)\pi(x)du.$$

$$(5.339)$$

The latter gives us that we can indeed represent all Gaussian martingales $\{\ell(\widetilde{M'}_\bullet(s,f_p)) : \ell \in Y^*, p \in \mathbb{N}\}$ according to the same vector W'.

Let $(Y_2, \|\cdot\|_{Y_2})$ a separable Banach space continuously embedded in $(Y_1, \|\cdot\|_{Y_1})$ such that $Y_2 \subseteq \widehat{\mathcal{D}}$. We now show our main result, which is the weak convergence of the process:

$$\epsilon^{-1/2}(V_\epsilon(s, \bullet) - \widehat{\Gamma}(s, \bullet)) \tag{5.340}$$

in a suitable topology, which is the space $\mathbb{D}_s^{Y_2 \to Y}$ defined similarly to $\mathbb{D}_s$, the following way:

$$\mathbb{D}_s^{Y_2 \to Y} := \{S \in \mathcal{B}(Y_2, Y)^{\mathbf{J}(s)} : Sf \in D(\mathbf{J}(s), Y) \quad \forall f \in Y_2\}. \tag{5.341}$$

We will need the following regularity assumption:

For $j \in \{2, 3\}$, there exists separable Banach spaces $(Y_j, \|\cdot\|_{Y_j})$ continuously embedded in Y_{j-1} such that $Y_2 \subseteq \widehat{\mathcal{D}}$ and such that the following hold true for $j \in \{2, 3\}$:

- Y_j contains a countable family which is dense in Y_{j-1}.

- $\widehat{\Gamma}$ is a regular Y_2-propagator with respect to the Banach spaces Y_2 and Y_3.

- $A_x(t) \in \mathcal{B}(Y_j, Y_{j-1})$ and $\sup_{u \in [0,t]} \|A_x(u)\|_{\mathcal{B}(Y_j, Y_{j-1})} < \infty \ \forall t \in J, \forall x \in \mathbb{J}$.

- $A'_x(t) \in \mathcal{B}(Y_2, Y)$ and $\sup_{u \in [0,t]} \|A'_x(u)\|_{\mathcal{B}(Y_2, Y)} < \infty \ \forall t \in J, \forall x \in \mathbb{J}$.

- $D_1^0(x, y) \in \mathcal{B}(Y_j, Y_{j-1}) \ \forall x, y \in \mathbb{J}$.

Remark: We previously introduced the notion of "regular" $Y-$propagator. This notion implicitly involves the Banach spaces Y and Y_1. Therefore by "$\widehat{\Gamma}$ is a regular Y_2-propagator with respect to the Banach spaces Y_2 and Y_3" we mean the same thing as our previously introduced notion of regularity, but with the roles of Y, Y_1 being respectively replaced by Y_2, Y_3.

A few comments: we need the Banach space $Y_2 \subseteq Y_1$ because the limiting Gaussian operator I_σ of our main result Theorem 5.4.7 takes value in $\mathbb{C}_s^{Y_2 \to Y}$, but not in $\mathbb{C}_s^{Y_1 \to Y}$, as shown in Lemma 5.4.5. We need the Banach space $Y_3 \subseteq Y_2$ for the same reason we needed $Y_1 \subseteq Y$ in the WLLN result Theorem 5.3.19: in order to get the well-posedness of a Cauchy problem and conclude on the unicity of the limit. This technical point is established in Lemma 5.4.6.

Lemma 5.4.5. *Let $s \in J$ and $(\Omega, \mathcal{F}, \mathbb{P})$ a probability space which supports $W = \{W^{(k)} : k = 1, 2, 3\}$ a vector of orthogonal martingale measures on $\mathbb{J} \times \mathbb{J} \times \mathbf{J}(s)$ with intensities $\frac{1}{\Pi m}P(x, y)\pi(x)dt$. With the definition of σ in Proposition 5.4.2, define:*

$$I_\sigma(s, t)f := \sum_{x, y \in \mathbb{J}} \int_s^t \sigma_s(x, y, u, \widehat{\Gamma}(u, t)f)^T W(x, y, du), \quad \forall f \in Y_2, t \in \mathbf{J}(s). \tag{5.342}$$

Under assumptions 5.3.1, 5.3.5, 5.4, $I_\sigma(s, \bullet)$ *is a* $\mathbb{C}_s^{Y_2 \to Y}$*-valued random variable. Further, for all* $T \in \mathbb{J}(s) \cap \mathbb{Q}_+$, *there exists a constant* C_T *such that for a.e.* $\omega \in \Omega$:

$$\sup_{t \in [s,T]} \|I_\sigma(s,t)\|_{\mathcal{B}(Y_2,Y)} \leq C_T \sum_{x,y \in \mathbb{J}, k=1..3} \sup_{t \in [s,T]} |W^{(k)}(x,y,t)|. \quad (5.343)$$

Proof. *First we recall the fact that if* $h \in C^1(\mathbb{R})$, *then by Ito's lemma, for every* $x, y \in \mathbb{J}$ *and* $k = 1, 2, 3$ *we have almost surely:*

$$\int_s^t h(u) W^{(k)}(x,y,du) = h(t) W^{(k)}(x,y,t) - h(s) W^{(k)}(x,y,s)$$

$$- \int_s^t h'(u) W^{(k)}(x,y,u) du \quad (5.344)$$

$$\Rightarrow \left| \int_s^t h(u) W^{(k)}(x,y,du) \right| \leq 3(1+t-s) \sup_{u \in [s,t]} (|h(u)| + |h'(u)|) |W^{(k)}(x,y,u)| \text{ a.s.} \quad (5.345)$$

In our case we have for $f \in Y_2$:

$$\frac{d}{du} \sigma_s^{(1)}(x,y,u,\widehat{\Gamma}(u,t)f) = \sigma(x)\widehat{\Gamma}(s,u)\widehat{A}(u)A_x(u)\widehat{\Gamma}(u,t)f$$

$$+ \sigma(x)\widehat{\Gamma}(s,u)A_x'(u)\widehat{\Gamma}(u,t)f - \sigma(x)\widehat{\Gamma}(s,u)A_x(u)\widehat{A}(u)\widehat{\Gamma}(u,t)f, \quad (5.346)$$

$$\frac{d}{du} \sigma_s^{(2)}(x,y,u,\widehat{\Gamma}(u,t)f) = \widehat{\Gamma}(s,u)\widehat{A}(u)D_1^0(x,y)\widehat{\Gamma}(u,t)f$$

$$- \sum_{y \in \mathbb{J}} P(x,y)\widehat{\Gamma}(s,u)\widehat{A}(u)D_1^0(x,y)\widehat{\Gamma}(u,t)f - \widehat{\Gamma}(s,u)D_1^0(x,y)\widehat{A}(u)\widehat{\Gamma}(u,t)f$$

$$+ \sum_{y \in \mathbb{J}} P(x,y)\widehat{\Gamma}(s,u)D_1^0(x,y)\widehat{A}(u)\widehat{\Gamma}(u,t)f. \quad (5.347)$$

For the derivative of $\sigma^{(3)}$, *denote:*

$$g_1(y,u) := (\widehat{\Gamma}(u,t)f)_1(y,u) = \Pi m(P + \Pi - I)^{-1}[\widehat{A}(u)\widehat{\Gamma}(u,t)f - a(\bullet,u)\widehat{\Gamma}(u,t)f](y), \quad (5.348)$$

so that:

$$g_1'(y,u) = \Pi m(P + \Pi - I)^{-1}[\widehat{A}'(u)\widehat{\Gamma}(u,t)f - \widehat{A}(u)^2\widehat{\Gamma}(u,t)f$$

$$- a'(\bullet,u)\widehat{\Gamma}(u,t)f + a(\bullet,u)\widehat{A}(u)\widehat{\Gamma}(u,t)f](y), \quad (5.349)$$

$$a'(x,u) = \frac{1}{\Pi m} m_1(x) A_x'(u), \quad (5.350)$$

$$\widehat{A}'(u) = \Pi a'(\bullet,u), \quad (5.351)$$

where we recall that:

$$a(x,t) := \frac{1}{\Pi m}\left(m_1(x)A_x(t) + PD_1^0(x,\bullet)(x)\right). \tag{5.352}$$

With these notations we have:

$$\frac{d}{du}\sigma_s^{(3)}(x,y,u,\widehat{\Gamma}(u,t)f) = \widehat{\Gamma}(s,u)\widehat{A}(u)g_1(y,u) + \widehat{\Gamma}(s,u)g_1'(y,u)$$
$$- \sum_{y\in\mathbb{J}} P(x,y)(\widehat{\Gamma}(s,u)\widehat{A}(u)g_1(y,u) + \widehat{\Gamma}(s,u)g_1'(y,u)). \tag{5.353}$$

By (5.345) and assumption 5.4, for every $t \leq T \in \mathbf{J}(s)$, there exists a constant C_T such that for every $\ell \in Y^*$, every $f \in Y_2$ we have for a.e. $\omega \in \Omega$:

$$\left|\sum_{x,y\in\mathbb{J}}\int_s^t \ell(\sigma_s(x,y,u,\widehat{\Gamma}(u,t)f))^T W(x,y,du)\right| \leq ||\ell||C_T||f||_{Y_2}$$

$$\sum_{x,y\in\mathbb{J},k=1..3}\sup_{t\in[s,T]}|W^{(k)}(x,y,t)|. \tag{5.354}$$

By [13], the unit ball of the dual of every separable Banach space Y contains a norming subset, namely a countable subset $S \subseteq Y^*$ such that for every $f \in Y$, $||f|| = \sup_{\ell\in S}|\ell(f)|$. In our case and using Definition 5.4.3, it means that we get for every $t \leq T \in \mathbf{J}(s)$, every $f \in Y_2$ and a.e. $\omega \in \Omega$:

$$||I_\sigma(s,t)f|| = \left|\left|\sum_{x,y\in\mathbb{J}}\int_s^t \sigma_s(x,y,u,\widehat{\Gamma}(u,t)f)^T W(x,y,du)\right|\right| \tag{5.355}$$

$$= \sup_{\ell\in S}\left|\sum_{x,y\in\mathbb{J}}\int_s^t \ell(\sigma_s(x,y,u,\widehat{\Gamma}(u,t)f))^T W(x,y,du)\right| \leq C_T||f||_{Y_2}$$

$$\sum_{x,y\in\mathbb{J},k=1..3}\sup_{t\in[s,T]}|W^{(k)}(x,y,t)|. \tag{5.356}$$

Lemma 5.4.6. *Let $s \in \mathbf{J}$ and fix $\{f_p\}$ a countable dense family of Y_3, as well as a probability space $(\Omega,\mathcal{F},\mathbb{P})$ supporting the orthogonal martingale measures W of Theorem 5.4.4. Consider the following Cauchy problem on $(\Omega,\mathcal{F},\mathbb{P})$, for a $\mathbb{C}_s^{Y_2\to Y}$–valued random variable Z_s, where σ was defined in Proposition 5.4.2:*

$$Z_s(t)f_p - \int_s^t Z_s(u)\widehat{A}(u)f_p du = \sum_{x,y\in\mathbb{J}}\int_s^t \sigma_s(x,y,u,f_p)^T W(x,y,du),$$

$$\forall t \in \mathbf{J}(s), \forall p, \quad a.s. \tag{5.357}$$

If we assume that the previous problem has at least one solution Z_s, and that assumptions 5.3.1, 5.3.5, 5.4 hold true, then we have $Z_s(\bullet) = I_\sigma(s,\bullet)$.

Proof. *First, let's prove uniqueness. Let Z_1, Z_2 two solutions and let $Z :=$
$Z_1 - Z_2$. Then we have:*

$$Z(\bullet)f_p - \int_s^{\bullet} Z(u)\widehat{A}(u)f_p du = 0, \quad \forall p, \quad a.s. \tag{5.358}$$

*Consider the following Cauchy problem, very similar to Theorem 5.1.10,
where $G_s \in \mathcal{B}(Y_2, Y)$, G is a an operator $G : \mathbf{J}(s) \to \mathcal{B}(Y_2, Y)$ which is
Y_2-strongly continuous and $Y - \frac{d}{dt}$ represents the derivative with respect to the
Y-norm (unless specified otherwise, all norms, limits, derivatives are taken
with respect to Y, here I prefer to specify it explicitly as we are dealing with
several Banach spaces):*

$$\begin{cases} Y - \frac{d}{dt}G(t)f = G(t)\widehat{A}(t)f & \forall t \in \mathbf{J}(s), \forall f \in Y_3. \\ G(s) = G_s \in \mathcal{B}(Y_2, Y) \end{cases} \tag{5.359}$$

*Because by assumption $\widehat{\Gamma}$ is a regular Y_2-propagator with respect to the
Banach spaces Y_2 and Y_3, we can prove - similarly to the proof of Theorem 5.1.10 - that the previous Cauchy problem has a unique solution $G(t) =
G_s\widehat{\Gamma}(s,t)$. In our context, let Ω_0 the subset of probability one on which we
have:*

$$Z(\bullet)f_p - \int_s^{\bullet} Z(u)\widehat{A}(u)f_p du = 0, \quad \forall p. \tag{5.360}$$

*By density of $\{f_p\}$ in Y_3 and because $Z(t)$ takes value in $\mathcal{B}(Y_2, Y)$ and
$\widehat{A}(t) \in \mathcal{B}(Y_3, Y_2)$ for all t, we can extend the previous equality for all $f \in Y_3$
on Ω_0. Observing that Z takes value in $\mathbb{C}_s^{Y_2 \to Y}$, we can use the unicity of the
solution to the Cauchy problem discussed above, and we get $Z(\bullet) = 0$ on Ω_0.
This proves uniqueness.*

*It will be proved below that $I_\sigma(s, \bullet)$ (using the notation of Lemma 5.4.5) is
the unique solution to our Cauchy problem. Indeed, Lemma 5.4.5 gives us that
$I_\sigma(s, \bullet)$ takes value in $\mathbb{C}_s^{Y_2 \to Y}$. It remains to show that it satisfies (5.357).
Let $\ell \in Y^*$. Since $f \to \sigma_s(x, y, u, f) \in \mathcal{B}(Y_1, Y)$ and since $\widehat{\Gamma}$ is a regular
Y_2-propagator by assumption, we get for all $f \in Y_2$:*

$$Y - \frac{d}{dt}\ell(\sigma_s(x, y, u, \widehat{\Gamma}(u, t)f)) = \ell(\sigma_s(x, y, u, \widehat{\Gamma}(u, t)\widehat{A}(t)f)). \tag{5.361}$$

We get for all p, using the stochastic Fubini theorem:

$$\ell(I_\sigma(s, t)f_p) - \sum_{x,y \in \mathbb{J}} \int_s^t \ell(\sigma_s(x, y, u, f_p))^T W(x, y, du) \tag{5.362}$$

$$= \sum_{x,y \in \mathbb{J}} \int_s^t \ell(\sigma_s(x, y, u, \widehat{\Gamma}(u, t)f_p))^T W(x, y, du)$$

$$- \sum_{x,y \in \mathbb{J}} \int_s^t \ell(\sigma_s(x, y, u, f_p))^T W(x, y, du) \tag{5.363}$$

$$= \sum_{x,y \in \mathbb{J}} \int_s^t \int_s^r \frac{d}{dr} \ell(\sigma_s(x,y,u,\widehat{\Gamma}(u,r)f_p))^T W(x,y,du)dr \tag{5.364}$$

$$= \sum_{x,y \in \mathbb{J}} \int_s^t \int_s^r \ell(\sigma_s(x,y,u,\widehat{\Gamma}(u,r)\widehat{A}(r)f_p))^T W(x,y,du)dr \tag{5.365}$$

$$= \int_s^t \ell(I_\sigma(s,r)\widehat{A}(r)f_p)dr. \tag{5.366}$$

By [13], the unit ball of the dual of every separable Banach space Y contains a norming subset, namely a countable subset $S \subseteq Y^*$ such that for every $f \in Y$, $\|f\| = \sup_{\ell \in S} |\ell(f)|$. We therefore obtain:

$$I_\sigma(s,t)f_p - \int_s^t I_\sigma(s,r)\widehat{A}(r)f_p dr = \sum_{x,y \in \mathbb{J}} \int_s^t \sigma_s(x,y,u,f_p)^T W(x,y,du),$$

$$\forall t \in \mathbf{J}(s), \forall p, \text{ a.s.}, \tag{5.367}$$

which completes the proof.

We now state the main result of this section:

Theorem 5.4.7. *Let $s \in \mathbf{J}$. Assume that assumptions 5.3.1, 5.3.3, 5.3.5, 5.4 hold true, that $\epsilon^{-1/2}(V_\epsilon(s,\bullet) - \widehat{\Gamma}(s,\bullet))$ is Y_2-relatively compact and that for all $T \in \mathbf{J}(s) \cap \mathbb{Q}_+$, the sequence of real-valued random variables $\{M_{s,T}^\epsilon\}_{\epsilon \in (0,1]}$ is tight, where:*

$$M_{s,T}^\epsilon := \sup_{t \in [s,T]} \epsilon^{-1/2} \|V_\epsilon(s,t) - \widehat{\Gamma}(s,t)\|_{\mathcal{B}(Y_2,Y)}. \tag{5.368}$$

Then we have the following weak convergence in $(\mathbb{D}_s^{Y_2 \to Y}, d)$:

$$\epsilon^{-1/2}(V_\epsilon(s,\bullet) - \widehat{\Gamma}(s,\bullet)) \Rightarrow I_\sigma(s,\bullet), \tag{5.369}$$

where $I_\sigma(s,\bullet)$ is the $\mathbb{C}_s^{Y_2 \to Y}$-valued random variable defined by:

$$I_\sigma(s,t)f := \sum_{x,y \in \mathbb{J}} \int_s^t \sigma_s(x,y,u,\widehat{\Gamma}(u,t)f)^T W(x,y,du), \quad \forall f \in Y_2, t \in \mathbf{J}(s), \tag{5.370}$$

where $W = \{W^{(k)} : k = 1,2,3\}$ is a vector of orthogonal martingale measures on $\mathbb{J} \times \mathbb{J} \times \mathbf{J}(s)$ with intensity $\frac{1}{\Pi m} P(x,y)\pi(x)dt$ and σ was defined in Proposition 5.4.2.

Proof. Denote $W_\epsilon(s,\bullet) := \epsilon^{-1/2}(V_\epsilon(s,\bullet) - \widehat{\Gamma}(s,\bullet))$. Take $\{f_p\}$ a countable family of Y_3 that is dense in both Y_3 and Y_2. Marginal relative compactness implies countable relative compactness and therefore, $\{W_\epsilon(s,\bullet)f_p\}_{p \in \mathbb{N}}$ is relatively compact in $D(\mathbf{J}(s), Y)^\infty$. Actually, $\{W_\epsilon(s,\bullet)f_p\}_{p \in \mathbb{N}}$ is C-relatively compact in $D(\mathbf{J}(s), Y)^\infty$, and therefore in $D(\mathbf{J}(s), Y^\infty)$. This is because the jumps of $W_\epsilon(s, \bullet)$ are the jumps of $\epsilon^{-1/2} V_\epsilon(s, \bullet)$ which are uniformly bounded by $C_T \epsilon^{1/2}$ on each interval $[s, T]$ for some constant C_T which depends only on T (see proof of Lemma 5.3.12). Take a weakly converging sequence ϵ_n. By the proof of Theorem 3.4.1, we can find a probability space $(\Omega', \mathcal{F}', \mathbb{P}')$ and $\mathbb{D}_s^{Y_2 \to Y}$−valued random variables $W'_{\epsilon_n}(s, \bullet)$, $W'_0(s, \bullet)$ on it, such that on Ω' we have the sure convergence:

$$W'_{\epsilon_n}(s, \bullet) \to W'_0(s, \bullet) \text{ in } (\mathbb{D}_s^{Y_2 \to Y}, d), \tag{5.371}$$

$$W'_{\epsilon_n}(s, \bullet) \overset{d}{=} W_{\epsilon_n}(s, \bullet) \text{ in } (\mathbb{D}_s^{Y_2 \to Y}, d), \forall n. \tag{5.372}$$

We can extend the probability space $(\Omega', \mathcal{F}', \mathbb{P}')$ so that it supports a Markov renewal process $(x'_k, T'_k)_{k \geq 0}$ such that:

$$(W'_{\epsilon_n}(s, \bullet), (x'_k, T'_k)_{k \geq 0}) \overset{d}{=} (W_{\epsilon_n}(s, \bullet), (x_k, T_k)_{k \geq 0}) \text{ in } (\mathbb{D}_s^{Y_2 \to Y}, d) \times (\mathbb{J} \times \mathbb{R}_+)^\infty, \forall n. \tag{5.373}$$

By continuity of $\widehat{\Gamma}$, $\widehat{A}$ we have the following representation of the Riemann sum, for all $f \in Y_3$ (actually, for all $f \in Y_1$):

$$\int_s^\bullet \widehat{\Gamma}(s, u) \widehat{A}(u) f \, du = \epsilon \Pi m \sum_{k=1}^{N_s\left(t^{\frac{1}{\epsilon}, s}\right)} \widehat{\Gamma}(s, T_k^{\epsilon_n, s}(s)) \widehat{A}(T_k^{\epsilon_n, s}(s)) f + O(\epsilon). \tag{5.374}$$

Therefore by Lemma 5.3.17 we have the representation, for all $f \in Y_3$ (actually, for all $f \in Y_1$):

$$\epsilon_n^{-1/2} \widetilde{M}_\bullet^{\epsilon_n}(s) f = W_{\epsilon_n}(s, \bullet) f - \epsilon_n \Pi m \sum_{k=1}^{N_s\left(t^{\frac{1}{\epsilon}, s}\right)} W_{\epsilon_n}(s, T_k^{\epsilon_n, s}(s)) \widehat{A}(T_k^{\epsilon_n, s}(s)) f$$
$$+ O(\epsilon_n^{1/2}) \text{ a.e.} \tag{5.375}$$

As in the proof of Theorem 5.3.19, using continuity of W'_0, we have the following convergence in the Skorokhod topology on Ω', for all $f \in Y_3$:

$$\epsilon_n \sum_{k=1}^{N'_s\left(t^{\frac{1}{\epsilon_n}, s}\right)} W'_{\epsilon_n}(s, T_k^{\epsilon_n, s}(s)') \widehat{A}(T_k^{\epsilon_n, s}(s)') f \to \frac{1}{\Pi m} \int_s^\bullet W'_0(s, u) \widehat{A}(u) f \, du. \tag{5.376}$$

Similarly to what was done in the proof of Theorem 5.4.4, we can extend the probability space $(\Omega', \mathcal{F}', \mathbb{P}')$ to a probability space $(\Omega'', \mathcal{F}'', \mathbb{P}'')$ which supports the orthogonal martingale measures $W^{(k)}(x, y, dt)$ of Theorem 5.4.4. Since we have:

$$(W'_{\epsilon_n}(s, \bullet), (x'_k, T'_k)_{k \geq 0}) \overset{d}{=} (W_{\epsilon_n}(s, \bullet), (x_k, T_k)_{k \geq 0}) \text{ in } (\mathbb{D}_s^{Y_2 \to Y}, d) \times (\mathbb{J} \times \mathbb{R}_+)^\infty, \forall n, \tag{5.377}$$

using (5.375) and Theorem 5.4.4 we get for all p, say on some set of probability one $\Omega_0'' \subseteq \Omega''$, that:

$$W_0'(s, \bullet)f_p - \int_s^\bullet W_0'(s, u)\widehat{A}(u)f_p du = \sum_{x,y\in\mathbb{J}} \int_s^\bullet \sigma_s(x, y, u, f_p)^T W(x, y, du).$$

$$(5.378)$$

The result of Lemma 5.4.6 concludes the proof.

Bibliography

[1] J. Conway. *A Course in Functional Analysis*. Springer, 2007.

[2] N. El Karoui and S. Méléard. Martingale measures and stochastic calculus. *Probab. Th. Rel. Fields*, 84(2):83-101, 1990.

[3] S. Ethier and T. Kurtz. *Markov Processes: Characterization and Convergence*. John Wiley, 1986.

[4] D. Filipovic. Time-inhomogeneous affine processes. *Stochastic Processes and Their Applications*, 115:639-659, 2005.

[5] P. Gassiat, F. Gozzi, and H. Pham. Investment/consumption problem in illiquid mar- kets with regime-switching. *SIAM Journal on Control and Optimization*, 52(3):1761-1786, 2014.

[6] R. Griego and R. Hersh. Random evolutions, Markov chains, and systems of partial differential equations. *Proc. National Acad. Sci.*, 62:305-308, 1969.

[7] A. Gulisashvili and J. van Casteren. *Non Autonomous Kato Classes and Feynman-Kac Propagators*. World Scientific Publishing Co. Pte. Ltd., 1986.

[8] R. Hersh. Random evolutions: A survey of results and problems. *Rocky Mountain J. Math.*, 4:443-475, 1972.

[9] R. Hersh and M. Pinsky. Random evolutions are asymptotically Gaussian. *Comm. Pure Appl. Math.*, XXV:33-44, 1972.

[10] J. Jacod and A. Shiryaev. *Limit Theorems for Stochastic Processes*. Springer, 2003.

[11] I. Karatzas and S. Shreve. *Brownian Motion and Stochastic Calculus*. Springer, 1998.

[12] W. Kluge. *Time-Inhomogeneous Lévy Processes in Interest Rate and Credit Risk Models.* Ph.D. Thesis, 2005.

[13] M. Ledoux and M. Talagrand. *Probability in Banach Spaces: Isoperimetry and Processes.* Springer-Verlag, 1991.

[14] R. Lyons. Strong laws of large numbers for weakly correlated random variables. *Michigan Math J.*, 35(3), 1988.

[15] P. Mathé. Numerical integration using V-uniformly ergodic Markov chains. *J. Appl. Probab.*, 41(4):1104-1112, 2004.

[16] A. Pazy. *Semigroups of Linear Operators and Applications to Partial Differential Equations.* Springer-Verlag, 1983.

[17] D. Revuz. *Markov Chains.* Elsevier Science Publishers B.V., 1984.

[18] M. Riedle and O. van Gaans. Stochastic integration for Lévy processes with values in Banach spaces. *Stochastic Processes and their Applications*, 119:1952-1974, 2009.

[19] L. Ruschendorf, A. Schnurr, and V. Wolf. *Comparison of Time-Inhomogeneous Markov Processes.* Preprint, 2014.

[20] K. Sato. *Lévy Processes and Infinitely Divisible Distributions.* Cambridge University Press, 1999.

[21] A. V. Skorokhod. *Random Linear Operators.* Reidel Publishing Company, Dordrecht, 1984.

[22] A. Swishchuk. *Random Evolutions and their Applications.* Kluwer Academic Publishers, 1997.

[23] A. Swishchuk and V. Korolyuk. *Evolutions of Systems in Random Media.* CRC Press, 1995.

[24] D.H. Thang. Transforming random operators into random bounded operators. *Random Operators and Stochastic Eqs.*, 16:293302, 2008.

[25] N. Vadori. *Semi-Markov Driven Models: Limit Theorems and Financial Applications.* PhD Thesis, University of Calgary, Calgary, AB, Canada, 2015.

[26] N. Vadori and A. Swishchuk. Strong law of large numbers and central limit theorems for functionals of inhomogeneous semi-Markov processes. *Stochastic Analysis and Applications*, 33(2):213?243, 2015.

[27] N. Vadori and A. Swishchuk. Convergence of random bounded operators in the Skorokhod space. *Random Operators and Stochastic Equations*, 2019, 27 (3): 1-13 (https://doi.org/10.1515/rose-2019-2011).

[28] N. Vadori and A. Swishchuk. Inhomogeneous random evolutions: limit theorems and financial applications. *Mathematics*, 2019, v. 7, N. 447; doi:10.3390/math7050447. Special Issue: *New Trends in Random Evolutions and their Applications.*

[29] J. Watkins. A central limit theorem in random evolution. *Ann. Prob.*, 12(2):480-513, 1984.

[30] J. Watkins. Limit theorems for stationary random evolutions. *Stoch. Proc. Appl.*, 19:189-224, 1985.

[31] J. Watkins. A stochastic integral representation for random evolution. *Ann. Prob.*, 13(2):531-557, 1985.

Part III

Applications of Inhomogeneous Random Evolutions

6

Applications of IHREs: Inhomogeneous Lévy-Based Models

In this chapter we present financial applications to illiquidity modeling using regime-switching time-inhomogeneous Lévy price dynamics and regime-switching Lévy driven diffusion based price dynamics. We also present a generalized version of the multi-asset model of price impact from distressed selling introduced in the recent article [1], for which we retrieve (and generalize) their diffusion limit result for the price process (see [11], [12]).

6.1 Regime-Switching Inhomogeneous Lévy-Based Stock Price Dynamics and Application to Illiquidity Modeling

We would like to mention that for some specific examples in this section, we state all the proofs related to regularity of the propagator, compact containment criterion, etc. It is worth noting that we focus below on the Banach space $Y = C_0(\mathbb{R}^d)$, but that in the article [6], other L^p-type Banach spaces denoted by $\bar{L}_\rho^p(\nu)$ are considered (in addition to $C_0(\mathbb{R}^d)$), for which a suitable weighted sup-norm is introduced (see their Lemma 2.6).

We will first give a brief overview of the proposed model and the intuition behind it, and then we will go into more formal details, presenting precisely all the various variables of interest, checking assumptions we have made up to this point, etc. We will consider a regime-switching inhomogeneous Lévy-based stock price model, very similar in the spirit to the recent article [4]. In short, an inhomogeneous Lévy process differs from a classical Lévy process in the sense that it has time-dependent (and absolutely continuous) characteristics. We let $\{L^x\}_{x \in \mathbb{J}}$ a collection of such $\mathbb{R}^d$−valued inhomogeneous Lévy processes with characteristics $(b_t^x, c_t^x, \nu_t^x)_{x \in \mathbb{J}}$, and we define:

$$\Gamma_x(s,t)f(z) := \mathbb{E}[f(L_t^x - L_s^x + z)], \quad z \in \mathbb{R}^d, \, x \in \mathbb{J}, \tag{6.1}$$

$$D^\epsilon(x,y)f(z) := f(z + \epsilon\alpha(x,y)), \quad z \in \mathbb{R}^d, \, x,y \in \mathbb{J}, \tag{6.2}$$

for some bounded function α. We will give below a financial interpretation of this function α, as well as reasons why we consider a regime-switching model. In this setting, f represents a contingent claim on a (d–dimensional) risky asset S having regime-switching inhomogeneous Lévy dynamics driven by the processes $\{L^x\}_{x\in\mathbb{J}}$: on each random time interval $[T_k^{\epsilon,s}(s), T_{k+1}^{\epsilon,s}(s))$, the risky asset is driven by the process $L^{x_k(s)}$. Indeed, we have the following representation, for $\omega' \in \Omega$ (to make clear that the expectation below is taken w.r.t. ω embedded in the process L and not ω'):

$$V_\epsilon(s,t)(\omega')f(z) =$$

$$\mathbb{E}\left[f\left(z + \sum_{k=1}^{N_s\left(t^{\frac{1}{\epsilon},s}\right)(\omega')+1} \Delta L_k + \sum_{k=1}^{N_s\left(t^{\frac{1}{\epsilon},s}\right)(\omega')} \epsilon\alpha(x_{k-1}(s)(\omega'), x_k(s)(\omega'))\right)\right],$$

(6.3)

where we have denoted for clarity:

$$\Delta L_k = \Delta L_k(\epsilon,\omega') := L_{T_k^{\epsilon,s}(s)(\omega')\wedge t}^{x_{k-1}(s)(\omega')} - L_{T_{k-1}^{\epsilon,s}(s)(\omega')}^{x_{k-1}(s)(\omega')}.$$

(6.4)

The random evolution $V_\epsilon(s,t)f$ represents in this case the present value of the contingent claim f of maturity t on the risky asset S, *conditionally on the regime switching process* $(x_n, T_n)_{n\geq 0}$: indeed, remember that $V_\epsilon(s,t)f$ is random, and that its randomness (only) comes from the Markov renewal process. Our main results from Theorems 5.3.19 and 5.4.7 allow us to approximate the impact of the regime-switching on the present value $V_\epsilon(s,t)f$ of the contingent claim. Indeed, we get the following normal approximation, for small ϵ:

$$V_\epsilon(s,t)f \approx \underbrace{\widehat{\Gamma}(s,t)f}_{\substack{1^{st}\text{ order} \\ \text{regime-switching approx.}}} + \underbrace{\sqrt{\epsilon}I_\sigma(s,t)f}_{\substack{\text{noise due to} \\ \text{regime-switching.}}}$$

(6.5)

The above approximation allows us to quantify the risk inherent to regime-switchings occurring at a high frequency governed by ϵ. The parameter ϵ reflects the frequency of the regime-switchings and can therefore be calibrated to market data by the risk manager. For market practitioners, because of the computational cost, it is often convenient to have asymptotic formulas that allow them to approximate the present value of a given derivative, and by extent the value of their whole portfolio. In addition, the asymptotic normal form of the regime-switching cost allows the risk manager to derive approximate confidence intervals for his portfolio, as well as other quantities of interest such as reserve policies linked to a given model.

In the recent article [4], an application to illiquidity is presented: they interpret the state process $\{x_n\}$ as a *"proxy for market liquidity with states (or regimes) representing the level of liquidity activity."* In their context however, $x(t)$ is taken to be a continuous-time Markov chain, which is a specific case of semi-Markov process with exponentially distributed sojourn times $T_{n+1} - T_n$.

With creative choices of the Markov renewal kernel, one could choose alternative distributions for the "waiting times" between regimes, thus generalizing the setting of [4]. In [4], the equivalent of our quantities $\alpha(x, y)$ are the constants γ_{xy} which represent an impact on the stock price at each regime-switching time. As they write it: *"Typically, there is a drop of the stock price after a liquidity breakdown."* This justifies the use of the operators D.

In addition to the liquidity modeling of [4], another application of our framework could be to view the different states as a reflection of *uncertainty on market data* induced by, for example, a significant bid-ask spread on the observed market data. The different characteristics $(b_t^x, c_t^x, \nu_t^x)_{x \in J}$ would, in this context, reflect this uncertainty induced by the observables. The random times $\{T_n\}$ would then correspond to times at which the observed market data (e.g. call option prices, or interest rates) switch from one value to another. As these switching times are typically very small ($\sim ms$), an asymptotic expansion as $\epsilon \to 0$ would make perfect sense in order to get an approximation of the stock price at a scale $\gg ms$, e.g. $\sim min$.

Now, let us formalize the model a bit more precisely. We let $Y :=$ $C_0(\mathbb{R}^d, \mathbb{R})$, $Y_1 := C_0^m(\mathbb{R}^d, \mathbb{R})$ for some $m \geq 2$. Let $(\Omega, \mathcal{F}, \mathbb{P})$ a probability space and let's begin with $(L_t)_{t \in J}$ a $\mathbb{R}^d$−valued (time-inhomogeneous) Markov process on it with transition function $p_{s,t}(z, A)$ ($z \in \mathbb{R}^d$, $A \in Bor(\mathbb{R}^d)$) and starting point $z_0 \in \mathbb{R}^d$ (in the sense of [7], II.10.2). Formally, $\forall (s, u), (u, t) \in \Delta_J$, $\forall A \in Bor(\mathbb{R}^d)$, p satisfies:

1. $\forall z \in \mathbb{R}^d$, $p_{s,t}(z, \cdot)$ is a probability measure.

2. $z \to p_{s,t}(z, A)$ is $Bor(\mathbb{R}^d) - Bor([0, 1])$ measurable.

3. $p_{s,s}(z, A) = 1_A(z)$.

4. $\int_{\mathbb{R}^d} p_{s,u}(z, dy) p_{u,t}(y, A) = p_{s,t}(z, A)$ (Chapman-Kolmogorov).

Example 1. One example of such time-inhomogeneous Markov processes are additive processes (in the sense of [7], I.1.6). This class of processes differs from Lévy processes only on the fact that increments are not assumed to be stationary. Additive processes are *translation invariant* (or *spatially homogeneous*) and we have (by [7], II.10.4) $p_{s,t}(z, A) = \mathbb{P}(L_t - L_s \in A - z)$. We will focus on additive processes in the following as they are rich enough to cover many financial applications. We define for $(s, t) \in \Delta_J$ and $f \in B_{\mathbb{R}}^b(\mathbb{R}^d)$:

$$\Gamma(s, t) f(z) := \mathbb{E}[f(L_t - L_s + z)] = \int_{\mathbb{R}^d} p_{s,t}(z, dy) f(y). \qquad (6.6)$$

It can be proved (see Chapter 6) that Γ is a regular $\mathcal{B}(Y)$−contraction propagator and $\forall n \in \mathbb{N}$ we have $\Gamma(s, t) \in \mathcal{B}(C_0^n(\mathbb{R}^d, \mathbb{R}))$ and $\|\Gamma(s, t)\|_{\mathcal{B}(C_0^n(\mathbb{R}^d, \mathbb{R}))} \leq 1$. The Lévy-Khintchine representation of such a process $(L_t)_{t \in J}$ (see [2], 14.1) ensures that there exists unique $(B_t)_{t \in J} \subseteq \mathbb{R}^d$, $(C_t)_{t \in J}$ a family of $d \times d$ symmetric nonnegative-definite matrices and $(\bar{\nu}_t)_{t \in J}$

a family of measures on $\mathbb{R}^d$ such that:

$$\mathbb{E}[e^{i\langle u, L_t\rangle}] = e^{\psi(u,t)}, \text{ with:} \tag{6.7}$$

$$\psi(u,t) := i\langle u, B_t\rangle - \frac{1}{2}\langle u, C_t u\rangle + \int_{\mathbb{R}^d}(e^{i\langle u,y\rangle} - 1 - i\langle u, y\rangle 1_{|y|\leq 1})\bar{\nu}_t(dy). \tag{6.8}$$

$(B_t, C_t, \bar{\nu}_t)_{t\in J}$ are called the *spot characteristics* of L. They satisfy the following regularity conditions:

- $\forall t \in J$, $\bar{\nu}_t\{0\} = 0$ and $\int_{\mathbb{R}^d}(|y|^2 \wedge 1)\bar{\nu}_t(dy) < \infty$.

- $(B_0, C_0, \bar{\nu}_0) = (0, 0, 0)$ and $\forall(s,t) \in \Delta_J$: $C_t - C_s$ is symmetric nonnegative-definite and $\bar{\nu}_s(A) \leq \bar{\nu}_t(A) \,\forall A \in Bor(\mathbb{R}^d)$.

- $\forall t \in J$ and as $s \to t$, $B_s \to B_t$, $\langle u, C_s u\rangle \to \langle u, C_t u\rangle \,\forall u \in \mathbb{R}^d$ and $\bar{\nu}_s(A) \to \bar{\nu}_t(A) \,\forall A \in Bor(\mathbb{R}^d)$ such that $A \subseteq \{z \in \mathbb{R}^d : |z| > \epsilon\}$ for some $\epsilon > 0$.

If $\forall t \in J$, $\int_{\mathbb{R}^d}|y|1_{|y|\leq 1}\bar{\nu}_t(dy) < \infty$, we can replace B_t by B_t^0 in the Lévy-Khintchine representation of L, where $B_t^0 := B_t - \int_{\mathbb{R}^d}y1_{|y|\leq 1}\bar{\nu}_t(dy)$. We denote by $(B_t^0, C_t, \bar{\nu}_t)_{t\in J}^0$ this other version of the spot characteristics of L.

Let $J = [0, T_\infty]$. We can consider a specific case of additive processes, called processes with independent increments and absolutely continuous characteristics (PIIAC), or inhomogeneous Lévy processes in [5], for which there exists $(b_t)_{t\in J} \subseteq \mathbb{R}^d$, $(c_t)_{t\in J}$ a family of $d \times d$ symmetric nonnegative-definite matrices and $(\nu_t)_{t\in J}$ a family of measures on $\mathbb{R}^d$ satisfying:

$$\int_0^{T_\infty}\left(|b_s| + ||c_s|| + \int_{\mathbb{R}^d}(|y|^2 \wedge 1)\nu_s(dy)\right)ds < \infty \tag{6.9}$$

$$B_t = \int_0^t b_s ds \tag{6.10}$$

$$C_t = \int_0^t c_s ds \tag{6.11}$$

$$\bar{\nu}_t(A) = \int_0^t \nu_s(A)ds \quad \forall A \in Bor(\mathbb{R}^d), \tag{6.12}$$

where $||c_t||$ denotes any norm on the space of $d \times d$ matrices. $(b_t, c_t, \nu_t)_{t\in J}$ are called the *local characteristics* of L. We can notice by [5] that PIIAC are semi-martingales, which is not the case of all additive processes. By [5], we have the following representation for L:

$$L_t = \int_0^t b_s ds + \int_0^t \sqrt{c_s}dW_s + \int_0^t\int_{\mathbb{R}^d}y1_{|y|\leq 1}(N - \bar{\nu})(dsdy) + \sum_{s\leq t}\Delta L_s 1_{|\Delta L_s|>1}, \tag{6.13}$$

where for $A \in Bor(\mathbb{R}^d)$: $\bar{\nu}([0,t] \times A) := \bar{\nu}_t(A)$ and N the Poisson measure of L ($N - \bar{\nu}$ is then called the compensated Poisson measure of L). $(W_t)_{t \in J}$ is a d-dimensional Brownian motion on $\mathbb{R}^d$, independent from the jump part. $\sqrt{c_t}$ here stands for the unique symmetric nonnegative-definite square root of c_t. Sometimes it is convenient to write a Cholesky decomposition $c_t = h_t h_t^T$ and replace $\sqrt{c_t}$ by h_t in the previous representation (in this case, the Brownian motion W would have to be replaced by another Brownian motion $\widetilde{W}$, the point being that both $\int_0^t \sqrt{c_s}dW_s$ and $\int_0^t h_s d\widetilde{W}_s$ are processes with independent Gaussian increments with mean 0 and variance $\int_0^t c_s ds$). It can be shown - see [6] - that the infinitesimal generator of the propagator Γ is given by:

$$
\begin{aligned}
A(t)f(z) = &\sum_{j=1}^{d} b_t(j) \frac{\partial f}{\partial z_j}(z) + \frac{1}{2} \sum_{j,k=1}^{d} c_t(j,k) \frac{\partial^2 f}{\partial z_j \partial z_k}(z) \\
&+ \int_{\mathbb{R}^d} \left(f(z+y) - f(z) - \sum_{j=1}^{d} \frac{\partial f}{\partial z_j}(z) y(j) 1_{|y| \le 1} \right) \nu_t(dy),
\end{aligned}
\tag{6.14}
$$

and that $C_0^2(\mathbb{R}^d, \mathbb{R}) \subseteq \mathcal{D}(A(t)) = \mathcal{D}(A)$. And if $b_t^0 := b_t - \int_{\mathbb{R}^d} y 1_{|y| \le 1} \nu_t(dy)$ is well-defined:

$$
\begin{aligned}
A(t)f(z) = &\sum_{j=1}^{d} b_t^0(j) \frac{\partial f}{\partial z_j}(z) + \frac{1}{2} \sum_{j,k=1}^{d} c_t(j,k) \frac{\partial^2 f}{\partial z_j \partial z_k}(z) \\
&+ \int_{\mathbb{R}^d} (f(z+y) - f(z)) \nu_t(dy).
\end{aligned}
\tag{6.15}
$$

As a first example, we can consider the *inhomogeneous Poisson process*, for which $d = 1$, local characteristics $(b_t^0, c_t, \nu_t)^0 = (0, 0, \lambda(t)\delta_1)$, where the intensity function $\lambda \in B_{\mathbb{R}_+^*}(J)$ s.t. $\int_0^{T_\infty} \lambda(s)ds < \infty$. We have, letting $\Lambda_{st} := \int_s^t \lambda(u)du$:

$$
p_{s,t}(z, z+n) = \frac{1}{n!} \Lambda_{st}^n e^{-\Lambda_{st}}, \text{ for } z, n \in \mathbb{R} \times \mathbb{N} \text{ and } p_{s,t}(z, z+n) = 0 \text{ if } n \notin \mathbb{N}
\tag{6.16}
$$

$$
A(t)f(z) = \lambda(t)(f(z+1) - f(z)).
\tag{6.17}
$$

Example 2. As a second example, we can consider a *risk process* L (used in insurance for example):

$$
dL_t = r_t dt - dZ_t
\tag{6.18}
$$

$$
Z_t = \sum_{i=1}^{N_t} Z_i,
\tag{6.19}
$$

where $(N_t)_{t \in J}$ is an inhomogeneous Poisson process, $(r_t)_{t \in J} \subseteq \mathbb{R}^{d+}$ is the premium intensity function and the Z_i's are iid random variables, independent from N with common law ν that represent the random payments that an insurance company has to pay. In this case, the local characteristics $(b_t^0, c_t, \nu_t)^0 = (r_t, 0, \lambda(t)\nu)$. We get:

$$A(t)f(z) = \sum_{j=1}^{d} r_t(j) \frac{\partial f}{\partial z_j}(z) + \lambda(t) \int_{\mathbb{R}^d} (f(z-y) - f(z))\nu(dy). \qquad (6.20)$$

We can also consider a *Brownian motion with time-dependent variance-covariance matrix* c_t *and drift* r_t (local characteristics $(b_t^0, c_t, \nu_t)^0 = (r_t, c_t, 0)$):

$$dL_t = r_t dt + \sqrt{c_t} dW_t \qquad (6.21)$$

$$A(t)f(z) = \sum_{j=1}^{d} r_t(j) \frac{\partial f}{\partial z_j}(z) + \frac{1}{2} \sum_{j,k=1}^{d} c_t(j,k) \frac{\partial^2 f}{\partial z_j \partial z_k}(z). \qquad (6.22)$$

To define the corresponding random evolution, we let L be an inhomogeneous Lévy process taking value in $\mathbb{R}^{d \times |J|}$, independent of the semi-Markov process $(x(t))_{t \in \mathbb{R}_+}$. In this case, the $\mathbb{R}$−valued Lévy processes $\{L^{j,x} : j = 1..d, x \in J\}$ are not necessarily independent but they have independent increments over non-overlapping time-intervals, which will be used in the proofs. Denote $\{L^x : x \in J\}$ the corresponding $\mathbb{R}^d$−valued inhomogeneous Lévy processes, with local characteristics (b_t^x, c_t^x, ν_t^x). Define for $z \in \mathbb{R}^d$, $(s,t) \in \Delta_J$, $x \in J$:

$$\Gamma_x(s,t)f(z) := \mathbb{E}[f(L_t^x - L_s^x + z)] = \int_{\mathbb{R}} p_{s,t}^x(z,dy)f(y). \qquad (6.23)$$

We define the jump operators, for $z \in \mathbb{R}^d$ by:

$$D^\epsilon(x,y)f(z) := f(z + \epsilon\alpha(x,y)), \qquad (6.24)$$

where $\alpha \in B_{\mathbb{R}^d}^b(J^2)$, so that $D_1^\epsilon(x,y)f(z) = \sum_{j=1}^{d} \alpha_j(x,y) \frac{\partial f}{\partial z_j}(z + \epsilon\alpha(x,y))$ and $Y_1 \subseteq \mathcal{D}(D_1)$. Assume that the $\mathbb{R}$−valued inhomogeneous Lévy processes $\{L^{j,x} : j = 1..d, x \in J\}$ admit a second moment, namely they satisfy:

$$\int_0^t \int_{|y| \geq 1} y^2 \nu_s^{j,x}(dy)ds < \infty, \quad \forall t \in J, \ j = 1..d, \ x \in J, \qquad (6.25)$$

then it can be proved (see below) that the compact containment criterion is satisfied: $\{V_\epsilon\} \in Y_1$-CCC (see Chapters 1 and 3). It is clear that $\hat{A}$ is the generator of the (inhomogeneous) Lévy propagator with "weighted" local characteristics $(\widehat{b}_t, \widehat{c}_t, \widehat{\nu}_t)$ given by:

$$\widehat{b}_t = \frac{1}{\Pi m} \sum_{x \in J} (m_1(x) b_t^x + P\alpha(x, \bullet)(x)) \, \pi(x) \tag{6.26}$$

$$\widehat{c}_t = \frac{1}{\Pi m} \sum_{x \in J} m_1(x) \pi(x) c_t^x \tag{6.27}$$

$$\widehat{\nu}_t(dy) = \frac{1}{\Pi m} \sum_{x \in J} m_1(x) \pi(x) \nu_t^x(dy). \tag{6.28}$$

In particular we notice that $\widehat{\nu}_t$ is indeed a Lévy measure on $\mathbb{R}^d$.

6.1.1   Proofs for Section 6.1

Proof that Γ is a propagator and that $\forall n \in \mathbb{N}$ we have $\Gamma(s,t) \in \mathcal{B}(C_0^n(\mathbb{R}^d, \mathbb{R}))$ and $\|\Gamma(s,t)\|_{\mathcal{B}(C_0^n(\mathbb{R}^d, \mathbb{R}))} \leq 1$.
Γ satisfies the propagator equation because of the Chapman-Kolmogorov equation. Now let's show that $\forall n \in \mathbb{N}$ we have $\Gamma(s,t) \in \mathcal{B}(C_0^n(\mathbb{R}^d, \mathbb{R}))$ and $\|\Gamma(s,t)\|_{\mathcal{B}(C_0^n(\mathbb{R}^d,\mathbb{R}))} \leq 1$. Let $Y := C_0(\mathbb{R}^d, \mathbb{R})$. Let's first start with $C_0(\mathbb{R}^d, \mathbb{R})$. Let $f \in C_0(\mathbb{R}^d, \mathbb{R})$. By ([7]), we get the representation $\Gamma(s,t)f(z) = \int_{\mathbb{R}^d} \mu_{s,t}(dy)f(z+y)$, where $\mu_{s,t}$ is the distribution of $L_t - L_s$, i.e. $\mu_{s,t}(A) = \mathbb{P}(L_t - L_s \in A)$ for $A \in Bor(\mathbb{R}^d)$. Let $z \in \mathbb{R}^d$ and take any sequence $(z_n)_{n \in \mathbb{N}} \subseteq \mathbb{R}^d \colon z_n \to z$ and denote $g_n := y \to f(z_n + y) \in C_0(\mathbb{R}^d, \mathbb{R})$ and $g := y \to f(z+y) \in C_0(\mathbb{R}^d, \mathbb{R})$. By continuity of f, $g_n \to g$ pointwise. Further, $\|g_n\| = \|f\| \in L_{\mathbb{R}}^1(\mathbb{R}^d, Bor(\mathbb{R}^d), \mu_{s,t})$. Therefore by Lebesgue dominated convergence theorem, we get $\lim_{n \to \infty} \Gamma(s,t)f(z_n) = \int_{\mathbb{R}^d} \lim_{n \to \infty} \mu_{s,t}(dy)f(z_n + y) = \int_{\mathbb{R}^d} \mu_{s,t}(dy)f(z+y) = \Gamma(s,t)f(z)$. Therefore $\Gamma(s,t)f \in C(\mathbb{R}^d, \mathbb{R})$. By the same argument but now taking any sequence $(z_n)_{n \in \mathbb{N}} \subseteq \mathbb{R}^d \colon |z_n| \to \infty$, we get $\lim_{|z| \to \infty} \Gamma(s,t)f(z) = 0$ and therefore $\Gamma(s,t)f \in C_0(\mathbb{R}^d, \mathbb{R})$. Further, we get:

$$|\Gamma(s,t)f(z)| = \left| \int_{\mathbb{R}^d} \mu_{s,t}(dy)f(z+y) \right| \leq \int_{\mathbb{R}^d} \mu_{s,t}(dy)|f(z+y)|$$

$$\leq \int_{\mathbb{R}^d} \mu_{s,t}(dy)\|f\| = \|f\|. \tag{6.29}$$

and therefore $\|\Gamma(s,t)\|_{\mathcal{B}(C_0(\mathbb{R}^d,\mathbb{R}))} \leq 1$. Let $z \in \mathbb{R}^d$, $f \in C_0^q(\mathbb{R}^d, \mathbb{R})$ $(q \geq 1)$. Take any sequence $(h_n)_{n \in \mathbb{N}} \subseteq \mathbb{R}^* \colon h_n \to 0$, $j \in [\![1,d]\!]$ and $f_{n,j}(z) := f(z_1, ..., z_j + h_n, ..., z_d)$. Then:

$$\frac{1}{h_n}((\Gamma(s,t)f)_{n,j}(z) - \Gamma(s,t)f(z)) = \int_{\mathbb{R}^d} \mu_{s,t}(dy)\frac{1}{h_n}(f_{n,j}(z+y) - f(z+y)). \tag{6.30}$$

Let $g_n := y \to \frac{1}{h_n}(f_{n,j}(z+y) - f(z+y)) \in C_0(\mathbb{R}^d, \mathbb{R})$ and $g := y \to \frac{\partial f}{\partial z_j}(z+y) \in C_0(\mathbb{R}^d, \mathbb{R})$. We have $g_n \to g$ pointwise since $f \in C_0^1(\mathbb{R}^d, \mathbb{R})$. By the Mean

Value theorem, $\exists \gamma_j(n,z,y) \in [-|h_n|,|h_n|] : g_n(y) = \frac{\partial f}{\partial z_j}(z_1 + y_1, ..., z_j + y_j +$

$\gamma_j(n,z,y), ..., z_d + y_d)$, and therefore $|g_n(y)| \leq \left\|\frac{\partial f}{\partial z_j}\right\| \in L^1_{\mathbb{R}}(\mathbb{R}^d, Bor(\mathbb{R}^d), \mu_{s,t})$.

Therefore by Lebesgue dominated convergence theorem, we get:

$$\frac{\partial \Gamma(s,t)f}{\partial z_j}(z) = \int_{\mathbb{R}^d} \mu_{s,t}(dy) \frac{\partial f}{\partial z_j}(z+y). \tag{6.31}$$

Using the same argument as for $C_0(\mathbb{R}^d, \mathbb{R})$, we get that $\Gamma(s,t)f \in C_0^1(\mathbb{R}^d, \mathbb{R})$ since $\frac{\partial f}{\partial z_j} \in C_0^1(\mathbb{R}^d, \mathbb{R}) \ \forall j \in [[1,d]]$. Repeating this argument by computing successively every partial derivative up to order q by the relationship $\partial^\alpha \Gamma(s,t)f(z) = \int_{\mathbb{R}^d} \mu_{s,t}(dy) \partial^\alpha f(z+y) \ \forall \alpha \in \mathbb{N}_{d,q}$, we get $\Gamma(s,t)f \in C_0^q(\mathbb{R}^d, \mathbb{R})$. Further, the same way we got $\|\Gamma(s,t)f\| \leq \|f\|$ for $f \in C_0(\mathbb{R}^d, \mathbb{R})$, we get for $f \in C_0^q(\mathbb{R}^d, \mathbb{R})$: $\|\partial^\alpha \Gamma(s,t)f\| \leq \|\partial^\alpha f\| \ \forall \alpha \in \mathbb{N}_{d,q}$. Therefore $\max_{\alpha \in \mathbb{N}_{d,q}} \|\partial^\alpha \Gamma(s,t)f\| \leq \max_{\alpha \in \mathbb{N}_{d,q}} \|\partial^\alpha f\| \Rightarrow \|\Gamma(s,t)f\|_{C_0^q(\mathbb{R}^d, \mathbb{R})} \leq \|f\|_{C_0^q(\mathbb{R}^d, \mathbb{R})} \Rightarrow \|\Gamma(s,t)\|_{\mathcal{B}(C_0^q(\mathbb{R}^d, \mathbb{R}))} \leq 1$.

Proof that Γ is $(Y_1, \|\cdot\|_{Y_1})$–strongly s-continuous, Y–strongly t-continuous.

Let $(s,t) \in \Delta_J$, $f \in Y_1$ and $h \in [-s, t-s]$:

$$\|\Gamma(s+h,t)f - \Gamma(s,t)f\|_{Y_1} = \max_{\alpha \in \mathbb{N}_{d,m}} \|\partial^\alpha \Gamma(s+h,t)f - \partial^\alpha \Gamma(s,t)f\| \tag{6.32}$$

$$= \max_{\alpha \in \mathbb{N}_{d,m}} \|\Gamma(s+h,t)\partial^\alpha f - \Gamma(s,t)\partial^\alpha f\|. \tag{6.33}$$

Let $\alpha \in \mathbb{N}_{d,m}$ and $\{h_n\}_{n \in \mathbb{N}} \subseteq [-s, t-s]$ any sequence such that $h_n \to 0$. Let $S_n := L_t - L_{s+h_n}$ and $S := L_t - L_s$. We have $S_n \overset{P}{\to} S$ by stochastic continuity of additive processes. By the Skorokhod's representation theorem, there exists a probability space $(\Omega', \mathcal{F}', \mathbb{P}')$ and random variables $\{S'_n\}_{n \in \mathbb{N}}$, S' on it such that $S_n \overset{D}{=} S'_n$, $S \overset{D}{=} S'$ and $S'_n \overset{a.e.}{\to} S'$. Let $z \in \mathbb{R}^d$, we therefore get:

$$|\Gamma(s+h_n,t)\partial^\alpha f(z) - \Gamma(s,t)\partial^\alpha f(z)| = |\mathbb{E}'[\partial^\alpha f(z+S'_n) - \partial^\alpha f(z+S')]| \tag{6.34}$$

$$\leq \mathbb{E}'|\partial^\alpha f(z+S'_n) - \partial^\alpha f(z+S')| \tag{6.35}$$

$$\Rightarrow \|\Gamma(s+h_n,t)\partial^\alpha f - \Gamma(s,t)\partial^\alpha f\| \leq \sup_{z \in \mathbb{R}^d} \mathbb{E}'|\partial^\alpha f(z+S'_n) - \partial^\alpha f(z+S')| \tag{6.36}$$

$$\leq \mathbb{E}'[\sup_{z \in \mathbb{R}^d} |\partial^\alpha f(z+S'_n) - \partial^\alpha f(z+S')|]. \tag{6.37}$$

Further, since $\partial^\alpha f \in Y$, $\partial^\alpha f$ is uniformly continuous on $\mathbb{R}^d$ and $\forall \epsilon > 0$, $\exists \delta > 0: |z - y| < \delta \Rightarrow |\partial^\alpha f(z) - \partial^\alpha f(y)| < \epsilon$. And because $S'_n \overset{a.e.}{\to} S'$, for a.e. $\omega' \in \Omega'$, $\exists N(\omega') \in \mathbb{N} : n \geq N(\omega') \Rightarrow |S'_n(\omega') - S'(\omega')| = |z + S'_n(\omega') - S'(\omega') -$

$z| < \delta \quad \forall z \in \mathbb{R}^d \Rightarrow |\partial^\alpha f(z + S'_n(\omega')) - \partial^\alpha f(z + S'(\omega'))| < \epsilon \quad \forall z \in \mathbb{R}^d$.
Therefore we have that $g_n := \sup_{z \in \mathbb{R}^d} |\partial^\alpha f(z + S'_n) - \partial^\alpha f(z + S')| \overset{a.e.}{\rightarrow} 0$.
Further $|g_n| \leq 2||\partial^\alpha f|| \in L^1_{\mathbb{R}}(\Omega', \mathcal{F}', \mathbb{P}')$. By Lebesgue dominated convergence theorem we get:

$$\lim_{n \to \infty} \mathbb{E}'[\sup_{z \in \mathbb{R}^d} |\partial^\alpha f(z + S'_n) - \partial^\alpha f(z + S')|] = 0. \tag{6.38}$$

We can notice that the proof strongly relies on the uniform continuity of f, and therefore on the topological properties of the space $C_0(\mathbb{R}^d, \mathbb{R})$ (which $C_b(\mathbb{R}^d, \mathbb{R})$ doesn't have). We prove that Γ is Y-strongly t-continuous exactly the same way, but now considering $Y_n := L_{t+h_n} - L_s$ and any sequence $\{h_n\}_{n \in \mathbb{N}} \subseteq K$ such that $h_n \to 0$, where $K := [s - t, T_\infty - t]$ if $J = [0, T_\infty]$ and $K := [s - t, 1]$ if $J = \mathbb{R}_+$.

Proof that Γ is regular.
By Taylor's theorem we get $\forall f \in Y_1, t \in J, z, y \in \mathbb{R}^d$:

$$\left| f(z + y) - f(z) - \sum_{j=1}^d \frac{\partial f}{\partial z_j}(z)y(j) \right| \leq \frac{1}{2}|y|^2 \sum_{j,k=1}^d \left\| \frac{\partial^2 f}{\partial z_j \partial z_k} \right\| \leq \frac{d^2}{2}|y|^2||f||_{Y_1} \tag{6.39}$$

$$\Rightarrow ||A_\Gamma(t)f|| \leq ||f||_{Y_1}$$

$$\left[\sum_{j=1}^d |b_t(j)| + \frac{1}{2} \sum_{j,k=1}^d |c_t(j, k)| + 2\nu_t\{|y| > 1\} + \frac{d^2}{2} \int_{|y| \leq 1} |y|^2 \nu_t(dy) \right] \tag{6.40}$$

$$\Rightarrow ||A_\Gamma(t)||_{\mathcal{B}(Y_1,Y)} \leq$$

$$\sum_{j=1}^d |b_t(j)| + \frac{1}{2} \sum_{j,k=1}^d |c_t(j, k)| + 2\nu_t\{|y| > 1\} + \frac{d^2}{2} \int_{|y| \leq 1} |y|^2 \nu_t(dy), \tag{6.41}$$

observing that $\nu_t\{|y| > 1\} + \int_{|y| \leq 1} |y|^2 \nu_t(dy) < \infty$ by assumption. Therefore by integrability assumption on the local characteristics and Theorem 5.1.8, we get the propagator evolution equations and therefore the regularity of Γ.

Proof that $V_\epsilon \in Y_1$-CCC.
We showed V_ϵ is a $\mathcal{B}(Y_1)$-contraction (see Chapter 3), so it remains to show the uniform convergence to 0 at infinity. We have the following representation, for $\omega' \in \Omega$ (to make clear that the expectation is w.r.t. ω and not ω'):

$$V_\epsilon(s, t)(\omega')f(z) =$$

$$\mathbb{E}\left[f\left(z + \sum_{k=1}^{N_s\left(t^{\frac{1}{\epsilon},s}\right)(\omega')+1} \Delta L_k + \sum_{k=1}^{N_s\left(t^{\frac{1}{\epsilon},s}\right)(\omega')} \epsilon\alpha(x_{k-1}(s)(\omega'), x_k(s)(\omega')) \right) \right],$$

where we have denoted for clarity $\Delta L_k = \Delta L_k(\epsilon, \omega') := L^{x_{k-1}(s)(\omega')}_{T_k^{\epsilon,s}(s)(\omega')\wedge t} - L^{x_{k-1}(s)(\omega')}_{T_{k-1}^{\epsilon,s}(s)(\omega')}$. In the following we drop the index ω'. Denote the d components of the inhomogeneous Lévy process L^x by $L^{(j),x}$ $(j = 1..d)$. The multivariate version of Chebyshev's inequality yields for any d-vector of L^2 integrable random variables X_j:

$$\mathbb{P}\left[\sum_{j=1}^d (X_j - \mathbb{E}(X_j))^2 \geq \delta^2\right] \leq \frac{1}{\delta^2}\sum_{j=1}^d var(X_j). \qquad (6.42)$$

We apply this inequality to the d-vector $\displaystyle\sum_{k=1}^{N_s\left(t^{\frac{1}{\epsilon},s}\right)+1} \Delta L_k$ having components

$$X_j := \sum_{k=1}^{N_s\left(t^{\frac{1}{\epsilon},s}\right)+1} \Delta L_k^{(j)},$$ where $\Delta L_k^{(j)}$ is the jth component of ΔL_k. If we denote by

$(b_t^{(j),x}, c_t^{(j),x}, \nu_t^{(j),x})$ the local characteristics of $L^{(j),x}$, we have ([2], Proposition 3.13):

$$\mathbb{E}(X_j) = \sum_{k=1}^{N_s\left(t^{\frac{1}{\epsilon},s}\right)+1} \mathbb{E}(\Delta L_k^{(j)}) = \sum_{k=1}^{N_s\left(t^{\frac{1}{\epsilon},s}\right)+1} \int_{T_{k-1}^{\epsilon,s}(s)}^{T_k^{\epsilon,s}(s)\wedge t} \left[b_u^{(j),x_{k-1}(s)} + \int_{|y|\geq 1} y\nu_u^{(j),x_{k-1}(s)}(dy)\right] du, \qquad (6.43)$$

and so:

$$|\mathbb{E}(X_j)| \leq \int_s^t \left[b_u^* + \int_{|y|\geq 1} |y|\nu_u^*(dy)\right] du, \qquad (6.44)$$

where $b_u^* := \sum_{j=1..d, x\in J} |b_u^{(j),x}|$ and $\nu_u^* := \sum_{j=1..d, x\in J} \nu_u^{(j),x}$. Similarly, we have:

$$var(\Delta L_k^{(j)}) = \int_{T_{k-1}^{\epsilon,s}(s)}^{T_k^{\epsilon,s}(s)\wedge t} \left[c_u^{(j),x_{k-1}(s)} + \int_{\mathbb{R}} y^2\nu_u^{(j),x_{k-1}(s)}(dy)\right] du \qquad (6.45)$$

$$\leq \int_{T_{k-1}^{\epsilon,s}(s)}^{T_k^{\epsilon,s}(s)\wedge t} \left[c_u^* + \int_{\mathbb{R}} y^2\nu_u^*(dy)\right] du, \qquad (6.46)$$

where $c_u^* := \sum_{j=1..d, x\in J} c_u^{(j),x}$. Now, since L is a $\mathbb{R}^{d\times|J|}$ valued inhomogeneous Lévy process, the $\mathbb{R}$ valued inhomogeneous Lévy processes $L^{(j),x}$ are not necessarily independent but they have independent increments over non overlapping time intervals. This yields:

$$
var(X_j) = var\left(\sum_{k=1}^{N_s\left(t^{\frac{1}{\epsilon},s}\right)+1} \Delta L_k^{(j)} \right) = \sum_{k=1}^{N_s\left(t^{\frac{1}{\epsilon},s}\right)+1} var(\Delta L_k^{(j)})
$$

$$
\leq \int_s^t \left[c_u^* + \int_{\mathbb{R}} y^2 \nu_u^*(dy) \right] du. \tag{6.47}
$$

Overall we get that for every $\delta > 0$, there exists $k_\delta = k_{\delta,s,t,f} > 0$ such that:

$$
\mathbb{P}\left[\left| \sum_{k=1}^{N_s\left(t^{\frac{1}{\epsilon},s}\right)(\omega')+1} \Delta L_k \right| > k_\delta \right] < \frac{\delta}{2\|f\|}. \tag{6.48}
$$

Define the set:

$$
B := \left\{ \left| \sum_{k=1}^{N_s\left(t^{\frac{1}{\epsilon},s}\right)(\omega')+1} \Delta L_k \right| > k_\delta \right\}. \tag{6.49}
$$

We have:

$$
\left| \mathbb{E}\left[1_B f(z + \cdot) \right] \right| \leq \|f\| \mathbb{P}(B) < \frac{\delta}{2}. \tag{6.50}
$$

On the other hand, there exists $c_\delta > 0$: $|z| > c_\delta \Rightarrow |f(z)| < \frac{\delta}{2}$. Define, as in the traffic case, $A_\epsilon := \{\epsilon N_s\left(t^{\frac{1}{\epsilon},s}\right) \leq n_{00}\}$ so that $\mathbb{P}(A_\epsilon) \geq 1 - \Delta$. We have for $\omega' \in A_\epsilon$ and $|z| > C_\delta := c_\delta + \|\alpha\| n_{00} + k_\delta$:

$$
\left| \mathbb{E}\left[1_{B^c} f(z + \cdot) \right] \right| < \frac{\delta}{2} \mathbb{P}(B^c) \leq \frac{\delta}{2}, \tag{6.51}
$$

so that overall, for $\omega' \in A_\epsilon$ and $|z| > C_\delta$ (uniform in ω', ϵ) we have:

$$
|V_\epsilon(s,t)(\omega') f(z)| < \delta, \tag{6.52}
$$

which completes the proof.

6.2 Regime-Switching Lévy Driven Diffusion-Based Price Dynamics

We consider the same financial setting as in Section 6.1, with the exception that the dynamics of the stock price will be driven by a Lévy driven diffusion process, and not a time-inhomogeneous Lévy process. As in [6] (Section 4.1), consider for $x \in \mathbb{J}$ the function $\Phi^x \in C_b^{1,1}(\mathbb{R}^d \times \mathbb{R}_+, \mathbb{R}^d \times \mathbb{R}^d)$ and $\{L^x\}_{x \in \mathbb{J}}$ a collection of $\mathbb{R}^d$-valued classical Lévy processes with local characteristics (b^x, c^x, ν^x). Let Z^x a solution of the SDE [1]:

$$dZ_t^x = \Phi^x(Z_{t-}^x, t)dL_t^x, \tag{6.53}$$

and define as in Section 6.1:

$$\Gamma_x(s,t)f(z) := \mathbb{E}[f(Z_t^x)|Z_s^x = z]. \tag{6.54}$$

This general setting includes many popular models in finance, for example the Heston model. Time-inhomogeneity of Φ also enables one to consider the time-dependent Heston model, which is characterized by time-dependent parameters. The Delayed Heston model considered in [8] can be seen as a specific example of the class of time-dependent Heston models, and therefore fits into this framework. Indeed, let S_t, V_t be the price and variance process, respectively. The time-dependent Heston dynamics read:

$$dS_t = (r_t - q_t)S_t dt + S_t\sqrt{V_t}\sqrt{1 - \rho_t^2}dW_t^{(1)} + S_t\sqrt{V_t}\rho_t dW_t^{(2)}, \tag{6.55}$$

$$dV_t = \gamma_t(\theta_t^2 - V_t)dt + \delta_t\sqrt{V_t}dW_t^{(2)}, \tag{6.56}$$

where γ_t, ρ_t, r_t, q_t, δ_t, θ_t are deterministic processes and $W^{(1)}$, $W^{(2)}$ independent Brownian motions. Then, letting the $\mathbb{R}^3$-valued Lévy process $L_t := (t, W_t^{(1)}, W_t^{(2)})^T$ and $Z_t := (S_t, V_t, t)^T$, we have $dZ_t = \Phi(Z_{t-}, t)dL_t$, where for $z = (z_1, z_2, z_3)^T \in \mathbb{R}^3$:

$$\Phi(z,t) := \begin{pmatrix} (r_t - q_t)z_1 & z_1\sqrt{z_2}\sqrt{1 - \rho_t^2} & z_1\sqrt{z_2}\rho_t \\ \gamma_t(\theta_t^2 - z_2) & 0 & \delta_t\sqrt{z_2} \\ 1 & 0 & 0 \end{pmatrix}$$

Then, we can consider regime-switching extensions of the latter time-dependent Heston dynamics by allowing the various coefficients γ_t, ρ_t, r_t, q_t, δ_t, θ_t (and therefore the matrix Φ) to depend on the regime $x \in \mathbb{J}$. It also has to be noted that the previously defined Φ is not in $C_b^{1,1}$, which is required in [6] to prove the strong continuity of Γ, as mentioned below. Nevertheless, in practice, it is always possible to consider bounded approximations of the latter Φ.

Let $Y := C_0(\mathbb{R}^d, \mathbb{R})$ and $Y_1 := C_0^m(\mathbb{R}^d, \mathbb{R})$ for some $m \geq 2$ as in Section 6.1. Lemma 4.6, the corresponding proof and Remark 4.1 of [6] give us strong

---

[1] actually, for any initial condition $Z_0^x = z \in \mathbb{R}^d$, the SDE admits a unique strong solution, cf. [6], Lemma 4.6.

continuity of Γ and for $f \in Y_1$:

$$A_x(t)f(z) = \sum_{j=1}^{d} [b^x \Phi^x(z,t)](j) \frac{\partial f}{\partial z_j}(z) + \frac{1}{2} \sum_{j,k=1}^{d} [c^x \Phi^x(z,t)](j,k) \frac{\partial^2 f}{\partial z_j \partial z_k}(z)$$

$$+ \int_{\mathbb{R}^d} \left(f(z + \Phi^x(z,t)y) - f(z) - \sum_{j=1}^{d} \frac{\partial f}{\partial z_j}(z)[\Phi^x(z,t)y](j)1_{|y|\leq 1} \right) \nu^x(dy).$$

$$(6.57)$$

In particular, taking $L_t^x = t$ we retrieve the so-called *traffic* random evolution of [11], [10]. If $\nu^x = 0$ for all x, $\widehat{A}$ is the generator of the Lévy driven diffusion propagator with "weighted" local characteristics $(\widehat{b}, \widehat{c}, 0)$ and driver $\widehat{\Phi}$ implicitly defined by the following equations:

$$\widehat{b\Phi}(z,t) = \frac{1}{\Pi m} \sum_{x \in \mathbb{J}} (m_1(x)[b^x \Phi^x(z,t)] + P\alpha(x, \bullet)(x)) \pi(x) \qquad (6.58)$$

$$\widehat{c\Phi}(z,t) = \frac{1}{\Pi m} \sum_{x \in \mathbb{J}} m_1(x)\pi(x)[c^x \Phi^x(z,t)]. \qquad (6.59)$$

If $\Phi^x = \Phi$ doesn't depend on x, then $\widehat{A}$ is the generator of the Lévy driven diffusion propagator with driver Φ and "weighted" local characteristics $(\widehat{b}, \widehat{c}, \widehat{\nu})$ defined implicitly by the following equations:

$$\widehat{b}\Phi(z,t) = \frac{1}{\Pi m} \sum_{x \in \mathbb{J}} (m_1(x)b^x \Phi(z,t) + P\alpha(x, \bullet)(x)) \pi(x) \qquad (6.60)$$

$$\widehat{c}\Phi(z,t) = \frac{1}{\Pi m} \sum_{x \in \mathbb{J}} m_1(x)\pi(x)c^x \Phi(z,t) \qquad (6.61)$$

$$\widehat{\nu}(dy) = \frac{1}{\Pi m} \sum_{x \in \mathbb{J}} m_1(x)\pi(x)\nu^x(dy). \qquad (6.62)$$

6.3 Multi-Asset Model of Price Impact from Distressed Selling: Diffusion Limit

Take, again, $Y := C_0(\mathbb{R}^d, \mathbb{R})$. The setting of the recent article [1] fits exactly into our framework. In this article, they consider a discrete-time stock price model for d stocks $(S^i)_{1 \leq i \leq d}$. It is assumed that a large fund V holds α_i units of each asset i. Denoting the ith stock price and fund value at time t_k by respectively $S_k^i := S_{t_k}^i$ and $V_k := V_{t_k}$, we have $V_k = \sum_{j=1}^d \alpha_j S_k^j$, and the following dynamics are assumed for the stock prices:

$$\ln \left(\frac{S_{k+1}^i}{S_k^i} \right) = \epsilon^2 m_i + \epsilon Z_{k+1}^i + \ln \left(1 + g_i(\bar{V}_{k+1}^*) - g_i(\bar{V}_k) \right), \qquad (6.63)$$

$$g_i := \frac{\alpha_i}{D_i} g, \qquad m_i := \mu_i - \frac{1}{2}\sigma_{ii}, \qquad (6.64)$$

$$\bar{V}_{k+1}^* = \frac{1}{V_0} \sum_{j=1}^d \alpha_j S_k^j e^{\epsilon^2 m_j + \epsilon Z_{k+1}^j}, \qquad \bar{V}_k = \frac{V_k}{V_0}, \qquad (6.65)$$

where $\epsilon^2 = t_{k+1} - t_k$ is the constant time-step, $\{Z_k\}_{k \geq 1} := \{Z_k^i : 1 \leq i \leq d\}_{k \geq 1}$ are iid $\mathbb{R}^d$-valued centered random variables with covariance matrix σ (i.e. $\mathbb{E}(Z_k^i)=0$, $\mathbb{E}(Z_k^i Z_k^j) = \sigma_{ij} \forall k$); $g : \mathbb{R} \to \mathbb{R}$ is increasing, concave, equal to 0 on $[\beta_0, \infty)$ and $g(\beta_{liq}) = -1$, where $0 < \beta_{liq} < \beta_0$; $D_i > 0$ represents the depth of the market in asset i, and μ_i are constants. It is also assumed that:

$$\|g\|_\infty < \frac{1}{2} \min_i \frac{D_i}{|\alpha_i|}, \qquad (6.66)$$

so that the stock prices remain positive. The idea is that when the fund value V reaches a certain "low" level $\beta_0 V_0$, market participants will start selling the stocks involved in that fund, inducing a correlation between the stocks which "adds" to their fundamental correlations: this is captured by the function g (which, if equal to 0, gives us the "standard" Black & Scholes setting).

The above setting is a particular case of our framework, where $\Gamma_x = I$ for all $x \in \mathbb{J}$, and the operators D have the following form:

$$D^\epsilon(x,y)f(z) := f \circ \exp \left(z + \epsilon^2 m + \epsilon y + \phi(h(z + \epsilon^2 m + \epsilon y) - h(z)) \right),$$
$$x, y \in \mathbb{J}, \qquad (6.67)$$

where $\circ$ denotes composition of functions, $z \in \mathbb{R}^d$, m is the vector with coordinates m_i, $h : \mathbb{R}^d \to \mathbb{R}^d$ has coordinate functions h_i defined by:

$$h_i(z) = g_i \left(\frac{1}{V_0} \sum_{j=1}^d \alpha_j e^{z_j} \right), \qquad (6.68)$$

$\phi : \mathbb{R}^d \to \mathbb{R}^d$ has coordinate functions ϕ_i defined by:

$$\phi_i(z) = \ln(1 + z_i), \qquad (6.69)$$

and finally where we have extended the definition of $\exp : \mathbb{R}^d \to \mathbb{R}^d$ by the convention that $\exp(z)$ is the vector with coordinates $\exp(z_i)$. We notice that the operator D defined above does not depend on x, and so we let $D^\epsilon(y)f := D^\epsilon(x,y)f$. By iid assumption on the random variables $\{Z_k\}$, the process $\{x_n\}$ will be chosen such that it is a Markov chain with all rows of the transition matrix P equal to each other, meaning that x_{n+1} is independent of x_n: in this context the ergodicity of the Markov chain is immediate, and we have $P = \Pi$. It remains to choose the finite state space $\mathbb{J}$: we assume that the random variables Z_k^i can only take finitely many values, say M, and we denote $\{\gamma_p\}_{p=1..M}$ these possible values. In the paper [1], they consider that these random variables can have arbitrary distributions: here we approximate these distributions by finite state distributions. Note that in practice (e.g., on a computer), all distributions are in fact approximated by finite-state distributions, since the finite floating-point precision of the computer only allows finitely many values for a random variable. The parameter M will in fact play no role in the analysis below. We let $\mathbb{J} := \{\gamma_1, \gamma_2, ..., \gamma_{M-1}, \gamma_M\}^d$. We have, for $x := (x_1, x_2, ..., x_d) \in \mathbb{J}$ and $y := (y_1, y_2, ..., y_d) \in \mathbb{J}$:

$$\mathbb{P}[x_{n+1} = y | x_n = x] = \pi(y), \qquad x_n := (Z_n^1, Z_n^2, ..., Z_n^d). \qquad (6.70)$$

In [1], the times $T_n = n$ are deterministic (so that $T_{n+1} - T_n = 1$), but in our context we allow them to be possibly random, which can be seen for example as an "illiquidity extension" of the model [1], similarly to what we have done in Section 6.1. Also, our very general framework allows one to be creative with the choices of the operators D and Γ, leading to possibly interesting extensions of the model [1] presented above. For example, one could simply "enlarge" the Markov chain $\{x_n\}$ and consider the same model but with regime-switching parameters μ and σ, and one could carry on all the analysis below at almost no extra technical and computational cost. Since $\frac{\partial \phi_j}{\partial z_i} = \delta_{ij} \frac{1}{1+z_i}$, we let $\phi_j' := \frac{\partial \phi_j}{\partial z_j}$, and we get:

$$D_1^\epsilon(y) f(z) = \sum_{j=1}^{d} \frac{\partial \widetilde{f}}{\partial z_j} (s_z^\epsilon(y)) \frac{\partial [s_z^\epsilon(y)]_j}{\partial \epsilon} \qquad (6.71)$$

$$= \sum_{j=1}^{d} \frac{\partial \widetilde{f}}{\partial z_j} (s_z^\epsilon(y)) \left[2\epsilon m_j + y_j + \phi_j'(h(z + \epsilon^2 m + \epsilon y) - h(z)) \right] \qquad (6.72)$$

$$\times g_j' \left(\frac{1}{V_0} \sum_{p=1}^{d} \alpha_p e^{z_p + \epsilon^2 m_p + \epsilon y_p} \right) \frac{1}{V_0} \sum_{p=1}^{d} \alpha_p (2\epsilon m_p + y_p) e^{z_p + \epsilon^2 m_p + \epsilon y_p} \right], \qquad (6.73)$$

where $\widetilde{f} := f \circ \exp$, and:

$$s_z^\epsilon(y) := z + \epsilon^2 m + \epsilon y + \phi(h(z + \epsilon^2 m + \epsilon y) - h(z)). \qquad (6.74)$$

Since $\phi_j(0) = 0$ and $\phi_j'(0) = 1$, it results that:

$$D_1^0(y) f(z) = \sum_{j=1}^{d} \frac{\partial \widetilde{f}}{\partial z_j} (z) \left[y_j + g_j' \left(\frac{1}{V_0} \sum_{p=1}^{d} \alpha_p e^{z_p} \right) \frac{1}{V_0} \sum_{p=1}^{d} \alpha_p y_p e^{z_p} \right]. \qquad (6.75)$$

From the last expression, and remembering that each component Z_n^i of $x_n := (Z_n^1, Z_n^2, ..., Z_n^d)$ is centred, we get:

$$\widehat{A}(t) = \frac{1}{\Pi m} \sum_{x \in \mathbb{J}} \pi(x) P[D_1^0(x, \bullet) f](x) = \frac{1}{\Pi m} \sum_{y \in \mathbb{J}} D_1^0(y) f \pi(y)$$

$$= \frac{1}{\Pi m} \mathbb{E}[D_1^0(x_n) f] \tag{6.76}$$

$$= \frac{1}{\Pi m} \mathbb{E}\left[\sum_{j=1}^d \frac{\partial \widetilde{f}}{\partial z_j}(z) \left(Z_n^j + g_j' \left(\frac{1}{V_0} \sum_{p=1}^d \alpha_p e^{z_p} \right) \frac{1}{V_0} \sum_{p=1}^d \alpha_p Z_n^p e^{z_p} \right) \right] \tag{6.77}$$

$$= \frac{1}{\Pi m} \sum_{j=1}^d \frac{\partial \widetilde{f}}{\partial z_j}(z) \left(\mathbb{E}[Z_n^j] + g_j' \left(\frac{1}{V_0} \sum_{p=1}^d \alpha_p e^{z_p} \right) \frac{1}{V_0} \sum_{p=1}^d \alpha_p \mathbb{E}[Z_n^p] e^{z_p} \right) = 0. \tag{6.78}$$

Let us denote the vector of initial log-spot prices s_0 which elements are $\ln(S_0^i)$ $(1 \leq i \leq d)$, as well as the d-dimensional price process $S_t(z) : \mathbb{R}^d \to \mathbb{R}^d$ with elements:

$$S_t^i(z) := \exp\left[z_i + \sum_{k=1}^{N_0(t)} \ln\left(\frac{S_k^i}{S_{k-1}^i} \right) \right]. \tag{6.79}$$

In this context the spot price changes at each jump time of the Markov renewal process (x_n, T_n). As mentioned above, in [1], the times $T_n = n$ are deterministic, so that $N_0(t) = \lfloor t \rfloor$. In our context, the random evolution $V(0, t) f$ and its rescaled version $V_\epsilon(0, t) f$ are simply equal to a functional of the spot price:

$$V(0, t) f(s_0) = f(S_t(s_0)), \qquad V_\epsilon(0, t) f(s_0) = f(S_{\frac{t}{\epsilon}}(s_0)). \tag{6.80}$$

Because $\widehat{A} = 0$, the limit of $f(S_{\frac{t}{\epsilon}})$ is trivial: $f(S_{\frac{t}{\epsilon}}) \Rightarrow f$. Also, we get:

$$\sigma_0^{(1)}(x, y, t, f) = \sigma_0^{(3)}(x, y, t, f) = 0, \tag{6.81}$$

$$\sigma_0^{(2)}(x, y, t, f) = D_1^0(y) f, \tag{6.82}$$

so that by Theorem 5.4.7, we have the convergence in the Skorokhod topology:

$$\epsilon^{-1/2}(f(S_{\frac{t}{\epsilon}}) - f) \Rightarrow \sum_{y \in \mathbb{J}} \int_0^t D_1^0(y) f W(y, du), \tag{6.83}$$

where $\{W(y, dt) : y \in \mathbb{J}\}$ are orthogonal martingale measures on $\mathbb{J} \times \mathbb{J}$ with intensity $\frac{1}{\Pi m} \pi(y) dt$. This leads to the approximation:

$$f(S_{\frac{t}{\epsilon}}) \approx f + \sqrt{\epsilon} \sum_{y \in \mathbb{J}} \int_0^t D_1^0(y) f W(y, du). \tag{6.84}$$

Because the "1st order" limit is trivial, it calls to study the "2nd order" limit, or diffusion limit, i.e. the convergence of $f(S_{\frac{t}{\epsilon^2}})$. This is indeed what is done in [1]. A complete and rigorous treatment of the diffusion limit case for time-inhomogeneous random evolutions is beyond the scope of this thesis. This was done in the homogeneous case in [23] (Section 4.2.2), or in a simplified context in [13], [14]. Nevertheless, in the model we are presently focusing on, we will be able to characterize the diffusion limit. The martingale representation is trivial as $\widehat{A} = 0$. We define another rescaled random evolution $L_\epsilon(s, t)$ which is defined exactly as $V_\epsilon(s, t)$, but with $N_s\left(t^{\frac{1}{\epsilon}, s}\right)$ replaced by $N_s\left(t^{\frac{1}{\epsilon^2}, s}\right)$ in the product. In our context:

$$L_\epsilon(0, t)f(s_0) = f(S_{\frac{t}{\epsilon^2}}(s_0)). \tag{6.85}$$

Using the same techniques as in Lemma 5.3.17 (but going one more order in the "Taylor" expansion and writing $f^\epsilon(x, t) := f + \epsilon f_1(x, t) + \epsilon^2 f_2(x, t)$ for suitable function f_2), we can get a martingale representation of the form:

$$\text{Martingale } = L_\epsilon(s, t)f - f - \epsilon^2 \Pi m \sum_{k=1}^{N_s\left(t^{\frac{1}{\epsilon^2}, s}\right)} L_\epsilon(s, T_k^{\epsilon, s}(s))\widehat{B}\left(T_k^{\epsilon, s}(s)\right)f$$

$$+ O(\epsilon) \text{ a.e.,} \tag{6.86}$$

for a suitable operator $\widehat{B}$. In fact, after a few lines of computation, we get:

$$\widehat{B}(t) = \frac{1}{\Pi m}\Pi b(\bullet, t), \tag{6.87}$$

$$b(x, t) := \frac{1}{2}m_2(x)A_x^2(t) + m_1(x)PA_x(t)D_1^0(x, \bullet)(x) + \frac{1}{2}PD_2^0(x, \bullet)(x) \tag{6.88}$$

$$+ m_1(x)Pf_1'(\bullet, t)(x), \quad x \in \mathbb{J}. \tag{6.89}$$

In our present model, $\widehat{B}(t) = \widehat{B}$ is independent of time and the above reduces to:

$$\widehat{B}f = \frac{1}{2\Pi m}\sum_{y \in \mathbb{J}}\pi(y)D_2^0(y)f = \frac{1}{2\Pi m}\mathbb{E}[D_2^0(x_n)f], \quad f \in C_0^3(\mathbb{R}^d). \tag{6.90}$$

Denote:

$$k_j(\epsilon; y) := \frac{\partial[s_\epsilon^\epsilon(y)]_j}{\partial \epsilon} = 2\epsilon m_j + y_j + \phi_j'(h(z + \epsilon^2 m + \epsilon y) - h(z)) \tag{6.91}$$

$$\times g_j'\left(\frac{1}{V_0}\sum_{p=1}^d \alpha_p e^{z_p + \epsilon^2 m_p + \epsilon y_p}\right)\frac{1}{V_0}\sum_{p=1}^d \alpha_p(2\epsilon m_p + y_p)e^{z_p + \epsilon^2 m_p + \epsilon y_p}. \tag{6.92}$$

With this notation we have:

$$D_2^\epsilon(y)f(z) = \sum_{j=1}^d \frac{\partial \widetilde{f}}{\partial z_j}(s_z^\epsilon(y))k_j'(\epsilon; y) + \sum_{j,m=1}^d \frac{\partial^2 \widetilde{f}}{\partial z_j \partial z_m}(s_z^\epsilon(y))k_j(\epsilon; y)k_m(\epsilon; y),$$

$$\tag{6.93}$$

and therefore:

$$\mathbb{E}[D_2^0(x_n)f(z)] = \sum_{j=1}^{d} \frac{\partial \widetilde{f}}{\partial z_j}(z)\mathbb{E}[k_j'(0; x_n)]$$

$$+ \sum_{j,m=1}^{d} \frac{\partial^2 \widetilde{f}}{\partial z_j \partial z_m}(z)\mathbb{E}[k_j(0; x_n)k_m(0; x_n)]. \tag{6.94}$$

We have:

$$k_j(0; y)k_m(0; y) = \left[y_j + g_j' \left(\frac{1}{V_0} \sum_{p=1}^{d} \alpha_p e^{z_p} \right) \frac{1}{V_0} \sum_{p=1}^{d} \alpha_p y_p e^{z_p} \right] \tag{6.95}$$

$$\times \left[y_m + g_m' \left(\frac{1}{V_0} \sum_{p=1}^{d} \alpha_p e^{z_p} \right) \frac{1}{V_0} \sum_{p=1}^{d} \alpha_p y_p e^{z_p} \right]. \tag{6.96}$$

Denoting $g_j' := g_j' \left(\frac{1}{V_0} \sum_{p=1}^{d} \alpha_p e^{z_p} \right)$ for clarity, we get:

$$\mathbb{E}[k_j(0; x_n)k_m(0; x_n)] = \sigma_{jm} + \frac{1}{V_0} g_j' \sum_{p=1}^{d} \alpha_p \sigma_{mp} e^{z_p} \tag{6.97}$$

$$+ \frac{1}{V_0} g_m' \sum_{p=1}^{d} \alpha_p \sigma_{jp} e^{z_p} + \frac{1}{V_0^2} g_j' g_m' \sum_{p,q=1}^{d} \alpha_p \alpha_q \sigma_{pq} e^{z_p + z_q}. \tag{6.98}$$

At this point it can be noted that the above result should coincide with the quantity $c^{j,m}$ of [1] (Proposition 5.1), and it does! Note however that because the times $\{T_n\}$ are random in our context, the above quantity is multiplied by $(\Pi m)^{-1}$ in $\widehat{B}$, i.e. the inverse of a weighted average of the mean values of the inter-arrival times $T_n - T_{n-1}$. If $T_n = n$ as in [1], then $T_n - T_{n-1} = 1$ and so $\Pi m = 1$: our framework is therefore a generalization of [1]. One interesting information coming from the above result is that in case of zero "fundamental correlations" $\sigma_{ij} = 0$ for $i \neq j$, we get for $j \neq m$, denoting $\sigma_i^2 := \sigma_{ii}$:

$$\mathbb{E}[k_j(0; x_n)k_m(0; x_n)] = \frac{1}{V_0} g_j' \alpha_m \sigma_m^2 e^{z_m} + \frac{1}{V_0} g_m' \alpha_j \sigma_j^2 e^{z_j}$$

$$+ \frac{1}{V_0^2} g_j' g_m' \sum_{p=1}^{d} (\alpha_p \sigma_p e^{z_p})^2 \neq 0. \tag{6.99}$$

This is the point of the model: distressed selling (modeled by the function g) induces a correlation between assets even if their fundamental correlation are equal to zero. It now remains to compute the quantities $\mathbb{E}[k_j'(0; x_n)]$. After some tedious computations we get, denoting again $g_j' := g_j' \left(\frac{1}{V_0} \sum_{p=1}^{d} \alpha_p e^{z_p} \right)$

(and similarly for g_j''):

$$\mathbb{E}[k_j'(0; x_n)] = 2m_j + \frac{1}{V_0^2}(g_j'' - (g_j')^2) \sum_{p,q=1}^{d} \alpha_p \alpha_q \sigma_{pq} e^{z_p + z_q}$$

$$+ \frac{1}{V_0} g_j' \sum_{p=1}^{d} 2\alpha_p \mu_p e^{z_p}. \tag{6.100}$$

Denote now:

$$\widetilde{\mathbf{a}_{ij}}(z) := \frac{1}{\Pi m} \mathbb{E}[k_i(0; x_n) k_j(0; x_n)] \quad \widetilde{\mathbf{b}_i}(z) := \frac{1}{2\Pi m} \mathbb{E}[k_i'(0; x_n)]. \tag{6.101}$$

The generator $\widehat{B}$ is therefore given by:

$$\widehat{B}f(z) = \sum_{i=1}^{d} \widetilde{\mathbf{b}_i}(z) \frac{\partial \widetilde{f}}{\partial z_i}(z) + \frac{1}{2} \sum_{i,j=1}^{d} \widetilde{\mathbf{a}_{ij}}(z) \frac{\partial^2 \widetilde{f}}{\partial z_i \partial z_j}(z), \quad f \in C_0^3(\mathbb{R}^d) \tag{6.102}$$

$$= \sum_{i=1}^{d} e^{z_i} \underbrace{\left[\widetilde{\mathbf{b}_i}(z) + \frac{1}{2}\widetilde{\mathbf{a}_{ii}}(z)\right]}_{\mathbf{b}_i(z)} \frac{\partial f}{\partial z_i}(z) + \frac{1}{2} \sum_{i,j=1}^{d} \underbrace{e^{z_i + z_j} \widetilde{\mathbf{a}_{ij}}(z)}_{\mathbf{a}_{ij}(z)} \frac{\partial^2 f}{\partial z_i \partial z_j}(z). \tag{6.103}$$

We get in consequence, still denoting $g_i := g_i \left(\frac{1}{V_0} \sum_{p=1}^{d} \alpha_p e^{z_p}\right)$ for clarity:

$$\mathbf{a}_{ij}(z) = \frac{1}{\Pi m} e^{z_i + z_j} \sigma_{ij} + \frac{1}{\Pi m} e^{z_i + z_j} \frac{1}{V_0} g_i' \sum_{p=1}^{d} \alpha_p \sigma_{jp} e^{z_p} \tag{6.104}$$

$$+ \frac{1}{\Pi m} e^{z_i + z_j} \frac{1}{V_0} g_j' \sum_{p=1}^{d} \alpha_p \sigma_{ip} e^{z_p} + \frac{1}{\Pi m} e^{z_i + z_j} \frac{1}{V_0^2} g_i' g_j' \sum_{p,q=1}^{d} \alpha_p \alpha_q \sigma_{pq} e^{z_p + z_q}, \tag{6.105}$$

and:

$$\mathbf{b}_i(z) = \frac{1}{\Pi m} e^{z_i} \mu_i + \frac{1}{\Pi m} e^{z_i} \frac{1}{2V_0^2} g_i'' \sum_{p,q=1}^{d} \alpha_p \alpha_q \sigma_{pq} e^{z_p + z_q}$$

$$+ \frac{1}{\Pi m} e^{z_i} \frac{1}{V_0} g_i' \sum_{p=1}^{d} \alpha_p (\underbrace{\mu_p}_{(*)} + \sigma_{ip}) e^{z_p}, \tag{6.106}$$

which is exactly equal to the result of [1], Theorem 4.2, at the small exception that in their drift result for $\mathbf{b}_i(z)$, the μ_p above denoted by $(*)$ is replaced by a m_p, where we recall that $m_p = \mu_p - \frac{1}{2}\sigma_p^2$. I have checked my computations and don't seem to find a mistake, therefore to the best of my knowledge the coefficient in (∗) should indeed be μ_p and not m_p. I will leave this small issue for further analysis. By 6.86, any limiting operator $L(0, \bullet)$ of $L_\epsilon(0, \bullet)$ satisfies:

$$\text{Martingale} = L(0, \bullet)f - f - \int_0^\bullet L(0, s)\widehat{B}f ds \text{ a.e.} \qquad (6.107)$$

Because of the above specific form of $\widehat{B}$, it can be proved as in [1] (assuming $g \in C_0^3(\mathbb{R})$), using the results of [3], that the martingale problem related to $(\widehat{B}, \delta_{s_0})$ is well-posed, and therefore $(S_{\frac{t}{\epsilon^2}}(s_0))_{t \geq 0}$ converges weakly in the Skorokhod topology as $\epsilon \to 0$ to the d-dimensional process $(P_t)_{t \geq 0}$ solution of:

$$dP_t = \mathbf{b}(P_t)dt + \mathbf{c}(P_t)dW_t, \qquad P_0 = S_0, \qquad (6.108)$$

where W is a standard d-dimensional Brownian motion and $\mathbf{a}(z) = \mathbf{c}(z)\mathbf{c}(z)^T$. This result is a generalization of the one of [1] as in our context, the times $\{T_n\}$ at which the price jumps are random. The consequence of this randomness is that the driving coefficients $\mathbf{b}$ and $\mathbf{c}$ of the limiting diffusion process $(P_t)_{t \geq 0}$ are multiplied by respectively $(\Pi m)^{-1}$ and $(\Pi m)^{-1/2}$. If $T_n = n$ as in [1], then $\Pi m = 1$ and we retrieve exactly their result. As mentioned above, our very general framework allows one to be creative with the choices of the operators D and Γ, leading to possibly interesting extensions of the considered model. For example, one could simply "enlarge" the Markov chain $\{x_n\}$ and consider the same model but with regime-switching parameters μ and σ, leading to diffusion limit results at almost no extra technical and computational cost.

Bibliography

[1] R. Cont and L. Wagalath. Fire sale forensics: Measuring endogenous risk. *Mathematical Finance*, 26 (4), 2016, 835-866.

[2] R. Cont and P. Tankov. *Financial Modelling with Jump Processes*. CRC Press LLC, 2004.

[3] S. Ethier and T. Kurtz. *Markov Processes: Characterization and Convergence*. John Wiley, 1986.

[4] P. Gassiat, F. Gozzi, and H. Pham. Investment/consumption problem in illiquid markets with regime-switching. *SIAM Journal on Control and Optimization*, 52(3):1761-1786, 2014.

[5] W. Kluge. *Time-Inhomogeneous Lévy Processes in Interest Rate and Credit Risk Models*. Ph.D. Thesis, 2005.

[6] L. Ruschendorf, A. Schnurr, and V. Wolf. *Comparison of Time-Inhomogeneous Markov Processes.* Preprint, 2014.

[7] K. Sato. *Lévy Processes and Infinitely Divisible Distributions.* Cambridge University Press, 1999.

[8] A. Swishchuk and N. Vadori. Smiling for the delayed volatility swaps. *Wilmott Magazine,* December 2014, 62-72.

[9] A. Swishchuk. *Random Evolutions and Their Applications.* Kluwer Academic Publishers, 1997.

[10] A. Swishchuk and V. Korolyuk. *Evolutions of Systems in Random Media.* CRC Press, 1995.

[11] N. Vadori. *Semi-Markov Driven Models: Limit Theorems and Financial Applications.* Ph.D. Thesis, University of Calgary, Calgary, AB, Canada, 2015.

[12] N. Vadori and A. Swishchuk. Inhomogeneous random evolutions: Limit theorems and financial applications. *Mathematics,* 2019, v. 7, N. 447; doi:10.3390/math7050447. Special Issue: *New Trends in Random Evolutions and their Applications.*

[13] J. Watkins. A central limit theorem in random evolution. *Ann. Prob.,* 12(2):480-513, 1984.

[14] J. Watkins. A stochastic integral representation for random evolution. *Ann. Prob.,* 13(2):531-557, 1985.

7

Applications of IHRE in High-Frequency Trading: Limit Order Books and their Semi-Markovian Modeling and Implementations

This chapter introduces a *semi-Markovian modeling of the limit order book*, and presents the main probabilistic results obtained in the context of this semi-Markovian model (such as duration until the next price change, probability of price increase and characterization of the Markov renewal process driving the stock price process). Let

$$s_t = s_0 + \sum_{k=1}^{N(t)} X_k$$

be our mid-price process, i.e., $s_t = (s_t^a + s_t^b)/2$, where s_t^a and s_t^b are ask and bid prices, respectively, $s_0 := s$ is an initial mid-price value, $N(t)$ is the number of price changes up to time t, described by a Markov renewal process, $X_k = \{+\delta, -\delta\}$. Let us introduce the following operators $D(x)$ on $B = C_0(R)$:

$$D(x)f(s) := f(s + x).$$

Then our mid-price process s_t above can be expressed as IHRE in the following way:

$$V(t)f(s) = f(s_t) = \Pi_{k=1}^{N(t)} D(X_k)f(s), \quad s_0 = s, \quad f(s) \in B.$$

One of the ways to get LLN and FCLT for mid-price process s_t is to use our main results, WLLN and FCLT for INREs, obtained in Chapter 5.

In this case we use the following scales: t/ϵ (or, equivalently, $nt, n \to +\infty$) for LLN, and t/ϵ^2 (or, equivalently, $tn^2, n \to +\infty$) for FCLT, when $\epsilon \to 0$.

However, we would like to show in this chapter that even another approach produces the same results: in Section 7 we obtain LLN and FCLT for mid-price process s_t using our martingale methods from [18] and change of time method.

We note that R. Cont and A. de Larrard [5] introduced a tractable stochastic model for the dynamics of a limit order book, computing various quantities of interest such as the probability of a price increase or the diffusion limit of the price process. As suggested by empirical observations, we extend their framework to 1) arbitrary distributions for book events inter-arrival times (possibly non-exponential) and 2) both the nature of a new book event and its corresponding inter-arrival time depend on the nature of the previous book event. We do so by resorting to Markov renewal processes to model the dynamics of the bid and ask queues. We keep analytical tractability via explicit expressions for the Laplace transforms of various quantities of interest. Here we also deal with diffusion limit results for the stock price process (based on our results from Chapter 5). We justify and illustrate our approach by calibrating our model to the five stocks of Amazon, Apple, Google, Intel and Microsoft on June 21st, 2012. As in [5], the bid-ask spread remains constant equal to one tick, only the bid and ask queues are modeled (they are independent from each other and get reinitialized after a price change), and all orders have the same size.

7.1   Introduction

Recently, interest in the modeling of limit order markets has increased. Some research has focused on optimal trading strategies in high-frequency environments: for example [8] studies such optimal trading strategies in a context where the stock price follows a semi-Markov process, while market orders arrive in the limit order book via a point process correlated with the stock price itself. [2] develops an optimal execution strategy for an investor seeking to execute a large order using both limit and market orders, under constraints on the volume of such orders. [13] studies optimal execution strategies for the purchase of a large number of shares of a financial asset over a fixed interval of time.

On the other hand, another class of articles has aimed at modeling either the high-frequency dynamics of the stock price itself, or the various queues of outstanding limit orders appearing on the bid and the ask side of the limit order book, resulting in specific dynamics for the stock price. In [7], a semi-Markov model for the stock price is introduced: the price increments are correlated and equal to arbitrary multiples of the tick size. The correlation between these price increments occurs via their sign only, and not their (absolute) size. In [4], the whole limit order book is modeled (not only the ask and bid queues) via an integer-valued continuous-time Markov chain. Using a Laplace analysis, they compute various quantities of interest such as the probability that the mid-price increases, or the probability that an order is executed before the mid-price moves. A detailed section on parameter estimation is also presented. For a more thorough literature search on limit order markets, we refer to the above cited articles and the references thereof.

The starting point of the present manuscript is the article [5], in which a stochastic model for the dynamics of the limit order book is presented.

Only the bid and ask queues are modeled (they are independent from each other and get reinitialized after a price change), the bid-ask spread remains constant equal to one tick and all orders have the same size. Their model is analytically tractable and allows them to compute various quantities of interest such as the distribution of the duration between price changes, the distribution and autocorrelation of price changes, the probability of an upward move in the price and the diffusion limit of the price process. Among the various assumptions made in this article, we seek to challenge two of them while preserving analytical tractability:

1. the inter-arrival times between book events (limit orders, market orders, order cancellations) are assumed to be independent and exponentially distributed.

2. the arrival of a new book event at the bid or the ask is independent from the previous events.

Assumption 1) is relatively common among the existing literature ([9], [12], [3], [6], [11], [15], [4]). Nevertheless, as it will be shown later, when calibrating the empirical distributions of the inter-arrival times to the Weibull and Gamma distributions (Amazon, Apple, Google, Intel and Microsoft on June 21st, 2012), we find that the shape parameter is in all cases significantly different than 1 ($\sim$ 0.1 to 0.3), which suggests that the exponential distribution is typically not rich enough to capture the behavior of these inter-arrival times.

Regarding Assumption 2), we split the book events into two different types: limit orders that increase the size of the corresponding bid or ask queue, and market orders/order cancellations that decrease the the size of the corresponding queue. Assimilating the former to the type "+1" and the latter to the type "−1", we find empirically that the probability to get an event of type "±1" is not independent of the nature of the previous event. Indeed, we present below the estimated transition probabilities between book events at the ask and the bid for the stock Microsoft on June 21st, 2012. It is seen that the *unconditional* probabilities $P(1)$ and $P(-1)$ to obtain respectively an event of type "+1" and "−1" are relatively close to 1/2, as in [5]. Nevertheless, denoting $P(i,j)$ the *conditional* probability to obtain an event of type j given that the last event was of type i, we observe that $P(i,j)$ can significantly depend on the previous event i. For example, on the bid side, $P(1,1) = 0.63$ whereas $P(-1,1) = 0.36$.

Microsoft	Bid	Ask
$P(1,1)$	0.63	0.60
$P(-1,1)$	0.36	0.41
$P(-1,-1)$	0.64	0.59
$P(1,-1)$	0.37	0.40
$P(1)$	0.49	0.51
$P(-1)$	0.51	0.49

Estimated probabilities for book event arrivals. June 21st, 2012.

On another front, we will show that we can obtain diffusion limit results for the stock price without resorting to the strong symmetry assumptions of [5]. In particular, the assumption that price increments are iid, which is contrary to empirical observations, as shown in [7] for example.

The book is organized as follows: Section 2 introduces our semi-Markovian modeling of the limit order book, Section 3 presents the main probabilistic results obtained in the context of this semi-Markovian model (duration until the next price change, probability of price increase and characterization of the Markov renewal process driving the stock price process), Section 4 deals with diffusion limit results for the stock price process, and Section 5 presents some calibration results on real market data.

7.2   A Semi-Markovian Modeling of Limit Order Markets

Throughout this book and to make the reading more convenient, we will use - when appropriate - the same notations as [5], as it is the starting point of the present article. In this section we introduce our model, highlighting when necessary the mains differences with the model in [5].

Let s_t, s_t^a, s_t^b be respectively the mid, the ask and the bid price processes. Denoting δ the "tick size," these quantities are assumed to be linked by the following relations:

$$s_t = \frac{1}{2}(s_t^a + s_t^b), \qquad\qquad s_t^a = s_t^b + \delta.$$

We will also assume that s_0^b is deterministic and positive. In this context, $s_0^a = s_0^b + \delta$ and $s_0 = s_0^b + \frac{\delta}{2}$ are also deterministic and positive. As shown in [5], the assumption that the bid-ask spread $s_t^a - s_t^b$ is constant and equal to one tick does not exactly match the empirical observations, but it is a reasonable assumption as [5] shows that - based on an analysis of the stocks Citigroup, General Electric, General Motors on June 26th 2008 - more than 98% of the observations have a bid-ask spread equal to one tick. This corresponds to a situation where the order book contains no empty levels (also called "gaps").

The price process s_t is assumed to be piecewise constant: at random times $\{T_n\}_{n \geq 0}$ (we set $T_0 := 0$), it jumps from its previous value $s_{T_n^-}$ to a new value $s_{T_n} = s_{T_n^-} \pm \delta$. By construction, the same holds for the ask and bid price processes s_t^a and s_t^b. These random times $\{T_n\}$ correspond to the times at which either the bid or the ask queue get depleted, and therefore, the distribution of these times $\{T_n\}$ will be obtained as a consequence of the dynamics on which we will choose to model the bid and ask queues. Let us denote q_t^a and q_t^b the non-negative integer-valued processes representing the respective sizes of the ask and bid queues at time t, namely the number of outstanding limit orders at each one of these queues. If the ask queue gets depleted before the bid queue at time T_n - i.e. $q^a_{T_n} = 0$ and $q^b_{T_n} > 0$ - then the price goes up: $s_{T_n} = s_{T_n^-} + \delta$ and both queue values $(q^b_{T_n}, q^a_{T_n})$ are immediately reinitialized to a new value drawn according to the distribution f, independently from all other random variables. In this context, if n_b, n_a are positive integers, $f(n_b, n_a)$ represents the probability that, after a price increase, the new values of the bid and ask queues are respectively equal to n_b and n_a. On the other hand, if the bid queue gets depleted before the ask queue at time T_n - i.e. $q^a_{T_n} > 0$ and $q^b_{T_n} = 0$ - then the price goes down: $s_{T_n} = s_{T_n^-} - \delta$ and both queue values $(q^b_{T_n}, q^a_{T_n})$ are immediately reinitialized to a new value drawn according to the distribution $\tilde{f}$, independently from all other random variables. Following the previous discussion, one can remark that the processes q^b_t, q^a_t will never effectively take the value 0, because whenever $q^b_{T_n} = 0$ or $q^a_{T_n} = 0$, we "replace" the pair $(q^b_{T_n}, q^a_{T_n})$ by a random variable drawn from the distribution f or $\tilde{f}$. The precise construction of the processes (q^b_t, q^a_t) will be explained below.

Let $\tau_n := T_n - T_{n-1}$ the "sojourn times" between two consecutive price changes, $N_t := \sup\{n : T_n \leq t\} = \sup\{n : \tau_1 + ... + \tau_n \leq t\}$ the number of price changes up to time t, $X_n := s_{T_n} - s_{T_{n-1}}$ the consecutive price increments (which can only take the values $\pm\delta$). With these notations we have:

$$s_t = \sum_{k=1}^{N_t} X_k.$$

Let us now present the chosen model for the dynamics of the bid and ask queues. As mentioned in introduction, we seek to extend the model [5] in the two following directions, as suggested by our calibration results:

1. inter-arrival times between book events (limit orders, market orders, order cancellations) are allowed to have an arbitrary distribution.

2. the arrival of a new book event at the bid or the ask and its corresponding inter-arrival time are allowed to depend on the nature of the previous event.

In order to do so, we will use a Markov renewal structure for the joint process of book events and corresponding inter-arrival times occurring at the ask and bid sides. Formally, for the ask side, consider a family $\{R^{n,a}\}_{n \geq 0}$ of Markov renewal processes given by:

$$R^{n,a} := \{(V_k^{n,a}, T_k^{n,a})\}_{k \geq 0}.$$

For each n, $R^{n,a}$ will "drive" the dynamics of the ask queue on the interval $[T_n, T_{n+1})$ where the stock price remains constant. $\{V_k^{n,a}\}_{k \geq 0}$ and $\{T_k^{n,a}\}_{k \geq 0}$ represent respectively the consecutive book events and the consecutive inter-arrival times between these book events at the ask side on the interval $[T_n, T_{n+1})$. At time T_{n+1} where one of the bid or ask queues gets depleted, the stock price changes and the model will be reinitialized with an independent copy $R^{n+1,a}$ of $R^{n,a}$: it will therefore be assumed that the processes $\{R^{n,a}\}_{n\geq 0}$ are independent copies of the same Markov renewal process of kernel Q^a, namely for each n:

$$\mathbb{P}[V_{k+1}^{n,a} = j, T_{k+1}^{n,a} \leq t | T_p^{n,a}, V_p^{n,a} : p \leq k] = Q^a(V_k^{n,a}, j, t), \quad j \in \{-1, 1\}$$
$$\mathbb{P}[V_0^{n,a} = j] = v_0^a(j), \quad j \in \{-1, 1\}$$
$$\mathbb{P}[T_0^{n,a} = 0] = 1.$$

We recall that as mentioned earlier, we consider two types of book events $V_k^{n,a}$: events of type $+1$ which increase the ask queue by 1 (limit orders), and events of type -1 which decrease the ask queue by 1 (market orders and order cancellations). In particular, the latter assumptions constitute a generalization of [5] in the sense that for each n:

- $V_{k+1}^{n,a}$ depends on the previous queue change $V_k^{n,a}$: $\{V_k^{n,a}\}_{k\geq 0}$ is a Markov chain.

- the inter-arrival times $\{T_k^{n,a}\}_{k\geq 0}$ between book events can have arbitrary distributions. Further, they are not strictly independent anymore but they are independent conditionally on the Markov chain $\{V_k^{n,a}\}_{k\geq 0}$.

We use the same notations to model the bid queue - but with indexes ^{a} replaced by ^{b} - and we assume that the processes involved at the bid and at the ask are independent.

In [5], the kernel Q^a is given by (the kernel Q^b has a similar expression with indexes ^{a} replaced by ^{b}):

$$Q^a(i, 1, t) = \frac{\lambda^a}{\lambda^a + \theta^a + \mu^a}(1 - e^{-(\lambda^a + \theta^a + \mu^a)t}), \quad i \in \{-1, 1\}$$

$$Q^a(i, -1, t) = \frac{\theta^a + \mu^a}{\lambda^a + \theta^a + \mu^a}(1 - e^{-(\lambda^a + \theta^a + \mu^a)t}), \quad i \in \{-1, 1\}.$$

Given these chosen dynamics to model to ask and bid queues between two consecutive price changes, we now specify formally the "state process":

$$\widetilde{L}_t := (s_t^b, q_t^b, q_t^a)$$

which will keep track of the state of the limit order book at time t (stock price and sizes of the bid and ask queues). In the context of [5], this process $\widetilde{L}_t$ was proved to be Markovian. Here, we will need to "add" to this process the process (V_t^b, V_t^a) keeping track of the nature of the last book event at the bid and the ask to make it Markovian: in this sense we can view it as being semi-Markovian. The process:

$$L_t := (s_t^b, q_t^b, q_t^a, V_t^b, V_t^a)$$

constructed below will be proved to be Markovian.

The process L is piecewise constant and changes value whenever a book event occurs at the bid or at the ask. We will construct both the process L and the sequence of times $\{T_n\}_{n\geq 0}$ recursively on $n \geq 0$. The recursive construction starts from $n = 0$ where we have $T_0 = 0$, $s_0^b > 0$ deterministic, and $(q_0^b, q_0^a, V_0^b, V_0^a)$ is a random variable with distribution $f_0 \times v_0^b \times v_0^a$, where f_0 is a distribution on $\mathbb{N}^* \times \mathbb{N}^*$, and both v_0^b and v_0^a are distributions on the 2-point space $\{-1, 1\}$, that is $v_0^b(1) = \mathbb{P}[V_0^b = 1]$ is given and $v_0^b(-1) = 1 - v_0^b(1)$ (and similarly for the ask). We will need to introduce the following processes for the bid side (for the ask side, they are defined similarly):

$$\bar{T}_k^{n,b} := \sum_{p=0}^{k} T_p^{n,b}, \qquad N_t^{n,b} := \sup\{k : T_n + \bar{T}_k^{n,b} \leq t\}.$$

With these notations, the book events corresponding to the interval $[T_n, T_{n+1})$ occur at times $T_n + \bar{T}_k^{n,b}$ ($k \geq 0$) until one of the queues gets depleted, and $N_t^{n,b}$ counts the number of book events on the interval $[T_n, t]$, for $t \in [T_n, T_{n+1})$.

The joint construction of L and of the sequence of times $\{T_n\}_{n\geq 0}$ is done recursively on $n \geq 0$. The following describes the step n of the recursive construction:

- For each $T \in \{T_n + \bar{T}_k^{n,b}\}_{k\geq 1}$, the book event $v_{n,T}^b := V_{N_T^{n,b}}^{n,b}$ occurs at time T at the bid side. If $q_{T-}^b + v_{n,T}^b > 0$, there is no price change at time T and we have:

$$(s_T^b, q_T^b, q_T^a, V_T^b, V_T^a) = (s_{T-}^b, q_{T-}^b + v_{n,T}^b, q_{T-}^a, v_{n,T}^b, V_{T-}^a).$$

If on the other hand $q_{T-}^b + v_{n,T}^b = 0$, there is a price change at time T and the model gets reinitialized:

$$(s_T^b, q_T^b, q_T^a, V_T^b, V_T^a) = (s_{T-}^b - \delta, \tilde{x}_n^b, \tilde{x}_n^a, v_{0,n}^b, v_{0,n}^a),$$

where $\{(\tilde{x}_k^b, \tilde{x}_k^a)\}_{k\geq 0}$ are iid random variables, independent from all other random variables, with joint distribution $\tilde{f}$ on $\mathbb{N}^* \times \mathbb{N}^*$, and $\{v_{0,k}^b, v_{0,k}^a\}_{k\geq 0}$ are iid random variables, independent from all other random variables, with joint distribution $v_0^b \times v_0^a$ on the space $\{-1, 1\} \times \{-1, 1\}$. We then set $T_{n+1} = T$ and move from the step n of the recursion to the step $n + 1$.

- For each $T \in \{T_n + \bar{T}_k^{n,a}\}_{k\geq 1}$, the book event $v_{n,T}^a := V_{N_T^{n,a}}^{n,a}$ occurs at time T at the ask side. If $q_{T-}^a + v_{n,T}^a > 0$, there is no price change at time T and we have:

$$(s_T^b, q_T^b, q_T^a, V_T^b, V_T^a) = (s_{T-}^b, q_{T-}^b, q_{T-}^a + v_{n,T}^a, V_{T-}^b, v_{n,T}^a).$$

If on the other hand $q_{T-}^a + v_{n,T}^a = 0$, there is a price change at time T and the model gets reinitialized:

$$(s_T^b, q_T^b, q_T^a, V_T^b, V_T^a) = (s_{T-}^b + \delta, x_n^b, x_n^a, v_{0,n}^b, v_{0,n}^a),$$

where $\{(x_k^b, x_k^a)\}_{k\geq 0}$ are iid random variables, independent from all other random variables, with joint distribution f on $\mathbb{N}^* \times \mathbb{N}^*$, and $\{v_{0,k}^b, v_{0,k}^a\}_{k\geq 0}$ are the iid random variables defined above. We then set $T_{n+1} = T$ and move from the step n of the recursion to the step $n+1$.

It results from the above construction and the Markov renewal structure of the processes $\{R^{n,a}\}_{n\geq 0}$, $\{R^{n,b}\}_{n\geq 0}$ that the process L_t is Markovian.

Since the processes $\{\bar{R}^{n,a}\}_{n\geq 0}$ are independent copies of the same Markov renewal process of kernel Q^a, we will drop the index n when appropriate in order to make the notations lighter. Following this remark, we will introduce the following notations for the ask, for $i, j \in \{-1, 1\}$ (for the bid, they are defined similarly):

$$P^a(i,j) := \mathbb{P}[V_{k+1}^a = j | V_k^a = i],$$

$$F^a(i,t) := \mathbb{P}[T_{k+1}^a \leq t | V_k^a = i],$$

$$H^a(i,j,t) := \mathbb{P}[T_{k+1}^a \leq t | V_k^a = i, V_{k+1}^a = j],$$

$$h^a(i,j) := \int_0^\infty t H^a(i,j,dt),$$

$$h_1^a := h^a(1,1) + h^a(-1,-1), \quad h_2^a := h^a(-1,1) + h^a(1,-1),$$

$$m^a(s,i,j) := \int_0^\infty e^{-st} Q^a(i,j,dt), \quad s \in \mathbb{C},$$

$$M^a(s,i) := m^a(s,i,-1) + m^a(s,i,1) = \int_0^\infty e^{-st} F^a(i,dt), \quad s \in \mathbb{C}.$$

Throughout this book, we will use the following mild technical assumptions:

(A1) $0 < P^a(i,j) < 1$, $0 < P^b(i,j) < 1$, $\quad i,j \in \{-1,1\}$.

(A2) $F^a(i,0) < 1$, $F^b(i,0) < 1$, $\quad i \in \{-1,1\}$.

(A3) $\int_0^\infty t^2 H^a(i,j,dt) < \infty$, $\int_0^\infty t^2 H^b(i,j,dt) < \infty$, $\quad i,j \in \{-1,1\}$.

Some brief comments on these assumptions: **(A1)** implies that each state ± 1 is accessible from each state. **(A2)** means that each inter-arrival time between book events has a positive probability to be non-zero, and **(A3)** constitutes a second moment integrability assumption on the cumulative distribution functions H^a and H^b.

7.3   Main Probabilistic Results

Throughout this section and as mentioned earlier, since the processes $\{R^{n,a}\}_{n\geq 0}$ are independent copies of the same Markov renewal process of kernel Q^a, we will drop the index n when appropriate in order to make the notations lighter on the random variables $T_k^{n,a}$, $\bar{T}_k^{n,a}$, $V_k^{n,a}$ (and similarly for the bid side).

7.3.1 Duration until the Next Price Change

Given an initial configuration of the bid and ask queues $(q_0^b, q_0^a) = (n_b, n_a)$ (n_b, n_a integers), we denote σ_b the first time at which the bid queue is depleted:

$$\sigma_b = \bar{T}_{k^*}^b, \qquad k^* := \inf\{k : n_b + \sum_{m=1}^{k} V_m^b = 0\}.$$

Similarly we define σ_a the first time at which the ask queue is depleted. The duration until the next price move is thus:

$$\tau := \sigma_a \wedge \sigma_b.$$

In order to have a realistic model in which the queues always get depleted at some point, i.e., $\mathbb{P}[\sigma_a < \infty] = \mathbb{P}[\sigma_b < \infty] = 1$, we impose the conditions:

$$P^a(1,1) \leq P^a(-1,-1), \qquad P^b(1,1) \leq P^b(-1,-1).$$

These conditions correspond to the condition $\lambda \leq \theta + \mu$ in [5], and the proof of the proposition below shows that they are respectively equivalent to $\mathbb{P}[\sigma_a < \infty] = 1$ and $\mathbb{P}[\sigma_b < \infty] = 1$. Indeed, as $s \to 0$ ($s > 0$), the Laplace transform $\mathcal{L}^a(s) := \mathbb{E}[e^{-s\sigma_a}]$ of σ_a tends to $\mathbb{P}[\sigma_a < \infty]$. The proposition below shows that if $P^a(1,1) > P^a(-1,-1)$, this quantity is strictly less than 1, and if $P^a(1,1) \leq P^a(-1,-1)$, this quantity is equal to 1. We have the following result which generalizes the Proposition 1 in [5] (see also Remark 7.3.2 below):

Proposition 7.3.1. *The conditional law of σ_a given $q_0^a = n \geq 1$ has a regularly varying tail with:*

- *tail exponent 1 if $P^a(1,1) < P^a(-1,-1)$.*

- *tail exponent 1/2 if $P^a(1,1) = P^a(-1,-1)$.*

More precisely, we get: if $P^a(1,1) = P^a(-1,-1) = p_a$:

$$\mathbb{P}[\sigma_a > t|q_0^a = n] \overset{t\to\infty}{\sim} \frac{\alpha^a(n)}{\sqrt{t}}$$

with:

$$\alpha^a(n) := \frac{1}{p_a\sqrt{\pi}}(n + \frac{2p_a - 1}{p_a - 1}v_0^a(1))\sqrt{p_a(1 - p_a)}\sqrt{p_a h_1^a + (1 - p_a)h_2^a}.$$

If $P^a(1,1) < P^a(-1,-1)$, we get:

$$\mathbb{P}[\sigma_a > t | q_0^a = n] \overset{t \to \infty}{\sim} \frac{\beta^a(n)}{t}$$

with:

$$\beta^a(n) := v_0^a(1)u_1^a + v_0^a(-1)u_2^a + (n-1)u_3^a,$$

$$u_1^a := h^a(1,-1) + \frac{P^a(1,1)}{1 - P^a(1,1)}(u_3^a + h^a(1,1))$$

$$u_2^a := -h^a(1,1) + \frac{1 - P^a(-1,-1)}{1 - P^a(1,1)}(u_3^a + h^a(1,1))$$

$$+ P^a(-1,-1)h_1^a + (1 - P^a(-1,-1))h_2^a,$$

$$u_3^a := h^a(1,1) + \frac{1 - P^a(1,1)}{P^a(-1,-1) - P^a(1,1)}$$

$$(P^a(-1,-1)h_1^a + (1 - P^a(-1,-1))h_2^a).$$

Similar expressions are obtained for $\mathbb{P}[\sigma_b > t | q_0^b = n]$, with indexes [a] *replaced by* [b].

Remark 7.3.2. *We retrieve the results of [5]: if $P^a(1,1) = P^a(-1,-1)$, then within the context/notations of [5] we get $p_a = 1/2$ and:*

$$h^a(i,j) = \int_0^\infty 2t\lambda e^{-2\lambda t}dt = \frac{1}{2\lambda},$$

and so $\alpha^a(n) = \frac{n}{\sqrt{\pi\lambda}}$. For the case $P^a(1,1) < P^a(-1,-1)$ ($\lambda < \theta + \mu$ with their notations), we find:

$$\beta^a(n) = \frac{n}{\theta + \mu - \lambda},$$

which is different from the result of [5] that is $\beta^a(n) = \frac{n(\theta+\mu+\lambda)}{2\lambda(\theta+\mu-\lambda)}$. We believe that they made a small mistake in their Taylor expansion on page 10: in the case $\lambda < \theta + \mu$, they should find:

$$\mathcal{L}(s,x) \overset{s \to 0}{\sim} 1 - \frac{sx}{\theta + \mu - \lambda}.$$

Proof. *Let $s > 0$ and denote $\mathcal{L}(s,n,i) := \mathbb{E}[e^{-s\sigma_a} | q_0^a = n, V_0^a = i]$. We have:*

$$\sigma_a = \sum_{m=1}^{k^*} T_m^a, \qquad k^* := \inf\{k : n + \sum_{m=1}^k V_m^a = 0\}.$$

Therefore:

$$\mathcal{L}(s,n,i) = \mathbb{E}[e^{-sT_1^a}\mathbb{E}[e^{-s(\sigma_a - T_1^a)}|q_0^a = n, V_0^a = i, V_1^a, T_1^a]|q_0^a = n, V_0^a = i]$$

$$= \mathbb{E}[e^{-sT_1^a}\underbrace{\mathbb{E}[e^{-s(\sigma_a - T_1^a)}|q_{T_1^a}^a = n + V_1^a, V_0^a = i, V_1^a, T_1^a]}_{\mathcal{L}(s, n+V_1^a, V_1^a)}|q_0^a = n, V_0^a = i]$$

$$= \mathbb{E}[e^{-sT_1^a}\mathcal{L}(s, n + V_1^a, V_1^a)|q_0^a = n, V_0^a = i]$$

$$= \int_0^\infty e^{-st}\mathcal{L}(s, n+1, 1)Q^a(i, 1, dt) + \int_0^\infty e^{-st}\mathcal{L}(s, n-1, -1)Q^a(i, -1, dt)$$

$$= m^a(s, i, 1)\mathcal{L}(s, n+1, 1) + m^a(s, i, -1)\mathcal{L}(s, n-1, -1).$$

Denote for sake of clarity $a_n := \mathcal{L}(s, n, 1)$, $b_n := \mathcal{L}(s, n, -1)$. These sequences therefore solve the system of coupled recurrence equations:

$$a_{n+1} = m^a(s, 1, 1)a_{n+2} + m^a(s, 1, -1)b_n, \quad n \geq 0$$
$$b_{n+1} = m^a(s, -1, 1)a_{n+2} + m^a(s, -1, -1)b_n$$
$$a_0 = b_0 = 1.$$

Simple algebra (computing $a_{n+1} - m^a(s, -1, -1)a_n$ on the on hand and $m^a(s, 1, 1)b_{n+1} - b_n$ on the other hand) gives us that both a_n and b_n solve the same following recurrence equation (but for different initial conditions):

$$m^a(s, 1, 1)u_{n+2} - (1 + \Delta^a(s))u_{n+1} + m^a(s, -1, -1)u_n, \quad n \geq 1$$

with:

$$\Delta^a(s) := m^a(s, 1, 1)m^a(s, -1, -1) - m^a(s, -1, 1)m^a(s, 1, -1).$$

The parameter $\Delta^a(s)$ can be seen as a coupling coefficient and is equal to 0 when the random variable (V_k^a, T_k^a) doesn't depend on the previous state V_{k-1}^a, for example in the context of [5].

If we denote $R(X)$ the characteristic polynomial associated to the previous recurrence equation $R(X) := m^a(s, 1, 1)X^2 - (1 + \Delta^a(s))X + m^a(s, -1, -1)$, then simple algebra gives us:

$$R(1) = \underbrace{(M^a(s, 1) - 1)}_{<0}\underbrace{(1 - m^a(s, -1, -1))}_{>0} + \underbrace{m^a(s, 1, -1)}_{>0}\underbrace{(M^a(s, -1) - 1)}_{<0} < 0$$

Note that $M^a(s, i) < 1$ for $s > 0$ because $F^a(i, 0) < 1$. Since $m^a(s, 1, 1) > 0$, this implies that R has only one root < 1 (and an other root > 1):

$$\lambda^a(s) := \frac{1 + \Delta^a(s) - \sqrt{(1 + \Delta^a(s))^2 - 4m^a(s, 1, 1)m^a(s, -1, -1)}}{2m^a(s, 1, 1)}.$$

Because we have $a_n, b_n \leq 1$ for $s > 0$, then we must have for $n \geq 1$:

$$a_n = a_1\lambda^a(s)^{n-1} \qquad\qquad b_n = b_1\lambda^a(s)^{n-1}$$

The recurrence equations on a_n, b_n give us:

$$a_1 = \frac{m^a(s, 1, -1)}{1 - \lambda^a(s)m^a(s, 1, 1)} \qquad b_1 = \frac{m^a(s, -1, 1)a_1 + \Delta^a(s)}{m^a(s, 1, 1)}$$

Finally, letting $\mathcal{L}(s, n) := \mathbb{E}[e^{-s\sigma_a}|q_0^a = n]$, we obtain:

$$\mathcal{L}(s, n) = \sum_i \mathcal{L}(s, n, i)v_0^a(i) = a_n v_0^a(1) + b_n v_0^a(-1).$$

The behavior of $\mathbb{P}[\sigma_a > t|q_0^a = n]$ as $t \to \infty$ is obtained by computing the behavior of $\mathcal{L}(s, n)$ as $s \to 0$, together with Karamata's Tauberian theorem. By the second moment integrability assumption on $H^a(i, j, dt)$, we note that:

$$m^a(s, i, j) = \int_0^\infty e^{-st} Q^a(i, j, dt) = P^a(i, j) \int_0^\infty e^{-st} H^a(i, j, dt)$$

$$\overset{s \to 0}{\sim} P^a(i, j) - sP^a(i, j) \int_0^\infty t H^a(i, j, dt) = P^a(i, j) - sP^a(i, j)h^a(i, j).$$

Now, assume $P^a(1, 1) = P^a(-1, -1) = p_a$. A straightforward but tedious Taylor expansion of $\mathcal{L}(s, n)$ as $s \to 0$ gives us:

$$\mathcal{L}(s, n) \overset{s \to 0}{\sim} 1 - \sqrt{\pi}\alpha^a(n)\sqrt{s}.$$

The same way, if $P^a(1, 1) < P^a(-1, -1)$, a straightforward Taylor expansion of $\mathcal{L}(s, n)$ as $s \to 0$ gives us:

$$\mathcal{L}(s, n) \overset{s \to 0}{\sim} 1 - \beta^a(n)s.$$

We are interested in the asymptotic behavior of the law of τ, which is, by independence of the bid/ask queues:

$$\mathbb{P}[\tau > t|(q_0^b, q_0^a) = (n_b, n_a)] = \mathbb{P}[\sigma_a > t|q_0^a = n_a]\mathbb{P}[\sigma_b > t|q_0^b = n_b].$$

We get the following immediate consequence of Proposition 7.3.1:

Proposition 7.3.3. *The conditional law of τ given $(q_0^b, q_0^a) = (n_b, n_a)$ has a regularly varying tail with:*

- *tail exponent 2 if $P^a(1, 1) < P^a(-1, -1)$ and $P^b(1, 1) < P^b(-1, -1)$. In particular, in this case, $\mathbb{E}[\tau|(q_0^b, q_0^a) = (n_b, n_a)] < \infty$.*

- *tail exponent 1 if $P^a(1, 1) = P^a(-1, -1)$ and $P^b(1, 1) = P^b(-1, -1)$. In particular, in this case, $\mathbb{E}[\tau|(q_0^b, q_0^a) = (n_b, n_a)] = \infty$ whenever $n_b, n_a \geq 1$.*

- *tail exponent 3/2 otherwise. In particular, in this case, $\mathbb{E}[\tau|(q_0^b, q_0^a) = (n_b, n_a)] < \infty$.*

More precisely, we get: if $P^a(1,1) = P^a(-1,-1)$ *and* $P^b(1,1) = P^b(-1,-1)$:

$$\mathbb{P}[\tau > t | (q_0^b, q_0^a) = (n_b, n_a)] \overset{t \to \infty}{\sim} \frac{\alpha^a(n_a)\alpha^b(n_b)}{t}$$

if $P^a(1,1) < P^a(-1,-1)$ *and* $P^b(1,1) < P^b(-1,-1)$:

$$\mathbb{P}[\tau > t | (q_0^b, q_0^a) = (n_b, n_a)] \overset{t \to \infty}{\sim} \frac{\beta^a(n_a)\beta^b(n_b)}{t^2}$$

if $P^a(1,1) = P^a(-1,-1)$ *and* $P^b(1,1) < P^b(-1,-1)$:

$$\mathbb{P}[\tau > t | (q_0^b, q_0^a) = (n_b, n_a)] \overset{t \to \infty}{\sim} \frac{\alpha^a(n_a)\beta^b(n_b)}{t^{3/2}}$$

if $P^a(1,1) < P^a(-1,-1)$ *and* $P^b(1,1) = P^b(-1,-1)$:

$$\mathbb{P}[\tau > t | (q_0^b, q_0^a) = (n_b, n_a)] \overset{t \to \infty}{\sim} \frac{\beta^a(n_a)\alpha^b(n_b)}{t^{3/2}}$$

Proof. *Immediate using Proposition 7.3.1.*

It will be needed to get the full law of τ, which is, by independence of the bid/ask queues:

$$\mathbb{P}[\tau > t | (q_0^b, q_0^a) = (n_b, n_a)] = \mathbb{P}[\sigma_a > t | q_0^a = n_a]\mathbb{P}[\sigma_b > t | q_0^b = n_b].$$

We have computed explicitly the Laplace transforms of σ_a and σ_b (cf. the proof of Proposition 7.3.1 above). There are two possibilities: either it is possible to invert those Laplace transforms so that we can compute $\mathbb{P}[\sigma_a > t | q_0^a = n_a]$ and $\mathbb{P}[\sigma_b > t | q_0^b = n_b]$ in closed form and thus $\mathbb{P}[\tau > t | (q_0^b, q_0^a) = (n_b, n_a)]$ in closed form as in [5]. If not, we will have to resort to a numerical procedure to invert the characteristic functions of σ_a and σ_b. Below we give the characteristic functions of σ_a and σ_b:

Proposition 7.3.4. *Let* $\phi^a(t,n) := \mathbb{E}[e^{it\sigma_a} | q_0^a = n]$ $(t \in \mathbb{R})$ *the characteristic function of* σ_a *conditionally on* $q_0^a = n \geq 1$. *We have:*

if $m^a(-it, 1, 1) \neq 0$:

$$\phi^a(t,n) = (c^a(-it)v_0^a(1) + d^a(-it)v_0^a(-1))\lambda^a(-it)^{n-1},$$

$$c^a(z) = \frac{m^a(z, 1, -1)}{1 - \lambda^a(z)m^a(z, 1, 1)},$$

$$d^a(z) = \frac{m^a(z, -1, 1)c^a(z) + \Delta^a(z)}{m^a(z, 1, 1)},$$

$$\Delta^a(z) := m^a(z, 1, 1)m^a(z, -1, -1) - m^a(z, -1, 1)m^a(z, 1, -1),$$

$$\lambda^a(z) := \frac{1 + \Delta^a(z) - \sqrt{(1 + \Delta^a(z))^2 - 4m^a(z, 1, 1)m^a(z, -1, -1)}}{2m^a(z, 1, 1)}.$$

and if $m^a(-it, 1, 1) = 0$:

$$\phi^a(t, n) = (m^a(-it, 1, -1)v_0^a(1) + \widetilde{\lambda^a}(-it)v_0^a(-1))\widetilde{\lambda^a}(-it)^{n-1},$$

$$\widetilde{\lambda^a}(z) := \frac{m^a(z, -1, -1)}{1 - m^a(z, 1, -1)m^a(z, -1, 1)}.$$

The coefficient $\Delta^a(z)$ can be seen as a coupling coefficient and is equal to 0 when the random variable (V_k^a, T_k^a) doesn't depend on the previous state V_{k-1}^a, for example in the context of [5].

The characteristic function $\phi^b(t, n) := \mathbb{E}[e^{it\sigma_b}|q_0^b = n]$ has the same expression, with indexes ^{a} replaced by ^{b}.

Proof. *Similarly to the proof of Proposition 7.3.1, we obtain (using the same notations but denoting this time $a_n := \mathcal{L}(-it, n, 1)$, $b_n := \mathcal{L}(-it, n, -1)$):*

$$a_{n+1} = m^a(-it, 1, 1)a_{n+2} + m^a(-it, 1, -1)b_n, \quad n \geq 0$$
$$b_{n+1} = m^a(-it, -1, 1)a_{n+2} + m^a(-it, -1, -1)b_n$$
$$a_0 = b_0 = 1.$$

If $m^a(-it, 1, 1) = 0$, we can solve explicitly the above system to get the desired result. If $m^a(-it, 1, 1) \neq 0$, we get as in the proof of Prop 7.3.1 that both a_n and b_n solve the same following recurrence equation (but for different initial conditions):

$$m^a(-it, 1, 1)u_{n+2} - (1 + \Delta^a(-it))u_{n+1} + m^a(-it, -1, -1)u_n, \quad n \geq 1.$$

Because $|m^a(-it, j, -1) + m^a(-it, j, 1)| = |M^a(-it, j)| = \left|\int_0^\infty e^{its}F^a(j, ds)\right| \leq 1$, tedious computations give us that $|\lambda_+^a(-it)| > 1$ whenever $t \neq 0$, where:

$$\lambda_+^a(z) := \frac{1 + \Delta^a(z) + \sqrt{(1 + \Delta^a(z))^2 - 4m^a(z, 1, 1)m^a(z, -1, -1)}}{2m^a(z, 1, 1)}.$$

Since both $|a_n|, |b_n| \leq 1$ for all n, it must be that:

$$a_n = a_1\lambda^a(-it)^{n-1} \qquad b_n = b_1\lambda^a(-it)^{n-1},$$

with a_1, b_1 being given by the recurrence equations on a_n, b_n:

$$a_1 = \frac{m^a(-it, 1, -1)}{1 - \lambda^a(-it)m^a(-it, 1, 1)} \qquad b_1 = \frac{m^a(-it, -1, 1)a_1 + \Delta^a(-it)}{m^a(-it, 1, 1)}.$$

Finally we conclude by observing that:

$$\phi^a(t, n) = a_n v_0^a(1) + b_n v_0^a(-1).$$

7.3.2 Probability of Price Increase

Starting from an initial configuration of the bid and ask queues, $(q_0^b, q_0^a) = (n_b, n_a)$, the probability that the next price change is a price increase will be denoted $p_1^{up}(n_b, n_a)$. This quantity is equal to the probability that σ_a is less than σ_b:

$$p_1^{up}(n_b, n_a) = \mathbb{P}[\sigma_a < \sigma_b | q_0^b = n_b, q_0^a = n_a].$$

Since we know the characteristic functions of σ_a, σ_b (cf. Proposition 7.3.4), we can compute their individual laws up to the use of a numerical procedure. Since σ_a and σ_b are independent, the law of $\sigma_b - \sigma_a$ can be computed using the individual laws of σ_a, σ_b, and therefore $p_1^{up}(n_b, n_a)$ can be computed up to the use of numerical procedures to 1) invert the characteristic function and 2) compute an indefinite integral. Indeed, denoting $f_{n_a,a}$ the p.d.f of σ_a conditionally on $q_0^a = n_a$, and $F_{n_b,b}$ the c.d.f. of σ_b conditionally on $q_0^b = n_b$, we have:

$$p_1^{up}(n_b, n_a) = \mathbb{P}[\sigma_a < \sigma_b | q_0^b = n_b, q_0^a = n_a] = \int_0^\infty f_{n_a,a}(t)(1 - F_{n_b,b}(t))dt,$$

where $F_{n_b,b}$ and $f_{n_a,a}$ are obtained by the following inversion formulas:

$$f_{n_a,a}(t) = \frac{1}{2\pi} \int_{\mathbb{R}} e^{-itx} \phi^a(x, n_a)dx,$$

$$F_{n_b,b}(t) = \frac{1}{2} - \frac{1}{\pi} \int_0^\infty \frac{1}{x} Im\{e^{-itx}\phi^b(x, n_b)\}dx.$$

7.3.3 The Stock Price Seen as a Functional of a Markov Renewal Process

As mentioned earlier, we can write the stock price s_t as:

$$s_t = \sum_{k=1}^{N_t} X_k,$$

where $\{X_n\}_{n \geq 0}$ are the consecutive price increments taking value $\pm\delta$, $\{\tau_n\}_{n \geq 0}$ are the consecutive durations between price changes and $\{T_n\}_{n \geq 0}$ the consecutive times at which the price changes.

In this context, the distribution of the random variable τ_{n+1} will depend on the initial configuration of the bid and ask queues at the beginning T_n of the period $[T_n, T_{n+1})$, which itself depends on the nature of the previous price change X_n: if the previous price change is a price decrease, the initial configuration will be drawn from the distribution $\tilde{f}$, and if it is an increase, the initial configuration will be drawn from the distribution f. Because for each n the random variable (X_n, τ_n) only depends on the previous increment X_{n-1}, it can be seen that the process $(X_n, \tau_n)_{n \geq 0}$ is a Markov renewal process

([10], [18]), and the stock price can therefore be seen as a functional of this Markov renewal process. We obtain the following result.

Proposition 7.3.5. *The process* $(X_n, \tau_n)_{n \geq 0}$ *is a Markov renewal process. The law of the process* $\{\tau_n\}_{n \geq 0}$ *is given by:*

$$F(\delta, t) := \mathbb{P}[\tau_{n+1} \leq t | X_n = \delta] = \sum_{p=1}^{\infty} \sum_{n=1}^{\infty} f(n, p) \mathbb{P}[\tau \leq t | (q_0^b, q_0^a) = (n, p)],$$

$$F(-\delta, t) := \mathbb{P}[\tau_{n+1} \leq t | X_n = -\delta] = \sum_{p=1}^{\infty} \sum_{n=1}^{\infty} \tilde{f}(n, p) \mathbb{P}[\tau \leq t | (q_0^b, q_0^a) = (n, p)].$$

The Markov chain $\{X_n\}_{n \geq 0}$ *is characterized by the following transition probabilities:*

$$p_{cont} := \mathbb{P}[X_{n+1} = \delta | X_n = \delta] = \sum_{i=1}^{\infty} \sum_{j=1}^{\infty} p_1^{up}(i, j) f(i, j).$$

$$p'_{cont} := \mathbb{P}[X_{n+1} = -\delta | X_n = -\delta] = \sum_{i=1}^{\infty} \sum_{j=1}^{\infty} (1 - p_1^{up}(i, j)) \tilde{f}(i, j).$$

The generator of this Markov chain is thus (we assimilate the state 1 to the value δ *and the state 2 to the value* $-\delta$*):*

$$P := \begin{pmatrix} p_{cont} & 1 - p_{cont} \\ 1 - p'_{cont} & p'_{cont} \end{pmatrix}$$

Let $p_n^{up}(b, a) := \mathbb{P}[X_n = \delta | q_0^b = b, q_0^a = a]$. *We can compute this quantity explicitly:*

$$p_n^{up}(b, a) = \pi^* + (p_{cont} + p'_{cont} - 1)^{n-1} (p_1^{up}(b, a) - \pi^*),$$

$$\pi^* := \pi^*(\delta) := \frac{p'_{cont} - 1}{p_{cont} + p'_{cont} - 2},$$

where π^* *is the stationary distribution of the Markov chain* $\{X_n\}$:

$$\pi^* = \lim_{n \to \infty} \mathbb{P}[X_n = \delta | X_1].$$

Further:

$$\mathbb{E}[X_n | q_0^b = b, q_0^a = a] = \delta(2 p_n^{up}(b, a) - 1),$$

and the (conditional) covariance between two consecutive price moves:

$$cov[X_{n+1}, X_n | q_0^b = b, q_0^a = a] = 4\delta^2 p_n^{up}(b, a)(1 - p_n^{up}(b, a))(p_{cont} + p'_{cont} - 1).$$

Remark 7.3.6. *In particular, if* $p_{cont} = p'_{cont}$, *then* $\pi^* = 1/2$ *and we retrieve the results of [5]. We also note that the sign of the (conditional) covariance between two consecutive price moves does not depend on the initial configuration of the bid and ask queues and is given by the sign of* $p_{cont} + p'_{cont} - 1$. *We also note that the quantities* p_{cont}, p'_{cont} *can be computed up to the knowledge* of the quantities $p_1^{up}(n_b, n_a)$ which computation was discussed in the previous section. The quantities $F(\pm\delta, t)$ can be computed up to the knowledge of the law of τ, which is known up to the use of a numerical procedure to invert the characteristic functions of σ_a and σ_b, together with the results of Proposition 7.3.4.

Proof.

    The results follow from elementary calculations in a similar way to what is done in [5]. Indeed, we have:

$$\left(\begin{array}{cc} p_n^{up}(b,a) & 1 - p_n^{up}(b,a) \end{array} \right) = \left(\begin{array}{cc} p_1^{up}(b,a) & 1 - p_1^{up}(b,a) \end{array} \right)$$

$$\left(\begin{array}{cc} p_{cont} & 1 - p_{cont} \\ 1 - p'_{cont} & p'_{cont} \end{array} \right)^{n-1}$$

We also have:

$$\left(\begin{array}{cc} p_{cont} & 1 - p_{cont} \\ 1 - p'_{cont} & p'_{cont} \end{array} \right) = S \left(\begin{array}{cc} 1 & 0 \\ 0 & p_{cont} + p'_{cont} - 1 \end{array} \right) S^{-1}$$

with:

$$S = \left(\begin{array}{cc} 1 & -\frac{1 - p_{cont}}{1 - p'_{cont}} \\ 1 & 1 \end{array} \right).$$

7.4   The Mid-Price Process as IHRE

Let

$$s_t = s_0 + \sum_{k=1}^{N(t)} X_k$$

be our mid-price process introduced before, $s_0 := s$ is an initial mid-price value, $N(t)$ is the number of price changes up to time t, $X_k = \{+\delta, -\delta\}$. Let us introduce the following operators $D(x)$ on $B = C_0(R)$:

$$D(x)f(s) := f(s + x).$$

Then our mid-price process s_t above can be expressed as IHRE in the following way:

$$V(t)f(s) = f(s_t) = \Pi_{k=1}^{N(t)} D(X_k)f(s), \quad f(s) \in B.$$

One of the way to get LLN and FCLT for mid-price process s_t is to use our main results, WLLN and FCLT for INREs, obtained in Chapter 5.

In this case we use the following scales: t/ϵ (or, equivalently, nt, $n \to +\infty$) for LLN, and t/ϵ^2 (or, equivalently, tn^2, $n \to +\infty$) for FCLT, when $\epsilon \to 0$.

However, we would like to show even another approach that produces the same results: in Section 5 we obtain LLN and FCLT for mid-price process s_t using our martingale methods from [18] and change of time method.

7.5 Diffusion Limit of the Price Process

In [5] it is assumed that $f(i,j) = \tilde{f}(i,j) = f(j,i)$ in order to make the price increments X_n independent and identically distributed. In fact, this assumption can be entirely relaxed. Indeed, as we mentioned above, $(X_n, \tau_n)_{n \geq 0}$ is in fact a Markov renewal process and therefore we can use the related theory to compute the diffusion limit of the price process. The results of this section generalize the results of Section 4 in [5].

7.5.1 Balanced Order Flow Case: $P^a(1,1) = P^a(-1,-1)$ and $P^b(1,1) = P^b(-1,-1)$

Throughout this section we make the assumption:

(A4) Using the notations of Proposition 7.3.1, the following holds:

$$\sum_{n=1}^{\infty}\sum_{p=1}^{\infty} a^b(n)a^a(p)f(n,p) < \infty, \quad \sum_{n=1}^{\infty}\sum_{p=1}^{\infty} a^b(n)a^a(p)\tilde{f}(n,p) < \infty.$$

Using Proposition 7.3.3, we obtain the following result generalizing Lemma 1 in [5]:

Lemma 7.5.1. *Under assumption* (A4), *the following weak convergence holds as $n \to \infty$:*

$$\frac{1}{n\log(n)}\sum_{k=1}^{n}\tau_k \Rightarrow \tau^* := \sum_{n=1}^{\infty}\sum_{p=1}^{\infty} a^b(n)a^a(p)f^*(n,p), \text{ where } f^*(n,p)$$

$$:= \pi^* f(n,p) + (1-\pi^*)\tilde{f}(n,p).$$

Proof. *We have:*

$$\frac{1}{n\log(n)}\sum_{k=1}^{n}\tau_k = \sum_{i\in\{-\delta,\delta\}}\frac{N_i(n)}{n}\frac{\log(N_i(n))}{\log(n)}\frac{1}{N_i(n)\log(N_i(n))}\sum_{k=1}^{N_i(n)}\tau_{p(k,i)},$$

where for $i \in \{-\delta, \delta\}$, $N_i(n)$ represents the number of times that $X_{k-1} = i$ for $1 \le k \le n$; and $\{p(k,i) : k \ge 1\}$ the successive indexes for which $X_{k-1} = i$. By the standard theory of Markov chains, we have for $i \in \{-\delta, \delta\}$:

$$\frac{N_i(n)}{n} \overset{a.e.}{\to} \pi^*(i),$$

and therefore we have $\frac{\log(N_i(n))}{\log(n)} \overset{a.e.}{\to} 1$. We recall that $\pi^*(\delta) := \pi^*$, and $\pi^*(-\delta) = 1 - \pi^*$. For fixed $i \in \{-\delta, \delta\}$, the random variables $\{\tau_{p(k,i)} : k \ge 1\}$ are iid with distribution $F(i, \cdot)$, and with tail index equal to 1 (by Proposition 7.3.3). Using [5] (Lemma 1) together with Proposition 7.3.3, we get that:

$$\frac{1}{n \log(n)} \sum_{k=1}^{n} \tau_{p(k,\delta)} \Rightarrow \sum_{n=1}^{\infty} \sum_{p=1}^{\infty} \alpha^b(n) \alpha^a(p) f(n,p), \quad \frac{1}{n \log(n)} \sum_{k=1}^{n} \tau_{p(k,-\delta)}$$

$$\Rightarrow \sum_{n=1}^{\infty} \sum_{p=1}^{\infty} \alpha^b(n) \alpha^a(p) \tilde{f}(n,p).$$

The latter convergence holds in probability and we finally have:

$$\frac{1}{n \log(n)} \sum_{k=1}^{n} \tau_k \overset{P}{\to} \pi^* \sum_{n=1}^{\infty} \sum_{p=1}^{\infty} \alpha^b(n) \alpha^a(p) f(n,p) + (1-\pi^*) \sum_{n=1}^{\infty} \sum_{p=1}^{\infty} \alpha^b(n) \alpha^a(p) \tilde{f}(n,p).$$

Let $s^* := \delta(2\pi^* - 1)$. Using the previous Lemma 7.5.1, we obtain the following diffusion limit for the renormalized price process $s_{tn \log(n)}$:

Proposition 7.5.2. *Under assumption* **(A4)**, *the renormalized price process $s_{tn \log(n)}$ satisfies the following weak convergence in the Skorokhod topology ([11]):*

$$\left(\frac{s_{tn \log(n)}}{n}, t \ge 0 \right) \overset{n \to \infty}{\Rightarrow} \left(\frac{s^* t}{\tau^*}, t \ge 0 \right),$$

$$\left(\frac{s_{tn \log(n)} - N_{tn \log(n)} s^*}{\sqrt{n}}, t \ge 0 \right) \overset{n \to \infty}{\Rightarrow} \frac{\sigma}{\sqrt{\tau^*}} W,$$

where W is a standard Brownian motion and σ is given by:

$$\sigma^2 = 4\delta^2 \left(\frac{1 - p'_{cont} + \pi^*(p'_{cont} - p_{cont})}{(p_{cont} + p'_{cont} - 2)^2} - \pi^*(1 - \pi^*) \right).$$

Remark 7.5.3. *If $p'_{cont} = p_{cont} = \pi^* = \frac{1}{2}$ as in [5], we find $s^* = 0$ and $\sigma = \delta$ as in [5]. If $p'_{cont} = p_{cont} = p$, we have $\pi^* = \frac{1}{2}$, $s^* = 0$ and:*

$$\sigma^2 = \delta^2 \frac{p}{1-p}.$$

Proof. *Because* $m(\pm\delta) := \mathbb{E}[\tau_n|X_{n-1} = \pm\delta] = +\infty$ *by Proposition 7.3.3, we cannot directly apply the well-known invariance principle results for semi-Markov processes. Denote for* $t \in \mathbb{R}_+$:

$$R_n := \sum_{k=1}^{n}(X_k - s^*), \quad U_n(t) := n^{-1/2}\left[(1 - \lambda_{n,t})R_{\lfloor nt \rfloor} + \lambda_{n,t}R_{\lfloor nt \rfloor+1}\right],$$

where $\lambda_{n,t} := nt - \lfloor nt \rfloor$. *We can show, following a martingale method similar to [18] (Section 3), that we have the following weak convergence in the Skorokhod topology:*

$$(U_n(t), t \geq 0) \overset{n \to \infty}{\Rightarrow} \sigma W,$$

where W *is a standard Brownian motion, and* σ *is given by:*

$$\sigma^2 = \sum_{i \in \{-\delta,\delta\}} \pi^*(i)v(i),$$

where for $i \in \{-\delta, \delta\}$:

$$v(i) := b(i)^2 + p(i)(g(-i) - g(i))^2 - 2b(i)p(i)(g(-i) - g(i)),$$
$$b(i) := i - s^*,$$
$$p(\delta) := 1 - p_{cont}, \quad p(-\delta) := 1 - p'_{cont},$$

and (the vector) g *is given by:*

$$g = (P + \Pi^* - I)^{-1}b,$$

where Π^* *is the matrix with rows equal to* $(\pi^* \quad 1 - \pi^*)$. *After completing the calculations we get:*

$$\sigma^2 = 4\delta^2\left(\frac{1 - p'_{cont} + \pi^*(p'_{cont} - p_{cont})}{(p_{cont} + p'_{cont} - 2)^2} - \pi^*(1 - \pi^*)\right).$$

For the sake of exhaustivity we also give the explicit expression for g:

$$g(\delta) = \delta\frac{p'_{cont} - p_{cont} + 2(1 - \pi^*)}{p_{cont} + p'_{cont} - 2} - s^*,$$
$$g(-\delta) = \delta\frac{p'_{cont} - p_{cont} - 2\pi^*}{p_{cont} + p'_{cont} - 2} - s^*.$$

Indeed, to show the above convergence of U_n, *we observe that we can write* R_n *as the sum of a* $\mathcal{F}_n$-*martingale* M_n *and a bounded process:*

$$R_n = M_n + \underbrace{g(X_n) - g(X_0) + X_n - X_0}_{unif.bounded},$$

$$M_n := \sum_{k=1}^{n} b(X_{k-1}) - g(X_k) + g(X_{k-1}),$$

where $\mathcal{F}_n := \sigma(\tau_k, X_k : k \leq n)$ and $X_0 := 0$. The process M_n is a martingale because g is the unique solution of the following Poisson equation, since $\Pi^* b = 0$:

$$[P - I]g = b.$$

The rest of the proof for the convergence of U_n follows exactly [18] (Section 3).

We proved earlier (Lemma 7.5.1) that:

$$\frac{T_n}{n \log(n)} \Rightarrow \tau^*,$$

where $T_n := \sum_{k=1}^n \tau_k$. Since the Markov renewal process $(X_n, \tau_n)_{n \geq 0}$ is regular (because the state space is finite), we get $N_t \to \infty$ a.s. and therefore:

$$\frac{T_{N_t}}{N_t \log(N_t)} \Rightarrow \tau^*.$$

Observing that $T_{N_t} \leq t \leq T_{N_t+1}$ a.s., we get:

$$\frac{T_{N_t}}{N_t \log N_t} \leq \frac{t}{N_t \log N_t} \leq \frac{(N_t + 1) \log(N_t + 1)}{N_t \log N_t} \frac{T_{N_t+1}}{(N_t + 1) \log(N_t + 1)},$$

and therefore:

$$\frac{t}{N_t \log(N_t)} \Rightarrow \tau^*.$$

Let $t_n := tn \log(n)$. We would like to show as in [5], equation (17) that:

$$N_{t_n} \overset{P}{\sim} \frac{nt}{\tau^*}.$$

We have denoted by $A_n \overset{P}{\sim} B_n$ iff $P - \lim \frac{A_n}{B_n} = 1$. We denote as in [5] $\rho : (1, \infty) \to (1, \infty)$ to be the inverse function of $t \log(t)$, and we note that $\rho(t) \overset{t \to \infty}{\sim} \frac{t}{\log(t)}$. The first equivalence in [5], equation (17): $N_{t_n} \overset{P}{\sim} \rho\left(\frac{t_n}{\tau^*}\right)$ is not obvious. Indeed, we have $N_{t_n} \log(N_{t_n}) \overset{P}{\sim} \frac{t_n}{\tau^*}$, and we would like to conclude that $N_{t_n} = \rho(N_{t_n} \log(N_{t_n})) \overset{P}{\sim} \rho\left(\frac{t_n}{\tau^*}\right)$. The latter implication is not true for every function ρ, in particular if ρ was exponential. Nevertheless, in our case, it is true because $\rho(t) \overset{t \to \infty}{\sim} \frac{t}{\log(t)}$, and therefore for any functions f, g going to $+\infty$ as $t \to \infty$:

$$\frac{\rho(f(t))}{\rho(g(t))} \overset{t \to \infty}{\sim} \frac{f(t) \log(g(t))}{g(t) \log(f(t))}.$$

Therefore we see that if $f(t) \overset{t \to \infty}{\sim} g(t)$, then by property of the logarithm $\log(f(t)) \overset{t \to \infty}{\sim} \log(g(t))$ and therefore $\rho(f(t)) \overset{t \to \infty}{\sim} \rho(g(t))$. This allows us to conclude as in [5] that:

$$\frac{N_{t_n}}{n} \overset{P}{\sim} \frac{t}{\tau^*}.$$

Therefore, we can make a change of time as in [18], Corollary 3.19 (see also [1], Section 14), and denoting $\alpha_n(t) := \frac{N_{t_n}}{n}$, we obtain the following weak convergence in the Skorokhod topology:

$$(U_n(\alpha_n(t)), t \geq 0) \Rightarrow (\sigma W_{\frac{t}{\tau^*}}, t \geq 0),$$

that is to say:

$$\left(\frac{S_{tn \log(n)} - N_{tn \log(n)} s^*}{\sqrt{n}}, t \geq 0 \right) \Rightarrow \frac{\sigma}{\sqrt{\tau^*}} W.$$

The law of large numbers result comes from the fact that $\frac{N_{t_n}}{n} \overset{P}{\sim} \frac{t}{\tau^*}$, together with the following fact (strong law of large numbers for Markov chains):

$$\frac{1}{n} \sum_{k=1}^{n} X_k \to s^* \quad a.e.$$

7.5.2   Other Cases: Either $P^a(1,1) < P^a(-1,-1)$ or $P^b(1,1) < P^b(-1,-1)$

In this case, we know by Proposition 7.3.3 that the conditional expectations $\mathbb{E}[\tau_k | q_0^b = n_b, q_0^a = n_a]$ are finite. Denoting the conditional expectations $m(\pm\delta) := \mathbb{E}[\tau_k | X_{k-1} = \pm\delta]$, we have:

$$m(\delta) = \sum_{p=1}^{\infty} \sum_{n=1}^{\infty} \mathbb{E}[\tau_k | q_0^b = n, q_0^a = p] f(n,p), m(-\delta)$$

$$= \sum_{p=1}^{\infty} \sum_{n=1}^{\infty} \mathbb{E}[\tau_k | q_0^b = n, q_0^a = p] \tilde{f}(n,p).$$

Throughout this section we will need the following assumption:

(A5) Using the previous notations, the following holds:

$$m(\pm\delta) < \infty.$$

For example, the above assumption is satisfied if the support of the distributions f and $\tilde{f}$ are compact, which is the case in practice. We obtain the following diffusion limit result as a classical consequence of invariance principle results for semi-Markov processes (see e.g. [18], Section 3):

Proposition 7.5.4. *Under assumption* **(A5)**, *the renormalized price process* s_{nt} *satisfies the following convergence in the Skorokhod topology:*

$$\left(\frac{s_{nt}}{n}, t \geq 0\right) \overset{n\to\infty}{\to} \left(\frac{s^* t}{m_\tau}, t \geq 0\right) \quad a.e.,$$

$$\left(\frac{s_{nt} - N_{nt} s^*}{\sqrt{n}}, t \geq 0\right) \overset{n\to\infty}{\Rightarrow} \frac{\sigma}{\sqrt{m_\tau}} W,$$

where W is a standard Brownian motion, σ is given in Proposition 7.5.2 and:

$$m_\tau := \sum_{i \in \{-\delta, \delta\}} \pi^*(i) m(i) = \pi^* m(\delta) + (1 - \pi^*) m(-\delta).$$

Proof. *This is an immediate consequence of the strong law of large numbers and invariance principle results for Markov renewal processes satisfying $m(\pm\delta) < \infty$ (see e.g. [18] Section 3). In the previous article [18], the proof of the invariance principle is carried on using a martingale method similar to the one of the proof of Proposition 7.5.2.*

7.6 Numerical Results

In this section, we present calibration results which illustrate and justify our approach.

In [5], it is assumed that the queue changes V_k^b, V_k^a do not depend on their previous values V_{k-1}^b, V_{k-1}^a. Empirically, it is found that $\mathbb{P}[V_k^b = 1] \approx \mathbb{P}[V_k^b = -1] \approx 1/2$ (and similarly for the ask side). Here, we challenge this assumption by estimating and comparing the probabilities $P(-1, 1)$ vs. $P(1, 1)$ on the one side and $P(-1, -1)$ vs. $P(1, -1)$ on the other side to check whether or not they are approximately equal to each other, for both the ask and the bid. We also give - for both the bid and ask - the estimated probabilities $\mathbb{P}[V_k = 1]$, $\mathbb{P}[V_k = -1]$ that we call respectively $P(1)$, $P(-1)$, to check whether or not they are approximately equal to $1/2$ as in [5].

The results below correspond to the 5 stocks Amazon, Apple, Google, Intel and Microsoft on June 21st, 2012[1]. The probabilities are estimated using the strong law of large numbers. We also give for indicative purposes the average time between order arrivals (in milliseconds (ms)) as well as the average number of stocks per order.

---

[1]The data were taken from the webpage *https://lobster.wiwi.hu-berlin.de/info/ DataSamples.php*

Amazon		Apple		Google		Intel		Microsoft	
	Bid	Ask	Bid	Ask	Bid	Ask	Bid	Ask	Bid	Ask
Avg time btw. orders (ms)	910	873	464	425	1123	1126	116	133	130	113
Avg nb. of stocks per order	100	82	90	82	84	71	502	463	587	565

Average time between orders (ms) and average number of stocks per order. June 21st, 2012.

	Amazon		Apple		Google		Intel		Microsoft	
	Bid	Ask	Bid	Ask	Bid	Ask	Bid	Ask	Bid	Ask
$P(1,1)$	0.48	0.57	0.50	0.55	0.48	0.53	0.55	0.61	0.63	0.60
$P(-1,1)$	0.46	0.42	0.40	0.42	0.46	0.49	0.44	0.40	0.36	0.41
$P(-1,-1)$	0.54	0.58	0.60	0.58	0.54	0.51	0.56	0.60	0.64	0.59
$P(1,-1)$	0.52	0.43	0.50	0.45	0.52	0.47	0.45	0.39	0.37	0.40
$P(1)$	0.47	0.497	0.44	0.48	0.47	0.51	0.495	0.505	0.49	0.508
$P(-1)$	0.53	0.503	0.56	0.52	0.53	0.49	0.505	0.495	0.51	0.492

Estimated transition probabilities of the Markov chains V_k^b, V_k^a. June 21st, 2012.

Findings: First of all, we find as in [5] that for all stocks, $\mathbb{P}[V_k = 1] \approx \mathbb{P}[V_k = -1] \approx 1/2$, except maybe in the case of Apple Bid. It is worth mentioning that we always have $P(1) < P(-1)$ except in three cases: Google Ask, Intel Ask and Microsoft Ask. Nevertheless, in these cases, $P(1)$ and $P(-1)$ are very close to each other and so they could be considered to fall into the case $P(1) = P(-1)$ of [5]. These three cases also correspond to the only three cases where $P(1,1) > P(-1,-1)$, which is contrary to our assumption $P(1,1) \leq P(-1,-1)$. Nevertheless, in these three cases, $P(1,1)$ and $P(-1,-1)$ are very close to each other so we can consider them to fall into the case $P(1,1) = P(-1,-1)$.

More importantly, we notice that the probabilities $P(-1,1)$, $P(1,1)$ can be significantly different from each other - and similarly for the probabilities $P(-1,-1)$, $P(1,-1)$ - which justifies the use of a Markov Chain structure for the random variables $\{V_k^b\}, \{V_k^a\}$. This phenomenon is particularly visible for example on Microsoft (Bid+Ask), Intel (Bid+Ask), Apple (Bid+Ask) or Amazon Ask. Further, regarding the comparison of $P(1,1)$ and $P(-1,-1)$, it turns out that they are often very similar, except in the cases of Amazon Bid, Apple Bid and Google Bid.

The second assumption of [5] that we would like to challenge is the assumed exponential distribution of the order arrival times T_k^a, T_k^b. To this end, on the same data set as used to estimate the transition probabilities $P^a(i,j)$, $P^b(i,j)$, we calibrate the empirical c.d.f.'s $H^a(i,j,\cdot), H^b(i,j,\cdot)$ to the Gamma and Weibull distributions (which are generalizations of the exponential distribution). We recall that the p.d.f.'s of these distributions are given by:

$$f_{Gamma}(x) = \frac{1}{\Gamma(k)\theta^k} x^{k-1} e^{-\frac{x}{\theta}} 1_{x>0},$$

$$f_{Weibull}(x) = \frac{k}{\theta} \left(\frac{x}{\theta}\right)^{k-1} e^{-\left(\frac{x}{\theta}\right)^k} 1_{x>0}.$$

Here, $k > 0$ and $\theta > 0$ represent respectively the shape and the scale parameter. The variable k is dimensionless, whereas θ will be expressed in ms^{-1}. We perform a maximum likelihood estimation of the Weibull and Gamma parameters for each one of the empirical distributions $H^a(i,j,\cdot)$, $H^b(i,j,\cdot)$ (together with a 95 % confidence interval for the parameters). As we can see on the tables below, the shape parameter k is always significantly different than 1 ($\sim$ 0.1 to 0.3), which indicates that the exponential distribution is not rich enough to fit our observations. To illustrate this, we present below the empirical c.d.f. of $H(1,-1)$ in the case of Google Bid, and we see that Gamma and Weibull allow to fit the empirical c.d.f. in a much better way than Exponential.

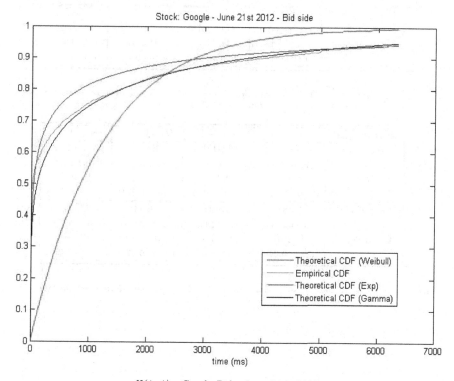

H(1,-1) - Google Bid - June 21st, 2012.

We summarize our calibration results in the tables below.

Amazon Bid	$H(1,1)$	$H(1,-1)$	$H(-1,-1)$	$H(-1,1)$
Weibull θ	99.1	185.5	87.7	87.0
	(90.2-109.0)	(171.3-200.8)	(80.1-96.0)	(78.7-96.1)
Weibull k	0.279	0.323	0.285	0.258
	(0.274-0.285)	(0.317-0.329)	(0.280-0.290)	(0.253-0.263)
Gamma θ	4927	4321	4712	5965
	(4618-5257)	(4075-4582)	(4423-5019)	(5589-6366)
Gamma k	0.179	0.215	0.179	0.165
	(0.174-0.184)	(0.209-0.220)	(0.175-0.184)	(0.161-0.169)

Amazon Bid: Fitted Weibull and Gamma parameters. 95 % confidence intervals in brackets. June 21st, 2012.

Amazon Ask	$H(1,1)$	$H(1,-1)$	$H(-1,-1)$	$H(-1,1)$
Weibull θ	80.8	197.8	57.9	137.0
	(74.4-87.7)	(181.9-215.1)	(52.8-63.4)	(124.2-151.2)
Weibull k	0.274	0.324	0.279	0.276
	(0.269-0.278)	(0.317-0.330)	(0.274-0.285)	(0.270-0.281)
Gamma θ	4732	4623	3845	5879
	(4475-5004)	(4345-4919)	(3609-4095)	(5502-6283)
Gamma k	0.174	0.215	0.173	0.181
	(0.170-0.178)	(0.209-0.221)	(0.168-0.177)	(0.176-0.186)

Amazon Ask: Fitted Weibull and Gamma parameters. 95 % confidence intervals in brackets. June 21st, 2012.

Apple Bid	$H(1,1)$	$H(1,-1)$	$H(-1,-1)$	$H(-1,1)$
Weibull θ	75.9	180.9	31.5	78.2
	(71.6-80.5)	(172.6-189.7)	(29.5-33.6)	(73.4-83.3)
Weibull k	0.317	0.400	0.271	0.300
	(0.313-0.321)	(0.394-0.405)	(0.267-0.274)	(0.296-0.304)
Gamma θ	2187	1860	2254	2711
	(2094-2284)	(1787-1935)	(2157-2355)	(2592-2835)
Gamma k	0.206	0.276	0.168	0.196
	(0.202-0.210)	(0.271-0.282)	(0.165-0.171)	(0.192-0.199)

Apple Bid: Fitted Weibull and Gamma parameters. 95 % confidence intervals in brackets. June 21st, 2012.

Apple Ask	$H(1,1)$	$H(1,-1)$	$H(-1,-1)$	$H(-1,1)$
Weibull θ	46.6	152.5	27.7	95.5
	(44.1-49.2)	(145.5-159.8)	(26.0-29.6)	(90.0-101.5)
Weibull k	0.298	0.394	0.271	0.308
	(0.294-0.301)	(0.388-0.399)	(0.267-0.275)	(0.303-0.312)
Gamma θ	2019	1666	1995	2740
	(1942-2099)	(1603-1732)	(1907-2087)	(2624-2861)
Gamma k	0.189	0.271	0.168	0.204
	(0.186-0.192)	(0.266-0.277)	(0.165-0.171)	(0.200-0.208)

Apple Ask: Fitted Weibull and Gamma parameters. 95 % confidence intervals in brackets. June 21st, 2012.

Google Bid	$H(1,1)$	$H(1,-1)$	$H(-1,-1)$	$H(-1,1)$
Weibull θ	113.9	158.5	67.9	56.8
	(102.8-126.2)	(143.4-175.3)	(60.6-76.0)	(50.5-63.8)
Weibull k	0.276	0.284	0.261	0.246
	(0.270-0.282)	(0.278-0.290)	(0.255-0.266)	(0.241-0.251)
Gamma θ	6720	6647	6381	7025
	(6263-7210)	(6204-7122)	(5913-6886)	(6517-7571)
Gamma k	0.174	0.185	0.160	0.151
	(0.169-0.179)	(0.180-0.191)	(0.155-0.165)	(0.147-0.156)

Google Bid: Fitted Weibull and Gamma parameters. 95 % confidence intervals in brackets. June 21st, 2012.

Google Ask	$H(1,1)$	$H(1,-1)$	$H(-1,-1)$	$H(-1,1)$
Weibull θ	196.7	271.6	38.1	57.0
	(180.6-214.2)	(248.5-296.8)	(33.8-43.0)	(51.3-63.3)
Weibull k	0.290	0.310	0.258	0.263
	(0.285-0.295)	(0.303-0.316)	(0.253-0.264)	(0.258-0.268)
Gamma θ	6081	6571	4304	4698
	(5734-6450)	(6165-7003)	(3971-4664)	(4380-5040)
Gamma k	0.195	0.209	0.156	0.164
	(0.190-0.200)	(0.203-0.215)	(0.151-0.161)	(0.159-0.168)

Google Ask: Fitted Weibull and Gamma parameters. 95 % confidence intervals in brackets. June 21st, 2012.

Intel Bid	$H(1,1)$	$H(1,-1)$	$H(-1,-1)$	$H(-1,1)$
Weibull θ	2.76	2.56	3.33	2.01
	(2.66-2.86)	(2.45-2.67)	(3.21-3.45)	(1.92-2.10)
Weibull k	0.227	0.226	0.267	0.209
	(0.226-0.229)	(0.225-0.228)	(0.265-0.269)	(0.208-0.211)
Gamma θ	1016	912	543	1093
	(991-1040)	(888-937)	(530-557)	(1063-1124)
Gamma k	0.129	0.130	0.151	0.120
	(0.128-0.130)	(0.129-0.131)	(0.150-0.152)	(0.119-0.121)

Intel Bid: Fitted Weibull and Gamma parameters. 95 % confidence intervals in brackets. June 21st, 2012.

Intel Ask	$H(1,1)$	$H(1,-1)$	$H(-1,-1)$	$H(-1,1)$
Weibull θ	1.33	5.46	4.63	5.15
	(1.28-1.38)	(5.21-5.73)	(4.45-4.80)	(4.90-5.41)
Weibull k	0.235	0.231	0.256	0.225
	(0.234-0.237)	(0.230-0.233)	(0.254-0.257)	(0.224-0.227)
Gamma θ	705	1219	884	1305
	(688-723)	(1183-1256)	(862-907)	(1266-1345)
Gamma k	0.126	0.137	0.146	0.133
	(0.125-0.127)	(0.136-0.139)	(0.144-0.147)	(0.132-0.135)

Intel Ask: Fitted Weibull and Gamma parameters. 95 % confidence intervals in brackets. June 21st, 2012.

Microsoft Bid	$H(1,1)$	$H(1,-1)$	$H(-1,-1)$	$H(-1,1)$
Weibull θ	0.79	2.98	2.68	2.64
	(0.76-0.82)	(2.83-3.13)	(2.59-2.78)	(2.50-2.78)
Weibull k	0.215	0.221	0.259	0.211
	(0.214-0.217)	(0.219-0.223)	(0.257-0.260)	(0.209-0.213)
Gamma θ	1012	1315	664	1488
	(987-1039)	(1274-1358)	(648-681)	(1440-1537)
Gamma k	0.112	0.125	0.142	0.120
	(0.111-0.113)	(0.124-0.127)	(0.141-0.143)	(0.118-0.121)

Microsoft Bid: Fitted Weibull and Gamma parameters. 95 % confidence intervals in brackets. June 21st, 2012.

Microsoft Ask	$H(1,1)$	$H(1,-1)$	$H(-1,-1)$	$H(-1,1)$
Weibull θ	0.85	1.57	2.07	1.43
	(0.82-0.89)	(1.50-1.64)	(2.00-2.15)	(1.36-1.50)
Weibull k	0.218	0.223	0.259	0.210
	(0.217-0.219)	(0.222-0.225)	(0.258-0.261)	(0.208-0.211)
Gamma θ	1004	1081	574	1138
	(980-1028)	(1051-1112)	(560-588)	(1105-1171)
Gamma k	0.113	0.121	0.140	0.116
	(0.112-0.114)	(0.120-0.122)	(0.139-0.141)	(0.115-0.117)

Microsoft Ask: Fitted Weibull and Gamma parameters. 95 % confidence intervals in brackets. June 21st, 2012.

Bibliography

[1] Billingsley, P. *Convergence of Probability Measures*, John Wiley & Sons, Inc., 1999.

[2] A. Cartea, and S. Jaimungal, *Optimal Execution with Limit and Market Orders*, Quantitative Finance, forthcoming.

[3] K. J. Cohen, R. M. Conroy and S. F. Maier, *Order flow and the quality of the market*, in Market Making and the Changing Structure of the Securities Industry (1985).

[4] R. Cont, S. Stoikov and R. Talreja, *A stochastic model for order book dynamics*, Operations Research, 58 (2010), pp. 549-563.

[5] R. Cont and A. de Larrard, *Price dynamics in a Markovian limit order book market*, SIAM Journal for Financial Mathematics, 4 (2013), No 1, pp. 1-25.

[6] I. Domowitz and J. Wang, *Auctions as algorithms*, J. Economic Dynamics and Control, 18 (1994), pp. 29-60.

[7] P. Fodra and H. Pham, *Semi Markov model for market microstructure*, ArXiv:1305v1 [q-fin.TR], 1 May (2013).

[8] P. Fodra and H. Pham, *High frequency trading and asymptotics for small risk aversion in a Markov renewal model*, ArXiv:1310.1765v2 [q-fin.TR], 4 Jan (2015).

[9] M. Garman, *Market microstructure*, J. Financial Economics, 3 (1976), pp. 257-275.

[10] N. Limnios and G. Oprisan, *Semi-Markov Processes and Reliability*, Birkhauser, 2001.

[11] H. Luckock, *A steady-state model of a continuous double auction*, Quant. Finance, 3 (2003), pp. 385-404.

[12] H. Mendelson, *Market behavior in a clearing house*, Econometrica, 50 (1982), pp. 1505-1524.

[13] S. Predoiu, G. Shaikhet and S. Shreve, *Optimal execution in a general one-sided limit order book*, SIAM J. Financial Math., 2(1) (2011), pp. 183-212.

[14] A. Skorokhod, *Studies in the Theory of Random Processes*, Addison-Wesley, Reading, Mass., 1965 (Reprinted by Dover Publications, NY).

[15] E. Smith, J.D. Farmer, L. Gillemot and S. Krishnamurthy, *Statistical theory of the continuous double auction*, Quant. Finance, 3 (2003), pp. 481-514.

[16] N. Vadori and A. Swishchuk, *Strong Law of Large Numbers and Central Limit Theorems for Functionals of Inhomogeneous Semi-Markov Processes*, Stochastic Analysis and Applications, 33:2 (2015), pp. 213-243.

8

Applications of IHREs in Insurance: Risk Model Based on General Compound Hawkes Process

In this chapter, we introduce a new model for the risk process based on general compound Hawkes process (GCHP) for the arrival of claims. We call it risk model based on general compound Hawkes process (RMGCHP). The Law of Large Numbers (LLN) and the Functional Central Limit Theorem (FCLT) are proved. If we take $B = C_0(R)$ and $f \in B$, define the following operators on B

$$\Gamma(t)f(u) := f(u + ct)$$

and

$$D(x)f(u) := f(u - a(x)),$$

where $u > 0$ and $c > 0$, and $a(x)$ is a bounded function, then the risk process $R(t)$, i.e., RMGCHP, can be expressed as IHRE $V(t)$ in the following way:

$$V(t)f(u) = \Gamma(t)\Pi_{k=1}^{N(t)}D(X_k)f(u), \quad f(u) \in B,$$

where $N(t)$ is a Hawkes process, X_k is a discrete-time Markov chain.

One of the ways to get LLN and FCLT for risk process $R(t)$ is to use our main results, WLLN and FCLT for INREs, obtained in Chapter 5.

In this case we use the following scales: t/ϵ (or, equivalently, nt, $n \to +\infty$) for LLN, and t/ϵ^2 (or, equivalently, tn^2, $n \to +\infty$) for FCLT, when $\epsilon \to 0$. We only need here one additional result for Hawkes process, namely, LLN for Hawkes process (see [3] or [4]):

$$N(t)/t \to_{t \to +\infty} \lambda/(1 - \hat{\mu}).$$

However, we would like to show in this chapter, at the same time, another approach that produce the same results: in Section 8 we obtain LLN and FCLT for RMGCHP using our martingale methods from [18] and change of time method.

8.1 Introduction

The Hawkes process ([8]) is a simple point process that has self-exciting property, clustering effect and long memory. It has been widely applied in seismology, neuroscience, DNA modeling and many other fields, including finance ([7] and insurance ([12]). In this chapter, we introduce a new model for the risk process, based on general compound Hawkes process (GCHP) for the arrival of claims. We call it *risk model based on general compound Hawkes process* (RMGCHP). To the best of the author's knowledge, this risk model is the most general relying on the existing literature. Compound Hawkes process and risk model based on it was introduced in [12]. In comparison to simple Poisson arrival of claims, GCHP model accounts for the risk of contagion and clustering of claims. We note that [12] were the first who replaced Poisson process by a simple Hawkes process in studying the classical problem of the probability of ruin. [5] considered the same ruin problem using marked mutually exciting process (dynamic contagion process). [9] implemented [5] to calculate insurance premiums and suggested higher premiums should be set up in general across different insurance product lines. Semi-Markov risk processes and their optimal control and stability were first introduced in [13] and studied and developed in [14]. Compound Hawkes processes were applied to Limit Order Books (LOB) in [17]. General compound Hawkes processes have also been applied to LOB in [16]. The general compound Hawkes process was first introduced in [15] to model a risk process in insurance. We also note that central limit theorem for nonlinear Hawkes processes was considered in [19].

The chapter is organized as follows. Section 8.2 is devoted to the description of Hawkes process. Section 8.3 describes RMGCHP as an IHRE. Section 8.4 contains Law of Large Numbers (LLN) and Functional Central Limit Theorem (FCLT) for RMGCHP. Section 8.5 contains applications of LLN and FCLT, including net profit condition, premium principle, ruin and ultimate ruin probabilities, and the probability density function of the time to ruin for RMGCHP. Section 8.6 describes applications of the results from Section 8.5 to the risk model based on compound Hawkes process (RMCHP). Section 8.6 contains the applications of the results from Section 8.6 to the classical risk model based on compound Poisson process (RMCPP), just for the completeness of the presentation. And Section 8.7 concludes the book.

8.2 Hawkes, General Compound Hawkes Process (GCHP) and Risk Model Based on GCHP

In this section we introduce Hawkes and general compound Hawkes processes and give some of their properties. We also introduce the risk model based on GCHP.

8.2.1 Hawkes Process

Definition 1 (Counting Process). A counting process is a stochastic process $N(t), t \geq 0$, taking positive integer values and satisfying: $N(0) = 0$. It is almost surely finite, and is a right-continuous step function with increments of size $+1$. (See, e.g., [4]).

Denote by $\mathcal{F}^N(t), t \geq 0$, the history of the arrivals up to time t, that is, $\{\mathcal{F}^N(t), t \geq 0\}$, is a filtration, (an increasing sequence of σ-algebras).

A counting process $N(t)$ can be interpreted as a cumulative count of the number of arrivals into a system up to the current time t.

The counting process can also be characterized by the sequence of random arrival times $(T_1, T_2, ...)$ at which the counting process $N(t)$ has jumped. The process defined by these arrival times is called a point process.

Definition 2 (Point Process). If a sequence of random variables $(T_1, T_2, ...)$, taking values in $[0, +\infty)$, has $P(0 \leq T_1 \leq T_2 \leq ...) = 1$, and the number of points in a bounded region is almost surely finite, then, $(T_1, T_2, ...)$ is called a point process. (See, e.g., [4]).

Definition 3 (Conditional Intensity Function). Consider a counting process $N(t)$ with associated histories $\mathcal{F}^N(t), t \geq 0$. If a non-negative function $\lambda(t)$ exists such that

$$\lambda(t) = \lim_{h \to 0} \frac{E[N(t+h) - N(t)|\mathcal{F}^N(t)]}{h}, \tag{1}$$

then it is called the conditional intensity function of $N(t)$. We note, that sometimes this function is called the hazard function.

Definition 4 (One-Dimensional Hawkes Process) ([8]). The one-dimensional Hawkes process is a point point process $N(t)$ which is characterized by its intensity $\lambda(t)$ with respect to its natural filtration:

$$\lambda(t) = \lambda + \int_0^t \mu(t-s)dN(s), \tag{2}$$

where $\lambda > 0$, and the response function $\mu(t)$ is a positive function and satisfies $\int_0^{+\infty} \mu(s)ds < 1$.

The constant λ is called the background intensity and the function $\mu(t)$ is sometimes also called the excitation function. We suppose that $\mu(t) \neq 0$ to avoid the trivial case, which is, a homogeneous Poisson process. Thus, the Hawkes process is a non-Markovian extension of the Poisson process.

The interpretation of equation (2) is that the events occur according to an intensity with a background intensity λ which increases by $\mu(0)$ at each new event then decays back to the background intensity value according to the function $\mu(t)$. Choosing $\mu(0) > 0$ leads to a jolt in the intensity at each new event, and this feature is often called a self-exciting feature, in other words, because an arrival causes the conditional intensity function $\lambda(t)$ in (1)-(2) to increase then the process is said to be self-exciting.

With respect to definitions of $\lambda(t)$ in (1) and $N(t)$ (2), it follows that

$$P(N(t+h) - N(t) = m | \mathcal{F}^N(t)) = \begin{cases} \lambda(t)h + o(h), & m = 1 \\ o(h), & m > 1 \\ 1 - \lambda(t)h + o(h), & m = 0. \end{cases}$$

We should mention that the conditional intensity function $\lambda(t)$ in (1)-(2) can be associated with the compensator $\Lambda(t)$ of the counting process $N(t)$, that is:

$$\Lambda(t) = \int_0^t \lambda(s)ds. \tag{3}$$

Thus, $\Lambda(t)$ is the unique $\mathcal{F}^N(t), t \geq 0$, predictable function, with $\Lambda(0) = 0$, and is non-decreasing, such that

$$N(t) = M(t) + \Lambda(t) \quad a.s.,$$

where $M(t)$ is an $\mathcal{F}^N(t), t \geq 0$, local martingale (this is the Doob-Meyer decomposition of N).

A common choice for the function $\mu(t)$ in (2) is one of exponential decay:

$$\mu(t) = \alpha e^{-\beta t}, \tag{4}$$

with parameters $\alpha, \beta > 0$. In this case the Hawkes process is called the Hawkes process with exponentially decaying intensity.

Thus, the equation (2) becomes

$$\lambda(t) = \lambda + \int_0^t \alpha e^{-\beta(t-s)} dN(s), \tag{5}$$

We note, that in the case of (4), the process $(N(t), \lambda(t))$ is a continuous-time Markov process, which is not the case for the choice (2).

With some initial condition $\lambda(0) = \lambda_0$, the conditional density $\lambda(t)$ in (5) with the exponential decay in (4) satisfies the SDE

$$d\lambda(t) = \beta(\lambda - \lambda(t))dt + \alpha dN(t), \quad t \geq 0,$$

which can be solved (using stochastic calculus) as

$$\lambda(t) = e^{-\beta t}(\lambda_0 - \lambda) + \lambda + \int_0^t \alpha e^{-\beta(t-s)} dN(s),$$

which is an extension of (5).

Another choice for $\mu(t)$ is a power law function:

$$\lambda(t) = \lambda + \int_0^t \frac{k}{(c + (t-s))^p} dN(s) \tag{6}$$

for some positive parameters c, k, p.

This power law form for $\lambda(t)$ in (6) was applied in the geological model called Omori's law, and used to predict the rate of aftershocks caused by an earthquake.

Many generalizations of Hawkes processes have been proposed. They include, in particular, multi-dimensional Hawkes processes, non-linear Hawkes processes, mixed diffusion-Hawkes models, Hawkes models with shot noise exogenous events and Hawkes processes with generation dependent kernels.

8.2.2 General Compound Hawkes Process (GCHP)

Definition 7 (General Compound Hawkes Process (GCHP)). Let $N(t)$ be any one-dimensional Hawkes process defined above. Let also X_n be ergodic continuous-time finite (or possibly infinite but countable) state Markov chain, independent of $N(t)$, with space state X, and $a(x)$ be any bounded and continuous function on X. The general compound Hawkes process is defined as

$$S_t = S_0 + \sum_{k=1}^{N(t)} a(X_k). \tag{7}$$

Some Examples of GCHP

1. Compound Poisson Process: $S_t = S_0 + \sum_{k=1}^{N(t)} X_k$, where $N(t)$ is a Poisson process and $a(X_k) = X_k$ are iidrv.

2. Compound Hawkes Process: $S_t = S_0 + \sum_{k=1}^{N(t)} X_k$, where $N(t)$ is a Hawkes process and $a(X_k) = X_k$ are i.i.d.r.v.

3. Compound Markov Renewal Process: $S_t = S_0 + \sum_{k=1}^{N(t)} a(X_k)$, where $N(t)$ is a renewal process and X_k is a Markov chain.

8.2.3 Risk Model Based on General Compound Hawkes Process

Definition 8 (RMGCHP: Finite State MC). We define the risk model $R(t)$ based on GCHP as follows:

$$R(t) := u + ct - \sum_{k=1}^{N(t)} a(X_k), \tag{8}$$

where u is the initial capital of an insurance company, c is the rate of at which premium is paid, X_k is continuous-time Markov chain in state space $X = \{1, 2, ..., n\}$, $N(t)$ is a Hawkes process, $a(x)$ is continuous and bounded function on X). $N(t)$ and X_k are independent.

Definition 8 (RMGCHP: Infinite State MC). We define the risk model $R(t)$ based on GCHP for infinite state but countable Markov chain as follows:

$$R(t) := u + ct - \sum_{k=1}^{N(t)} a(X_k). \qquad (8')$$

Here: $X = \{1, 2, ..., n, ...\}$-infinite but countable space of states for Markov chain X_k.

Some Examples of RMGCHP

1. Classical Risk Process (Cramer-Lundberg Risk Model): If $a(X_k) = X_k$ are i.i.d.r.v. and $N(t)$ is a homogeneous Poisson process, then $R(t)$ is a classical risk process also known as the Cramer-Lundberg risk model (see Asmussen and Albrecher (2010)). In the latter case we have compound Poisson process (CPP) for the outgoing claims.

Remark 1. Using this analogy, we call our risk process as a risk model based on general compound Hawkes process (GCHP).

2. Risk Model based on Compound Hawkes Process: If $a(X_k) = X_k$ are i.i.d.r.v. and $N(t)$ is a Hawkes process, then $R(t)$ is a risk process with non-stationary Hawkes claims arrival introduced in [12].

8.3   RMGCHP as IHRE

Let $B = C_0(R)$ and $f \in B$. Define the following operators on B:

$$\Gamma(t)f(u) := f(u + ct)$$

and

$$D(x)f(u) := f(u - a(x)),$$

where $u > 0$ and $c > 0$, and $a(x)$ is a bounded function (see (8)). Then risk process $R(t)$ in (8), i.e., RMGCHP, can be expressed as IHRE $V(t)$ in the following way:

$$V(t)f(u) = \Gamma(t)\Pi_{k=1}^{N(t)} D(X_k)f(u), \quad f(u) \in B.$$

One of the ways to get LLN and FCLT for risk process $R(t)$ is to use our main results, WLLN and FCLT for INREs, obtained in Chapter 5.

In this case we use the following scales: t/ϵ (or, equivalently, nt, $n \to +\infty$) for LLN, and t/ϵ^2 (or, equivalently, tn^2, $n \to +\infty$) for FCLT, when $\epsilon \to 0$.

We only need here one additional result for Hawkes process, namely, LLN for Hawkes process (see [3] or [4]):

$$N(t)/t \to_{t \to +\infty} \lambda/(1 - \hat{\mu}).$$

However, we would like to show even another approach that produce the same results: in Section 4 we obtain LLN and FCLT for RMGCHP using our martingale methods from [18] and change of time method.

8.4 LLN and FCLT for RMGCHP

In this section we present LLN and FCLT for RMGCHP.

8.4.1 LLN for RMGCHP

Theorem 1 (LLN for RMGCHP). Let $R(t)$ be the risk model (RMGCHP) defined above in (8), and X_k be an ergodic Markov chain with stationary probabilities π_n^*. We suppose that $0 < \hat{\mu} := \int_0^{+\infty} \mu(s)ds < 1$. Then

$$\lim_{t \to +\infty} \frac{R(t)}{t} = c - a^* \frac{\lambda}{1 - \hat{\mu}}, \tag{9}$$

where $a^* = \sum_{k \in X} a(k)\pi_k^*$.

Proof. From (8) we have

$$R(t)/t = u/t + c - \sum_{i=1}^{N(t)} a(X_k)/t. \tag{10}$$

The first term goes to zero when $t \to +\infty$. From the other side, w.r.t. the strong LLN for Markov chains (see, e.g., [10])

$$\frac{1}{n} \sum_{k=1}^{n} a(X_k) \to_{n \to +\infty} a^*, \tag{11}$$

where a^* is defined in (9).

Finally, taking into account (10) and (11), we obtain:

$$\sum_{i=1}^{N(t)} a(X_k)/t = \frac{N(t)}{t} \frac{1}{N(t)} \sum_{i=1}^{N(t)} a(X_k) \to_{t \to +\infty} a^* \frac{\lambda}{1 - \hat{\mu}},$$

and the result in (9) follows.

We note, that we have used above the result that $N(t)/t \to_{t \to +\infty} \lambda/(1-\hat{\mu})$. (See, e.g., [3] or [4]). Q.E.D.

Remark 2. When $a(X_k) = X_k$ are i.i.d.r.v., then $a^* = EX_k$.

Remark 3. When $\mu(t) = \alpha e^{-\beta t}$ is exponential, then $\hat{\mu} = \alpha/\beta$.

8.4.2   FCLT for RMGCHP

Theorem 2 (FCLT for RMGCHP). Let $R(t)$ be the risk model (RMGCHP) defined above in (8), and X_k be an ergodic Markov chain with stationary probabilities π_n^*. We suppose that $0 < \hat{\mu} := \int_0^{+\infty} \mu(s)ds < 1$ and $\int_0^{+\infty} s\mu(s)ds < +\infty$. Then

$$\lim_{t \to +\infty} \frac{R(t) - (ct - a^*N(t))}{\sqrt{t}} =^D \sigma\Phi(0,1), \tag{12}$$

(or in Skorokhod topology (see [11]))

$$\lim_{n \to +\infty} \frac{R(nt) - (cnt - a^*N(nt))}{\sqrt{n}} = \sigma W(t)) \tag{12'}$$

where $\Phi(\cdot, \cdot)$ is the standard normal random variable ($W(t)$ is a standard Wiener process),

$$\begin{aligned} \sigma &:= \sigma^*\sqrt{\lambda/(1-\hat{\mu})}, \\ (\sigma^*)^2 &:= \sum_{i \in X} \pi_i^* v(i), \end{aligned} \tag{13}$$

and

$$\begin{aligned} v(i) &= b(i)^2 \\ &+ \sum_{j \in X}(g(j) - g(i))^2 P(i,j) - 2b(i)\sum_{j \in S}(g(j) - g(i))P(i,j), \\ b &= (b(1), b(2), ..., b(n))', \\ b(i) : &= a^* - a(i), \\ g : &= (P + \Pi^* - I)^{-1}b, \\ a^* &:= \sum_{i \in X} \pi_i^* a(i), \end{aligned}$$

$$\tag{14}$$

P is a transition probability matrix for X_k,, i.e., $P(i,j) = P(X_{k+1} = j | X_k = i)$, Π^* denotes the matrix of stationary distributions of P and $g(j)$ is the jth entry of g.

Proof. From (8) it follows that

$$R(t)/\sqrt{t} = (u + ct - \sum_{i=1}^{N(nt)} a(X_k))/\sqrt{t},$$

and

$$R(t)/\sqrt{t} = (u + ct + \sum_{i=1}^{N(t)}(a^* - a(X_k)) - N(t)a^*)/\sqrt{t}, \tag{15}$$

where a^* is defined in (14)).

Therefore,

$$\frac{R(t) - (ct - N(t)a^*)}{\sqrt{t}} = \frac{u + \sum_{i=1}^{N(t)}(a^* - a(X_k))}{\sqrt{t}}. \tag{16}$$

As long as $\frac{u}{\sqrt{t}} \to_{t \to +\infty} 0$, we have to find the limit for

$$\frac{\sum_{i=1}^{N(t)}(a^* - a(X_k))}{\sqrt{t}}$$

when $t \to +\infty$.

Consider the following sums

$$R_n^* := \sum_{k=1}^{n}(a(X_k) - \hat{a}^*) \tag{17}$$

and

$$U_n^*(t) := n^{-1/2}[(1 - (nt - \lfloor nt \rfloor))R_{\lfloor nt \rfloor}^* + (nt - \lfloor nt \rfloor))R_{\lfloor nt \rfloor + 1}^*], \tag{18}$$

where $\lfloor \cdot \rfloor$ is the floor function.

Following the martingale method from [18], we have the following weak convergence in the Skorokhod topology (see [11]):

$$\hat{U}_n^*(t) \to_{n \to +\infty} \sigma^* W(t), \tag{19}$$

where σ^* is defined in (13).

We note again, that w.r.t LLN for Hawkes process $N(t)$ (see, e.g., Daley, D.J. and Vere-Jones, D. (1988)) we have:

$$\frac{N(t)}{t} \to_{t \to +\infty} \frac{\lambda}{1 - \hat{\mu}},$$

or

$$\frac{N(nt)}{n} \to_{n \to +\infty} \frac{t\lambda}{1 - \hat{\mu}}, \tag{20}$$

where $\hat{\mu}$ is defined in (13).

Using change of time in (19), $t \to N(t)/t$, we can find from (19) and (20):

$$U_n^*(N(nt)/n) \to_{n \to +\infty} \sigma W\left(t\lambda/(1 - \hat{\mu})\right),$$

or

$$U_n^*(N(nt)/n) \to_{n \to +\infty} \sigma\sqrt{\lambda/(1 - \hat{\mu})}W(t), \tag{21}$$

where $W(t)$ is the standard Wiener process, and σ^* and $\hat{\mu}$ are defined in (13). The result (12) now follows from (15)-(21). Q.E.D.

We note that LLN and FCLT for general compound Hawkes processes in limit order books were considered in [16].

Remark 4. When $a(X_k) = X_k \in \{+\delta, -\delta\}$ are independent and $P(1,2) = P(2,1) = \pi^* = 1/2$, then $a^* = 0$ and $\sigma^* = +\delta$.

Remark 5. When $a(X_k) = X_k \in \{+\delta, -\delta\}$ is two-state Markov chain, $P(1,2) = P(2,1) = p$, then $\pi^* = 1/2$, $a^* = 0$ and $(\sigma^*)^2 = \delta^2 p/(1-p)$.

Remark 6. When $a(X_k) = X_k \in \{+\delta, -\delta\}$ is two-state Markov chain and $P(1,1) = p', P(2,2) = p$, then $a^* = \delta(2\pi^* - 1)$ and

$$(\sigma^*)^2 = 4\delta^2 \left(\frac{1 - p' + \pi^*(p' - p)}{(p + p' - 2)^2} - \pi^*(1 - \pi^*) \right).$$

Remark 7. When $a(X_k) = X_k$ are i.i.d.r.v., then $\sigma^* = Var(X_k)$ in (13) and $\sigma = Var(X_k)\sqrt{\lambda/(1 - \hat{\mu})}$.

8.5 Applications of LLN and FCLT for RMGCHP

In this section we consider some applications of LLN and FCLT for RMGCHP that include net profit condition, premium principle and ruin and ultimate ruin probabilities.

8.5.1 Application of LLN: Net Profit Condition

From Theorem 1 (LLN for RMGCHP) follows that net profit condition has the following form:

Corollary 1 (NPC for RMGCHP).

$$c > a^* \frac{\lambda}{1 - \hat{\mu}}, \tag{22}$$

where $a^* = \sum_{k \in X} a(k)\pi_k^*$.

Corollary 2 (NPC for RMCHP). When $a(X_k) = X_k$ are i.i.d.r.v., then $a^* = EX_k$, and the net profit condition in this case has the form

$$c > \frac{\lambda}{1 - \hat{\mu}} \times E[X_k].$$

Corollary 3 (NPC for RMCPP). Of course, in the case of Poisson process $N(t)$ ($\hat{\mu} = 0$) we have well-known net profit condition:

$$c > \lambda \times E[X_k].$$

8.5.2 Application of LLN: Premium Principle

A premium principle is a formula for how to price a premium against an insurance risk. There many premium principles, and the following are three classical examples of premium principles ($S_t = \sum_{k=1}^{N(t)} a(X_k)$):

- The expected value principle: $c = (1 + \theta) \times E[S_t]/t$,

where the parameter $\theta > 0$ is the safety loading;
- The variance principle: $c = E[S_t]/t + \theta \times Var[S_t/t]$;
- The standard deviation principle: $c = E[S_t]/t + \theta \times \sqrt{Var[S_t/t]}$.

We present here the expected value principle as one of the premium principles (that follows from Theorem 1 (LLN for RMGCHP)):

Corollary 4 (Premium Principle for RMGCHP)

$$c = (1 + \theta)\frac{a^*\lambda}{1 - \hat{\mu}}, \tag{23}$$

where the parameter $\theta > 0$ is the safety loading.

8.5.3 Application of FCLT for RMGCHP: Ruin and Ultimate Ruin Probabilities

8.5.3.1 Application of FCLT for RMGCHP: Approximation of RMGCHP by a Diffusion Process

From Theorem 2 (FCLT for RMGCHP) it follows that risk process $R(t)$ can be approximated by the following jump-diffusion process $D(t)$:

$$R(t) \approx u + ct - N(t)a^* + \sigma W(t) := u + J(t),$$

where a^* and σ are defined above, $N(t)$ is a Hawkes process and $W(t)$ is a standard Wiener process.

Taking into account that $N(t) \approx [\lambda/(1 - \bar{\mu})]t$ (see Theorem 1), we conclude that our jump-diffusion process $J(t)$ has drift $(c - a^*\lambda/(1 - \hat{\mu}))$ and diffusion coefficient σ, i.e., $J(t)$ is $\Phi(c - a^*\lambda/(1 - \hat{\mu})t, \sigma^2 t)$-distributed, where Φ is standard normal distribution function. It means that $R(t)$ can be approximated by a diffusion process $D(t)$:

$$D(t) \approx u + ct - a^*[\lambda/(1 - \hat{\mu})] + \sigma W(t).$$

We use this diffusion approximation above for the RMGCHP to calculate the ruin probability in a finite time interval $(0, \tau)$.

We note that the rate of approximation is

$$E|R(t) - (ct - a^*N(t)) - \sigma W(t)| \le \frac{1}{\sqrt{t}}C(c, a^*, \sigma, \lambda, \hat{\mu}),$$

or

$$E|R(tn) - (cnt - a^*N(nt)) - \sigma W(t)| \le \frac{1}{\sqrt{n}}C(c, a^*, \sigma, \lambda, \hat{\mu}, T),$$

where $t \in [0, T]$, and $C(c, a^*, \sigma, \lambda, \hat{\mu})$ or $C(c, a^*, \sigma, \lambda, \hat{\mu}, T)$ are the constants that depend on all initial parameters of the model. See [14] for more details about the rates of approximation for random evolutions.

8.5.3.2   Application of FCLT for RMGCHP: Ruin Probabilities

The ruin probability up to time τ is given by (T_u is a ruin time)

$$
\begin{aligned}
\psi(u, \tau) &= 1 - \phi(u, \tau) = P(T_u < \tau) \\
&= P(\min_{0<t<\tau} R(t) < 0) \\
&= P(\min_{0<t<\tau} D(t) < -u).
\end{aligned}
$$

Applying now the result for ruin probabilities for diffusion process (see, e.g., [1] or [2]) we obtain the following

Theorem 3 (Ruin Probability for Our Diffusion Process):

$$
\begin{aligned}
\psi(u, \tau) &= \Phi\left(-\frac{u + (c - a^* \lambda/(1-\hat{\mu}))\tau}{\sigma\sqrt{\tau}}\right) \\
&+ e^{-\frac{2(c - a^* \lambda/(1-\hat{\mu}))}{\sigma^2}u} \Phi\left(-\frac{u - (c - a^* \lambda/(1-\hat{\mu}))\tau}{\sigma\sqrt{\tau}}\right),
\end{aligned}
\tag{24}
$$

where Φ is the standard normal distribution function.

8.5.3.3   Application of FCLT for RMGCHP: Ultimate Ruin Probabilities

Letting $\tau \to +\infty$ in Theorem 3 above, we obtain:

Corollary 5 (The Ultimate Ruin Probability for RMGCHP):

$$
\psi(u) = 1 - \phi(u) = P(T_u < +\infty) = e^{-\frac{2(c - a^* \lambda/(1-\hat{\mu}))}{\sigma^2}u},
\tag{25}
$$

where σ and $\hat{\mu}$ are defined in Theorem 2 (FCLT for RMGCHP).

8.5.4   Application of FCLT for RMGCHP: The Distribution of the Time to Ruin

From Theorem 3 and Corollary 5 follows:

Corollary 6 (The Distribution of the Time to Ruin). The distribution of the time to ruin, given that ruin occurs is:

$$
\begin{aligned}
\frac{\psi(u, \tau)}{\psi(u)} &= P(T_u < \tau | T_u < +\infty) \\
&= e^{\frac{2(c - a^* \lambda/(1-\hat{\mu}))}{\sigma^2}u} \Phi\left(-\frac{u + (c - a^* \lambda/(1-\hat{\mu}))\tau}{\sigma\sqrt{\tau}}\right) \\
&+ \Phi\left(-\frac{u - (c - a^* \lambda/(1-\hat{\mu}))\tau}{\sigma\sqrt{\tau}}\right).
\end{aligned}
$$

Differentiation in previous distribution by u gives the probability density function $f_{T_u}(\tau)$ of the time to ruin:

Corollary 7 (The Probability Density Function of the Time to Ruin):

$$f_{T_u}(\tau) = \frac{u}{\sigma\sqrt{2\pi}}\tau^{-3/2}e^{-\frac{(u-(c-a^*\lambda/(1-\hat{\mu}))\tau)^2}{2\sigma^2\tau}}, \quad \tau > 0. \tag{26}$$

Remark 8 (Inverse Gaussian Distribution): Substituting $u^2/\sigma^2 = a$ and $u/(c - a^*\lambda/(1 - \hat{\mu})) = b$ in the density function we obtain:

$$f_{T_u}(\tau) = (\frac{a}{2\pi\tau^3})^{1/2}e^{-\frac{a}{2\tau}(\frac{\tau-b}{\sigma})^2}, \quad \tau > 0,$$

which is the standard Inverse Gaussian distribution with expected value $u/(c - a^*\lambda/(1 - \hat{\mu}))$ and variance $u\sigma^2/(c - a^*\lambda/(1 - \hat{\mu}))$.

Remark 9 (Ruin Occurs with $P = 1$): If $c = a^*\lambda/(1 - \hat{\mu})$, then ruin occurs with $P = 1$ and the density function is obtained from Corollary 6 with $c = a^*\lambda/(1 - \hat{\mu})$, i.e.,

$$f_{T_u}(\tau) = \frac{u}{\sigma\sqrt{2\pi}}\tau^{-3/2}e^{-\frac{u^2}{2\sigma^2\tau}}, \quad \tau > 0.$$

The distribution function is :

$$F_{T_u}(\tau) = 2\Phi(-\frac{u}{\sigma\sqrt{\tau}}), \quad \tau > 0.$$

8.6 Applications of LLN and FCLT for RMCHP

In this section we list the applications of LLN and FCLT for a risk model based on a compound Hawkes process (RMCHP). The LLN and FCLT for RMCHP follow from Theorem 1 and Theorem 2 above, respectively. In this case $a(X_k) = X_k$ are i.i.d.r.v. and $a^* = EX_k$, and our risk model $R(t)$ based on compound Hawkes process $N(t)$ (RMCHP) has the following form:

$$R(t) = u + ct - \sum_{k=1}^{N(t)} X_k,$$

where $N(t)$ is a Hawkes process.

8.6.1 Net Profit Condition for RMCHP

From (22) it follows that net profit condition for RMCHP has the following form ($a^* = EX_k$):

$$c > \frac{\lambda EX_1}{1 - \hat{\mu}}.$$

8.6.2 Premium Principle for RMCHP

From (23) it follows that premium principle for RMCHP has the following form:

$$c = (1 + \theta)\frac{\lambda EX_1}{1 - \hat{\mu}},$$

where $\theta > 0$ is the safety loading parameter.

8.6.3 Ruin Probability for RMCHP

From (24) it follows that the ruin probability for RMCHP has the following form:

$$\psi(u, \tau) = \Phi\left(-\frac{u + (c - EX_1\lambda/(1 - \hat{\mu}))\tau}{\sigma\sqrt{\tau}}\right)$$

$$+ e^{-\frac{2(c - EX_1\lambda/(1 - \hat{\mu}))}{\sigma^2}u}\Phi\left(-\frac{u - (c - EX_1\lambda/(1 - \hat{\mu}))\tau}{\sigma\sqrt{\tau}}\right).$$

Remark 10. Here, $\sigma = Var(X_k)\sqrt{\lambda/(1 - \hat{\mu})}$ (see Remark 7.).

8.6.4 Ultimate Ruin Probability for RMCHP

From (25) it follows that the ultimate ruin probability for RMCHP has the following form:

$$\psi(u) = 1 - \phi(u) = P(T_u < +\infty) = e^{-\frac{2(c - EX_1\lambda/(1 - \hat{\mu}))}{\sigma^2}u}.$$

8.6.5 The Probability Density Function of the Time to Ruin

From (26) it follows that the probability density function of the time to ruin for RMCHP has the following form:

$$f_{T_u}(\tau) = \frac{u}{\sigma\sqrt{2\pi}}\tau^{-3/2}e^{-\frac{(u - (c - EX_1\lambda/(1 - \hat{\mu}))\tau)^2}{2\sigma^2\tau}}, \quad \tau > 0.$$

8.7 Applications of LLN and FCLT for RMCPP

In this section we list, just for completeness, the applications of LLN and FCLT for risk model based on compound Poisson process (RMCPP). The LLN and FCLT for RMCPP follow from Section 5 above. In this case $a(X_k) = X_k$ are i.i.d.r.v. and $a^* = EX_k$, and $\hat{\mu} = 0$ and our risk model $R(t)$ based on compound Poisson process $N(t)$ (RMCHP) has the following form:

$$R(t) = u + ct - \sum_{k=1}^{N(t)} X_k,$$

where $N(t)$ is a Poisson process.

Of course, all the results below are classical and well-known (see, e.g., [1]), and we list them just to show that they are followed from our results above.

8.7.1 Net Profit Condition for RMCPP

From (22) it follows that net profit condition for RMCPP has the following form ($a^* = EX_k$):

$$c > \lambda EX_1.$$

8.7.2 Premium Principle for RMCPP

From (23) it follows that premium principle for RMCPP has the following form:

$$c = (1 + \theta)\lambda EX_1,$$

where $\theta > 0$ is the safety loading parameter.

8.7.3 Ruin Probability for RMCPP

From (24) it follows that the ruin probability for RMCPP has the following form:

$$\psi(u, \tau) = \Phi\left(-\frac{u + (c - EX_1\lambda)\tau}{\sigma\sqrt{\tau}}\right)$$

$$+ \quad e^{-\frac{2(c - EX_1\lambda)}{\sigma^2}u}\Phi\left(-\frac{u - (c - EX_1\lambda)\tau}{\sigma\sqrt{\tau}}\right).$$

Remark 11. Here, $\sigma = Var(X_k)\sqrt{\lambda}$ because $\hat{\mu} = 0$ (see Remark 7.).

8.7.4 Ultimate Ruin Probability for RMCPP

From (25) it follows that the ultimate ruin probability for RMCPP has the following form:

$$\psi(u) = 1 - \phi(u) = P(T_u < +\infty) = e^{-\frac{2(c - EX_1\lambda)}{\sigma^2}u}.$$

8.7.5 The Probability Density Function of the Time to Ruin for RMCPP

From (26) it follows that the probability density function of the time to ruin for RMCPP has the following form:

$$f_{T_u}(\tau) = \frac{u}{\sigma\sqrt{2\pi}}\tau^{-3/2}e^{-\frac{(u - (c - EX_1\lambda)\tau)^2}{2\sigma^2\tau}}, \quad \tau > 0.$$

Bibliography

[1] Asmussen, S. 2000. *Ruin Probabilities.* Singapore: World Scientific.

[2] Asmussen, S. and Albrecher, H. 2010. *Ruin Probabilities.* 2nd edition. Singapore: World Scientific.

[3] Bacry, E., Mastromatteo, I. and Muzy, J.-F. 2015. Hawkes processes in finance. *Market Microstructure and Liquidity,* June, Vol. 01, No. 012.

[4] Daley, D.J. and Vere-Jones, D. 1988. *An Introduction to the Theory of Point Processes.* Springer.

[5] Dassios, A. and Zhao, HB. 2011. A dynamic contagion process. *Adv. in Appl. Probab,* 43(3), 814-846.

[6] Dassios, A. and Jang, J. 2012. A double shot noise process and its application in insurance. *J. Math. System Sci.,* 2, 82-93.

[7] Embrechts, P., Liniger, T. and Lin, L. 2011. Multivariate Hawkes processes: An application to financial data. *Journal of Applied Probability,* 48A, 367-378.

[8] Hawkes, A. G. 1971. Spectra of some self-exciting and mutually exciting point processes. *Biometrika,* 58, 83-90.

[9] Jang, J. and Dassios, A. 2013. A bivariate shot noise self-exciting process for insurance. *Insurance: Mathematics and Economics,* 53 (3), 524-532.

[10] Norris, J. R. 1997. *Markov Chains.* In: Cambridge Series in Statistical and Probabilistic Mathematics. UK: Cambridge University Press.

[11] Skorokhod, A. 1965. *Studies in the Theory of Random Processes*. Reading (Mass.): Addison-Wesley. (Reprinted by Dover Publications, NY).

[12] Stabile, G. and G. L. Torrisi. 2010. Risk processes with non-stationary Hawkes arrivals. *Methodol. Comput. Appl. Prob.*, 12, 415-429.

[13] Swishchuk A.V. and Goncharova S.Y. 1998. Optimal control of semi-Markov risk processes. *Nonlinear Oscillations*, No2, 122-131.

[14] Swishchuk, A. 2000. *Random Evolutions and Their Applications: New Trends*. Dordrecht: Kluwer AP.

[15] Swishchuk, A. 2017a. Risk model based on compound Hawkes process. Abstract, IME 2017, Vienna.

[16] Swishchuk, A. 2017b. General Compound Hawkes Processes in Limit Order Books. *Available on arXiv*: https://arxiv.org/submit/1929048

[17] Swishchuk, A., Chavez-Casillas, J., Elliott, R. and Remillard, B. 2017. Compound Hawkes processes in limit order books. *Available on SSRN*: https://papers.ssrn.com/sol3/papers.cfm?abstract_id=2987943.

[18] Vadori, N. and Swishchuk, A. 2015. Strong law of large numbers and central limit theorems for functionals of inhomogeneous semi-Markov processes. *Stochastic Analysis and Applications*, 13 (2), 213-243.

[19] Zhu, L. 2013. Central limit theorem for nonlinear Hawkes processes. *Journal of Applied Probability*, 50, 760-771.

Index

Printed in the United States
by Baker & Taylor Publisher Services